MANAGING TCP/IP NETWORKS

MANAGING TCP/IP NETWORKS: TECHNIQUES, TOOLS, AND SECURITY CONSIDERATIONS

Gilbert Held
4 Degree Consulting
Macon, Georgia, USA

JOHN WILEY & SONS, LTD
Chichester • New York • Weinheim • Brisbane • Singapore • Toronto

Baffins Lane, Chichester,
West Sussex, PO19 1UD, England

National 01243 779777
International (+44) 1234 779777
e-mail (for orders and customer service enquiries): cs-books@wiley.co.uk

Visit our Home Page on http://www.wiley.co.uk or http://www.wiley.com

Reprinted June 2000

Other Wiley Editorial Offices

John Wiley & Sons, Inc., 605 Third Avenue,
New York, NY 10158-0012, USA

WILEY-VCH Verlag GmbH
Pappelallee 3, D-69469 Weinheim, Germany

Jacaranda Wiley Ltd, 33 Park Road, Milton,
Queensland 4064, Australia

John Wiley & Sons (Asia) Pte Ltd, 2 Clementi Loop #02-01,
Jin Xing Distripark, Singapore 129809

John Wiley & Sons (Canada) Ltd, 22 Worcester Road
Rexdale, Ontario, M9W 1L1, Canada

Library of Congress cataloging-in-Publication Data

Held, Gilbert, 1943-
Managing TCP/IP networks: techniques, tools and security considerations/Gilbert Held.
p. cm.
ISBN 0-471-80003-1 (alk. paper)
1. TCP/IP (Computer network protocol) 2. Computer networks–Management. I. Title.
TK5105.585.H447 2000 99-44748
004.6'2—dc21 CIP

British Library Cataloguing in Publication Data
A catalogue record for this book is available from the British Library

ISBN 0 471 80003 1

Typeset in 10/12pt Bookman-Light by Dobbie Typesetting Limited
Printed and bound in Great Britain by Bookcraft (Bath) Ltd
This book is printed on acid-free paper responsibly manufactured from sustainable forestry, in which at least two trees are planted for each one used for paper production.

CONTENTS

PREFACE

Today we live in the era of the Internet, intranets, and extranets, with virtual networking being employed to maximize the use of the Internet. Each of these rapidly growing areas of communications technology is based upon the TCP/IP protocol suite, which has exploded in use over the past decade. Accompanying this growth is the need to manage TCP/IP networks, which is the focus of this book.

Because the management of TCP/IP networks requires detailed knowledge of the protocol suite, the first few chapters in this book are focused on this topic. Once this has been accomplished, we will proceed up the layers of the protocol stack by examining tools and techniques that can be used at each layer. In doing so, we will investigate the use of several diagnostic tools to discover the cause of network problems, recognize potential problems prior to their occurrence, and note corrective actions that can be taken to alleviate actual and potential problems.

Although this book is not titled 'SNMP and RMON,' any coverage of the TCP/IP protocol suite needs to recognize the importance of those management tools and appropriately cover these areas of communications technology. With the focus of this book on managing TCP/IP networks, coverage of SNMP and RMON is an integral part. Another key area of TCP/IP network management is network security, which is also covered in this book.

Recognizing that the size of TCP/IP networks can range in scope from a few hub-based LANs interconnected via a wide area network transmission facility to large mesh structured private networks and the mother of all networks, the Internet, this book is focused upon concepts that can be applied to all TCP/IP-based networks, regardless of their size.

As a professional author I highly value reader feedback. Your comments concerning topics presented in this book such as areas you believe require additional elaboration or other comments are welcome. You can write to me through my publisher, whose address is on the cover of this book, or you can contact me directly via email at gil_held@yahoo.com

Gilbert Held
Macon, GA

ACKNOWLEDGMENTS

The preparation of a book is a team effort, even though only the author's name is displayed. Thus, I would be remiss if I did not acknowledge the efforts of other people who had a significant impact upon the evolution of this book from an author's concept into the book you are reading.

Once again I would like to thank Ann-Marie Halligan, my editor at John Wiley & Sons, for backing another of my writing projects. I would also like to thank Sarah Lock and the members of the Wiley production department for the fine job they accomplished in producing this book.

As an old-fashioned author who frequently travels to locations where his electrical adapters never seem to work, many years ago I decided pen and paper provided a higher level of reliability than a four-hour lap top battery on a two-week trip. Working by hand in drafting a manuscript results in the need for an alert typist who can translate my writing and drawings into a professional manuscript. Thus, I am most fortunate to again be able to count on Mrs. Linda Hayes to convert my longhand manuscript into an acceptable text.

Last but not least, writing a book is a time-consuming effort that requires many nights and long weekends of effort. I am most appreciative to my wife Beverly for her understanding as I literally locked myself in my office and network laboratory for long periods of time as I experimented with different networking tools and techniques while working on this book.

1

INTRODUCTION

In less than thirty years the TCP/IP protocol suite has evolved from a Department of Defense research initiative into a ubiquitous transmission capability that is used by academia, government agencies, businesses, and home computer users. Networks constructed using the TCP/IP protocol suite range in scope from a small hub-based local area network in a home office to the giant network of interconnected networks known as the Internet. As the use of the TCP/IP protocol suite proliferated, so did its support of a range of new applications that only a few years ago were considered by many persons to represent science fiction. Today real time audio and video, as well as digitized voice and fax, can be transmitted over the Internet and private intranets. While the growth in the use of the TCP/IP protocol stack and its role as a mechanism to transport different types of data has been quite impressive, it has not been problem-free. In actuality, it has introduced a new set of problems that network managers and administrators must consider as they manage their networks. Thus, the need for network management has increased in tandem with the growth in the use of the TCP/IP protocol suite, as has its expanded role in transporting different types of data.

In this introductory chapter we will focus our attention upon the process of network management and how it relates to the TCP/IP protocol suite. Although no definition can be expected to be all-encompassing, we will commence our investigation of network management with one. This definition will form a base for describing the different and varied facets of network management, which can include techniques, tools, and systems. However, prior to actually examining what network management encompasses, let us first examine the rationale for this activity. Doing so will provide us with additional insight into the various components that constitute this functional area.

1.1 RATIONALE FOR NETWORK MANAGEMENT

As mentioned above, we are in the midst of an explosive growth in the use of the TCP/IP protocol suite with respect to both the quantity of data transmitted and applications transmitting data. Today many vendors depend greatly upon their online Web sites for sales that can easily exceed several million dollars per day, other vendors provide low cost fax transmission services anywhere in the world for hundreds of thousands of customers, and millions of businesses and tens of millions of consumers depend upon the delivery of electronic mail to expedite messaging rather than use what is referred to as snail mail when speaking about the various postal services of different countries. This growth in the use of the TCP/IP protocol suite makes both individuals and organizations highly dependent upon the use of TCP/IP-based networks to perform their normal day-to-day tasks.

1.1.1 Cost of service interruptions

As a result of the previously described dependence upon the use of TCP/IP-based networks, interruptions or small abnormal situations can have serious consequences. For example, the failure of an Internet connection not only can terminate the delivery of electronic mail to a business but, in addition, can terminate access to its online order catalogue if they also operate a Web site that provides that capability. For a merchant the loss of a communications circuit could result in the loss of thousands or even millions of dollars of sales during the outage. Thus, methods to predict or rapidly detect failures and alert personnel to take remedial action can produce benefits ranging from a reduction in customer inconvenience to alleviating a loss of revenue. Other areas of concern in today's communications environment are the size and complexity of networks, their operating costs and performance, and the ability to learn enough information to take advantage of the sophistication of the protocol suite.

1.1.2 Size and complexity of networks

As the need for communications expanded, the size, complexity, and operating cost of networks increased in tandem. This was a driving force for the development of systems to monitor network equipment and transmission facilities, provide technicians with the ability to implement configuration changes from a central site location, and generate alarms when predefined conditions occur. Within the TCP/IP protocol suite the development of the Simple Network Management Protocol (SNMP) and Remote Monitor (RMON) makes a network more manageable with fewer personnel. However, their effective utilization requires an understanding of the protocol suite and communications concepts. To paraphrase a great general, 'in network management there is no substitute for understanding communications concepts.'

1.1.3 Performance monitoring

Through the use of management systems it becomes possible to monitor the performance and capacity of TCP/IP networks. A related issue is the management of network costs, since, as a general rule, excellent performance occurs at a low utilization level, which can result in an excessive expenditure of funds for equipment and transmission facilities only partially used. Thus, network management can be expected to balance performance and capacity while attempting to minimize costs.

1.1.4 Coping with equipment sophistication

As the use of TCP/IP networks has proliferated, devices used in their construction and access have grown in complexity. For example, many routers now include a voice digitization capability. Coping with the sophistication of modern networking devices requires personnel to have a high degree of training, which must be considered as another vital aspect of the network management. Fortunately, many network management products hide the inner workings of communications products by displaying a graphic user interface with an easily accessible help capability in place of a command-driven interface that might cause administrators to use cryptic command line entries to perform different equipment operations. Thus, modern network management products can assist us in coping with network device sophistication. Table 1.1 summarizes the major reasons why TCP/IP-based networks must be managed, providing the rationale for network management.

1.2 THE NETWORK MANAGEMENT PROCESS

Network management as a process resembles many other common activities in that we are fairly certain about what it is, but would probably be hard-pressed to provide a definition. The following definition, while not all-inclusive, provides a base upon which we will expand:

> Network management is the process of using hardware and software by trained personnel to monitor the status of network components and transmission facilities, question end-users and communications carrier personnel, and implement or recommend actions to alleviate outages and/or improve communications performance as well as conduct administrative tasks associated with the operation of the network.

As indicated by the previous definition, network management first and foremost requires trained personnel. In a TCP/IP environment this means that personnel must be very familiar with the protocol suite, how packets are formed,

Table 1.1 Rationale for network management

- Dependence upon network availability
- Effect of network failure
- Network size and complexity
- Coping with network device sophistication
- Network performance and capacity planning balance
- Operating cost containment

the role, use, and composition of packet headers, and both common and specialized networking concepts. Concerning the latter, this could include latency tolerance if your organization uses or is investigating the use of a TCP/IP network for voice or fax transmission. Secondly, it involves the use of hardware and software to examine network components, such as bridges and routers, as well as transmission facilities and equipment, such as Data Service Units (DSUs) and Channel Service Units (CSUs), connected to these facilities. Note that personnel may be required to question both end-users and communications carrier personnel to develop knowledge about a situation to which they will apply their expertise. In addition, after acquiring knowledge concerning an activity or event, network personnel will either implement or recommend actions to alleviate a current outage or devise methods to improve communications performance. Here, methods to improve communications performance can include changing an existing network configuration or preparing a long range study of the communications requirements of the organization and their effect upon the network. Finally, the performance of administrative tasks can be considered as a catchall phrase to include tasks associated with a variety of functions that can include generating and monitoring the progress of trouble tickets, developing and implementing a charge back procedure for sharing network costs among users, and ensuring that only valid users use the network. While many of the previously mentioned tasks may be optional, again, ensuring that only valid users use the network, these tasks are the tip of the proverbial iceberg, as they represent a few of many security-related topics that network managers and LAN administrators must consider.

1.2.1 The OSI framework for network management

Based upon the preceding, we can subdivide the tasks associated with network management into several functional areas. In fact, this was done by the International Organization for Standardization (ISO) with the development of its Open System Interconnection (OSI) Reference Model. In developing the OSI Reference Model the ISO defined five network management functional areas or disciplines, which are indicated in Table 1.2.

Configuration/change management

Configuration or change management involves the process of keeping track of the various parameters of communications devices and facilities that make up a network. Parameters can be set, reset, or simply read and displayed.

Table 1.2 OSI framework for network management

- Configuration/change management
- Fault/problem management
- Performance/growth management
- Security/access management
- Accounting/cost management

For complex networks that have hundreds or thousands of devices and transmission facilities, the use of SNMP and RMON will more than likely be used to facilitate the control of the network from a single point or from a few management locations. However, the actual platform under which SNMP and RMON operate can range in scope from a PC-based network management system to minicomputer- and mainframe-based systems. Regardless of the actual platform, most systems will include the ability to autodiscover devices and display a geographical representation of the network in addition to providing the user with the ability to read and possibly change device parameters as well as display a variety of transmission line parameters. Unlike devices whose parameters can be displayed and reset, transmission facilities are controlled by one or more communications carriers, and adjustment of those parameters is normally beyond the control of the network end-user operation. In this situation the ability to rapidly display and understand the meaning of transmission parameters may enable potential problems to be alleviated prior to their occurrence or can enable alternate routing procedures to be implemented when an outage of a marginal or failed facility is reported and the circuit is removed from operation for carrier testing.

Although a network management system facilitates configuration management, most organizations do not have one ubiquitous system. This is because SNMP and RMON were primarily developed as a monitoring and alerting facility, and also because of some security limitations that are not integrated to allow parameters changes to be made to routers, DSUs, CSUs, and other network devices. Instead, many organizations may maintain several systems, some of which may be used to control equipment from one vendor, while an SNMP and RMON station may be used as a separate monitoring and alerting facility. In addition, some devices may simply be controlled from their front panel display.

In concluding our initial discussion of configuration or change management, it should be noted that this area of network management is dependent upon a database of parameter settings and knowledge of their meanings. This database can consist of information recorded on 3×5 inch index cards, typewritten sheets, or files stored on a computer. Regardless of the media used to store information, the database represents a repository of information that can be used to determine alternatives as well as implement changes in the operation and structure of the network.

Fault/problem management

Fault or problem management is the process by which the detection, logging and ticketing, problem isolation, tracking, and eventual resolution of abnormal conditions is accomplished. Since you must know that a problem exists, the first and one of the most important steps in fault management is to detect an abnormal situation. This can be accomplished in a variety of ways, ranging from the setting of thresholds on a network management system that generates different types of alerts or alarm conditions when exceeded to users and customers calling a technical control center to report problems. Once a problem has been detected, many organizations will have a predefined operating procedure whereby the situation is recorded in a log and, if determined to represent a legitimate problem, is assigned a trouble ticket that enables the problem resolution process to be tracked.

It is important to understand that many problem-related calls to a technical control center are immediately resolved. Such calls may require trained technical control center personnel to spend a few minutes to a few hours checking equipment settings, viewing graphic displays to examine the status of remote devices, and questioning the user concerning their hardware and software settings, or performing other functions that resolve the problem without further action. Other calls or alarms may result in the issuing of a trouble ticket that requires action on the part of the communications carrier or the assistance of vendor personnel. Regardless of the extent of the problem, the initial logging involves an attempt to identify the cause of the abnormal situation and determine appropriate action for its correction.

Problem isolation can include a simple discussion with an end-user, diagnostic testing of equipment and transmission facilities, or extensive research. Once the cause of a problem is isolated it may be beyond the capability of an end-user's organization for correction, such as an unacceptable level of performance on a circuit or a failed device within the communications carrier network your organization is using. Thus, in addition to seeking appropriate assistance, another important step of the fault management process is to track progress of both internal and external personnel in their efforts towards correcting faults. Many times, fault management will require aged trouble tickets to be escalated to receive the attention they deserve. At other times, repetitive calls to a vendor or communications carrier to track the progress of a trouble ticket may reveal that the ticket was closed. Although we would logically hope that the carrier or vendor fixed the problem and inadvertently forgot to inform us of the problem resolution, we live in an imperfect world in which a trouble ticket can inadvertently be closed without resolving the problem. Thus, it is very important to track problems, including the status of trouble tickets.

While the resolution of an abnormal condition may appear to be the last task involved in the fault management process, in actuality it may require the performance of a configuration or change management task. For example, if an abnormal condition resulted in the implementation of alternative routing, the resolution of the problem could result in a configuration change in which routing reverts to its normal condition. This illustrates the interrelationship between each of the functional areas of network management.

Performance/growth management

Performance or growth management involves tasks required to evaluate the utilization of network equipment and transmission facilities and adjust them as required. Tasks performed can range from the visual observation of equipment indicators to the gathering of statistical information into a database that can be used to project utilization trends. Regardless of the method used, the objective of performance and growth management is to ensure that sufficient capacity exists to support end-user communications requirements. Thus, another term commonly used for performance or growth management is capacity planning.

One of the more interesting aspects of capacity planning concerns the reaction of end-users to capacity problems. If your organization's network has insufficient capacity, end-user complaints will commonly occur whenever response time increases or users encounter a busy signal when attempting to remotely access the network. Conversely, you will probably never encounter an end-user complaining that they always receive a good response time or never encounter a busy signal, and that the network has excessive capacity. This means that excessive capacity will more than likely require recognition by network management personnel, and it is incumbent upon such personnel to examine the potential for both network expansion and contraction.

A variety of tools can be used for the performance or growth management process, including communications carrier bills, network management systems, and such utility applications as Ping and Traceroute. Carrier bills may indicate the utilization of dial-in lines or leased lines connected to an Internet Service Provider. Network management systems may provide information about the use of local and remote networks and the operation and utilization of different network devices. The use of Ping, Traceroute, and other utility programs can indicate if a device is operational as well as the round-trip delay to the device.

Security/access management

Security or access management represents the set of tasks that ensures that only authorized personnel can use the network. In addition, some organizations may require the ability to hide the contents of data, especially when using the Internet as a virtual private network. Thus, tasks and functions associated with security management can include the authentication of users, encryption of data, the management and distribution of encryption keys, maintenance and examination of security logs, configuration of router access lists, and implementation of various firewall features to include proxy services and intrusion detection and alarm generation.

Allied with security access management are such tasks and functions as virus prevention measures, continuity of operation procedures, and the planning for and implementation, when necessary, of disaster recovery methods. Although a network manager cannot perform guard duty to ensure that personnel do not acquire or transmit suspicious files over the network, managers can and should publicize methods to test unknown software as well as procedures to be followed concerning the distribution of public domain software obtained from many shareware Web sites.

Accounting/cost management

In addition to the assurance of birth, death, and taxes, you can also expect the old adage 'there is no free lunch' to be essentially true. One of the processes of network management thus involves obtaining the right information at the right time, which provides a basis for establishing charges for the utilization of network resources. Tasks associated with accounting or cost management includes the issuing of equipment and transmission facility orders, the reconciliation and recording of invoices, the computation of depreciation and amortization charges, the assignment of personnel costs to network operations, the development of algorithms to prorate charges to users, and the periodic review of billing methods to ensure the fair and equitable assignment of costs based upon network usage.

The accounting management process may require the efforts of a team of specialists at large organizations. For small and medium size organizations, the effort involved in accounting management may still be substantial, especially when compared with the necessity to perform other network management functions. Many organizations therefore centrally fund the cost of communications or add a surcharge to the cost of using their data processing facilities. While this will certainly reduce the tasks associated with accounting management, other cost management functions, including budgeting, examining the effect of tariff changes upon the structure of the network, and verifying the correctness of vendor and communications carrier bills, must still continue. Accounting and cost management functions thus remain an important part of the network management process, regardless of whether or not the cost of the network is directly charged back to network users or organizational departments.

1.2.2 Other network management functions

Although the OSI framework for network management is comprehensive, it is not all-inclusive. Two key functional areas that are only partially covered within the OSI framework and are important enough to justify their identification as separate entities are asset management and planning or support management.

Asset management

Asset management is that set of tasks associated with the development and retrieval of records of equipment, facilities, and personnel. Equipment records can include one or more databases of information-covering devices used in the network, their parameter settings, manufacturer data and telephone numbers to call for maintenance, and similar information. Equipment records may reside on network management systems, may supplement information obtained from a network management system, or may be completely independent of a network management system.

Transmission facility records may simply include circuit numbers and carrier points of contact, or they can contain such additional information as the expected or guaranteed level of performance and the results of previous monitoring periods. If the latter is included, end-users may be able to note trends, which can include a deterioration in the quality of a circuit that, if unchecked, could result in a communications outage. The analysis of circuit record data may thus enable end-users to contact their communications carrier to request assistance prior to a degradation in service resulting in the failure of a circuit that would inhibit communications.

Unfortunately, personnel are often excluded from the asset management process even though they are your most valuable asset. Under the asset management process, you should consider developing records that indicate employee work experience, education, training, and certification levels. You can use this information to facilitate the assignment of employees to different networking projects. Similarly, information concerning prior education, training, and certification can be used in conjunction with organizational requirements to implement individual development plans that allocate training and travel funds to enable employees to receive appropriate training.

Planning/support management

Planning and support management includes those tasks that enable network managers and administrators to provide support for current users as well as to plan for the future. Support for current users can be viewed as a super-set of previously described network management functional areas. In fact, support as well as planning consistently draws upon other network management functional areas. Examples of support management functions can include adjusting network facilities to accommodate changes in the use of such facilities, ordering equipment and facilities to support new or expanded applications, and meeting with end-users to determine their degree of satisfaction or dissatisfaction with current communications methods.

Closely related to support management is the planning management process. During the planning process you may meet with end-users to determine their requirements as well as their satisfaction or dissatisfaction with existing communications. In addition, the planning process can involve the collection of data from other functional network management areas, which enables you to develop models to assist in the design of a new network structure or the optimization of an existing network structure. Finally, if the planning process results in a recommendation for a change in the structure of a network, upon approval those changes must be implemented. Thus, the planning process must include steps required to implement configuration or change management tasks.

Figure 1.1 summarizes the network management functional areas and tasks associated with each area. You should note that a valid case can be made for the inclusion of many tasks under two or more functional areas. Thus, you can view the tasks associated with the functional areas shown in Figure 1.1 as a guide to the primary areas in which certain tasks are performed and not as an all-inclusive example of where tasks are performed.

1.3 TOOLS AND SYSTEMS

Today there are a wide variety of tools whose utilization can provide a considerable level of assistance in managing a TCP/IP based network. Such

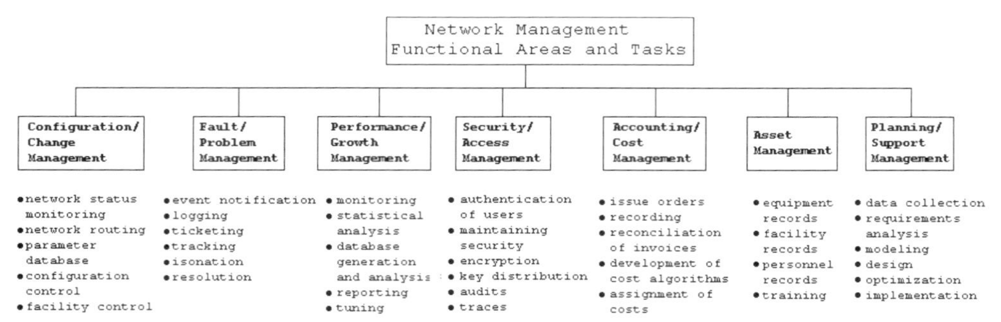

Figure 1.1 Network management functional areas and tasks

tools range in scope from the use of utility programs such as Ping, Traceroute, and NSLOOKUP to protocol analyzers and statistical reporting programs that provide an insight concerning the use of a network. In general, network management tools can be divided into three primary categories: monitoring tools, diagnostic tools, and computer-based management systems.

1.3.1 Monitoring tools

Monitoring tools provide you with the ability to observe the operation and performance of equipment and transmission facilities. Examples of monitoring tools include utility applications such as Ping that can inform you if a device is operational and reachable, as well as layer 2 and layer 3 software monitoring programs, such as EtherVision, EtherPeek, and other products that will be described and discussed in later chapters in this book.

1.3.2 Diagnostic tools

A diagnostic tool is typically used to detect problems with equipment or transmission facilities. Examples of diagnostic tools can also include Ping, as its use can provide information about the operational status of a device, as well as packet decoders that can shed light on reasons why communications devices are not operating properly.

1.3.3 Computer-based management systems

Computer-based management systems run the gamut from personal computer SNMP management platforms to more inclusive platforms that support SNMP as well as other vendor proprietary management hardware.

Figure 1.2 illustrates the general components of a computer-based management system. The management platform provides a control point for accessing devices either on the same network or located on a remote network. In a TCP/IP environment the management station uses SNMP as the communications protocol to access other devices to perform different management functions.

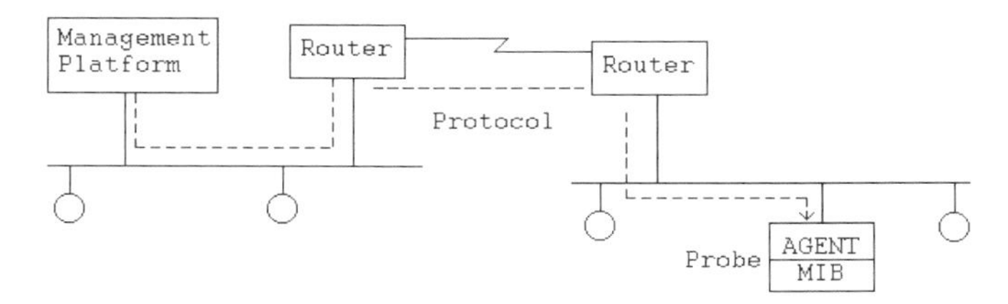

---- indicates data flow from management platform to a remote probe.

The key components of a network management system include a management platform, communications protocol to convey requests and responses, an agent that interprets requests and responds to the requests, and a Management Information Base (MIB) that represents a database of statistics and parameter values.

Figure 1.2 Key components of a network management system

The agent represents software that is responsible for interpreting and acting upon requests from the management platform. The third key component of a management system is the Management Information Base (MIB). The MIB represents a database of objects that represent either performance metrics maintained by a device or parameter values associated with a device that can be read and possibly reset. In a TCP/IP environment agents that perform remote monitoring occur through the use of RMON probes, with the term probe used to represent a remote agent and its MIB. Note that the agent and its MIB can reside as a module within a communications device, such as a router or CSU, or can operate as a stand-alone device connected to a network, such as a probe. Now that we have a general appreciation for network management functions, tools, and systems, we will conclude this chapter with a preview of the contents of succeeding chapters in this book.

1.4 BOOK PREVIEW

In the remainder of this chapter we will briefly preview the contents of succeeding chapters. This information can be used either as is, or in conjunction with the index of this book to locate items of particular interest. Although the chapters are ordered in a logical progression of information presentation, this author recognizes that the background of readers can considerably differ. Thus, although it is suggested that you should read the chapters in the book in their order, it is also recognized that for many reasons some individuals will require specific information. Thus, as much as possible each chapter was written to be as independent as possible with respect to preceding and succeeding chapters.

1.4.1 The TCP/IP protocol suite

In Chapter 2 we turn our attention to obtaining a detailed overview of the TCP/IP protocol suite. In this chapter we will first examine the evolution of the protocol suite and then focus our attention upon the ISO Reference Model. Next, we will use that model as a reference to examine the layers of the TCP/IP protocol suite, noting the data delivery mechanism by which an application data stream is converted into packets for delivery on LANs and over WANs.

1.4.2 The Internet Protocol

Because the TCP/IP protocol suite represents a layered communications protocol, we will commence our detailed examination of the suite at the network layer. In Chapter 3 we will examine the Internet Protocol (IP), which is a layer 3 protocol. We will examine both IP versions 4 and 6, to include their protocol header fields and addressing, data link delivery using the Address Resolution Protocol (ARP), and the method by which IP networks can be subdivided.

1.4.3 The transport protocols

Moving up the TCP/IP protocol stack, we will examine the two transport protocols supported by the TCP/IP protocol suite, the Transmission Control Protocol (TCP), and the User Datagram Protocol (UDP). In Chapter 4 we will examine the fields in the TCP and UDP headers, their data flow, and the tasks they perform.

1.4.4 DNS operations

The Domain Name System (DNS) provides the mechanism to convert host names into IP addresses used by communications equipment. In Chapter 5 we turn our attention to this topic. By understanding how the DNS operates, the different type of records that can be included on a server, and the use of the NSLOOKUP tool, we can many times understand the reason for host-IP address resolution problems and initiate appropriate actions to alleviate such problems.

1.4.5 Layer 2 management

Layer 2 represents the data delivery mechanism for transporting TCP/IP packets in LAN frames, and is not actually part of the protocol suite. However, it is important to understand layer 2 operations, as they effect the transmission and reception of TCP/IP packets. In Chapter 6 we turn our attention to this topic, examining in detail the operation and level of performance of different types of Ethernet and Token-Ring networks. To facilitate the layer 2 management process, we will also investigate the use of layer 2 monitoring tools.

1.4.6 Layer 3 and layer 4 management

Once again we will move up the protocol stack, examining tools and techniques to manage TCP/IP networks at layers 3 and 4 of the protocol stack. In Chapter 7 we will examine the use of two network monitoring and packet decoding programs to examine traffic at layers 3 and 4 of the protocol stack.

1.4.7 SNMP and RMON

In Chapter 8 we turn our attention to the structure, operation, and utilization of SNMP and RMON. In doing so we will first examine the SNMP protocol in detail as well as noting the differences between the first version of the protocol and SNMPv2 and SNMPv3. Once this is accomplished we will tour the MIB and examine the use of various object to determine the state of health of a TCP/IP network.

1.4.8 Management by utility program

Earlier in this chapter we noted the ability to use certain network utility programs to facilitate the state of a network. In Chapter 9 we turn our attention to this topic, examining the use of several application utility programs to determine the status of different devices in a network.

1.4.9 Security management

No book about TCP/IP management would be complete without discussing security. In Chapter 10, the concluding chapter of this book, we turn our attention to this topic. In this chapter we will examine network vulnerabilities as well as methods you can employ to reduce such vulnerabilities. Concerning the latter, we will examine the use of router access lists and the different functions that can be performed by firewalls.

2

THE TCP/IP PROTOCOL SUITE

This chapter represents the first of four that have been included to provide you with an in-depth examination of the operation of key areas of the TCP/IP protocol suite. In this chapter we will turn our attention to obtaining an appreciation of the TCP/IP protocol suite. In doing so we will briefly note the evolution of the protocol suite and then focus our attention on the International Organization for Standardization's (ISO) Open System Interconnection (OSI) Reference Model. We will use the layered structure of the OSI Reference Model as a foundation for examining the major components of the TCP/IP protocol suite to include how data flows from source to destination within a network as well as between networks.

2.1 EVOLUTION

The development of the TCP/IP protocol suite has its roots in a research project funded by the United States Department of Defense (DOD) as a mechanism to interconnect government agencies and academic research laboratories. The agency of the US DOD that funded this communications research project was the Advanced Research Projects Agency (ARPA), and the network developed as a result of the research effort was appropriately named ARPAnet.

During the development of ARPAnet, funds were provided to incorporate the evolving TCP/IP protocol suite into the University of California at Berkeley Unix operating system. This resulted in Berkeley Unix and TCP/IP becoming the standard operating system and transmission protocol of choice for use by many universities and government research agencies that required the use of engineering workstations for performing various research activities.

During the late 1960s through the mid 1970s, ARPAnet employed several networking technologies. Prior to the use of the TCP/IP protocol suite, a protocol referred to as the Network Control Program (NCP) was employed. Soon limitations of NCP resulted in its replacement by the Transmission

Control Program (TCP), which eventually formed the basis for the TCP/IP protocol suite. In fact, by 1983 all computers connected to the ARPAnet were restricted to using the TCP/IP protocol suite.

In 1983 the original ARPAnet was subdivided into two networks. One network, referred to as the Defense Data Network (DDN) and later as the military network (MILNET), was reserved for use by the US military. The second network, which consisted of the remainder of ARPAnet, was now referred to as ARPA Internet. In addition to those networks, other networks were formed by associations of colleges and universities and research associations that were also based upon the use of the TCP/IP protocol. These networks included the National Science Foundation network (NSFnet), New York State Educational Research Network (NYSERnet), California Educational Research Network (CERFnet), and the Southeastern University Research Association (SURAnet). Gradually, all of these networks were connected via the ARPAnet backbone to form a network of interconnected networks. By 1989, the original ARPAnet was taken out of service and the NSFnet, which provided a near mirror image of ARPAnet, became the backbone network.

Due to the growth in the popularity of the resulting Internet, the NSF realized that it could not continue to support its rapid expansion and commercialization. In place of the NSFnet, a series of Network Access Points (NAPs) were developed. NAPs were conceived as locations where companies that constructed their own networks could interconnect such networks via the concept of public peering. Today thousands of Internet Service Providers (ISPs) connect their networks to NAPs operated by approximately 20 companies, with NAPs considered to represent backbone networks that span major geographical areas.

2.2 GOVERNING BODIES

Although the Internet can be considered to represent a complex organization of thousands of interconnected networks, there are several governing bodies whose efforts permit it to operate and enable new technology to be applied to the TCP/IP protocol suite in an orderly manner. These bodies include the Internet Architecture Board (IAB), the Internet Assigned Number Authority (IANA), and the Internet Engineering Task Force (IETF).

2.2.1 The IAB

The TCP/IP protocol suite was originally governed by an organization known as the Internet Activities Board (IAB). In 1992 the Internet Activities Board was reorganized and renamed the Internet Architecture Board.

2.2.2 The IANA

The Internet Assigned Numbers Authority (IANA) is the central coordinator for the assignment of unique parameter values for Internet protocols. IANA is responsible for maintaining a registry of currently assigned values as well as responding to requests for the assignment of values to such parameters as protocols and ports. For example, the IANA maintains the assignment of TCP and UDP well-known ports.

2.2.3 The IETF

The Internet Engineering Task Force (IETF) consists of over 75 working groups that periodically develop memorandums referred to as Requests for Comments (RFCs).

2.2.4 RFCs

RFCs represent a mechanism by which technical standards governing the TCP/IP protocol suite evolve. Although the IETF is the major source of RFCs, they can also be submitted by individuals and companies. Once submitted, the proposed RFC becomes a draft document and is reviewed by the Internet Engineering Steering Group (IESG). The IESG consists of the chairperson of the IETF and other members of that group, and it performs an oversight and coordination function for the IETF. Once approved by the IESG, the RFC is sent to an RFC editor and assigned a number.

The time period between the submission of a draft RFC until its publication as a standard requires a minimum of 10 months to provide sufficient time for comments to be received and reviewed. Figure 2.1 illustrates the time track for the development of an Internet standard in the form of an RFC.

Once an RFC has been submitted and published, it does not change. Instead, RFCs are updated by new RFCs, resulting in the status of the earlier RFC becoming obsolete. There are currently over 3000 RFCs. You can retrieve RFCs from the Internet Network Information Center (InterNIC) at *http://ds.internic.net/rfc/rfc-index.txt*, as well as from numerous mirrored sites located throughout the world. By entering 'RFC' or a specific RFC number in a Web search engine, you can easily locate several sources for RFCs.

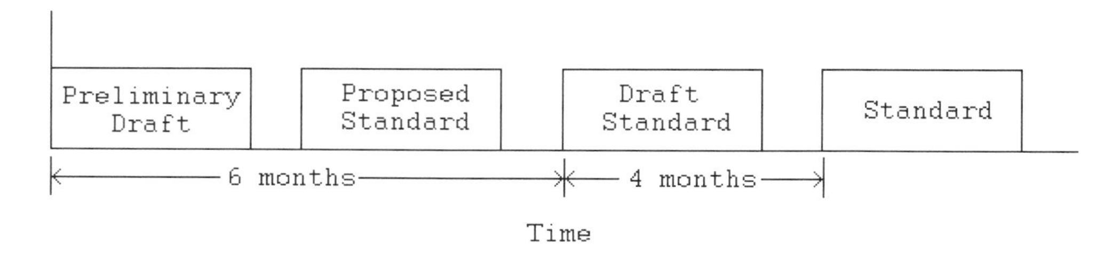

Figure 2.1 Internet standards time track

Now that we have an appreciation for the evolution of the Internet and the TCP/IP protocol suite, let us turn our attention to the structure of the protocol suite. However, since the TCP/IP protocol suite has a layered structure, we will first examine the ISO Reference Model and the subdivision of its second layer by the Institute of Electrical and Electronic Engineers (IEEE) to provide a standardized frame of reference.

2.3 THE ISO REFERENCE MODEL

The International Organization for Standardization is an agency of the United Nations headquartered in Geneva, Switzerland. The ISO is tasked with the development of worldwide standards to facilitate the international exchange of goods and services. The membership of the ISO consists of the national standards organization of most countries, with over 100 countries participating in its work. One of the most notable achievements of the ISO in the field of data communications was its development of the seven-layer Open Systems Interconnection (OSI) Reference Model. This model defines the communications process as a set of seven layers, with specific functions isolated and associated with each layer.

Figure 2.2 illustrates the seven layers of the ISO Reference Model. Each layer covers lower layer processes, effectively isolating them from higher layer functions. In this way, each layer performs a set of functions necessary to provide a set of services to the layer above it. Layer isolation permits the characteristics of a given layer to change without impacting the remainder of the model, provided that the supporting services remain the same. This layering was developed as a mechanism to enable users to mix and match OSI-conforming communications products to tailor their communications systems to satisfy a particular networking requirement. Although OSI-conforming communications products never gained a significant degree of acceptance, the OSI Reference Model provides a framework for comparing

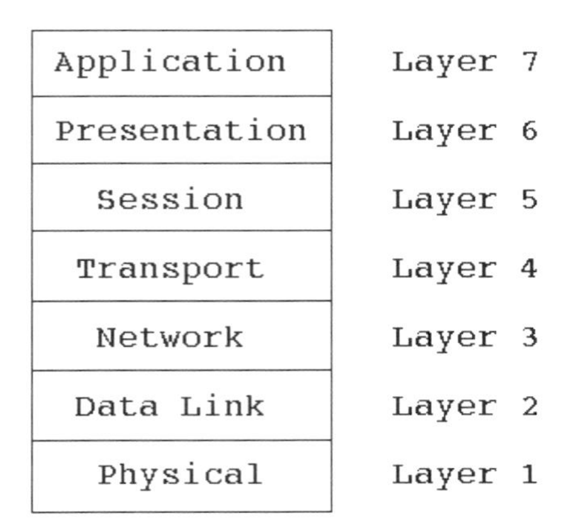

Figure 2.2 The International Organization for Standardization (ISO) Open System Interconnection (OSI) Reference Model

and contrasting the features and structure of other protocol suites. In addition, by understanding the structure of the model and the subdivision of its second layer by the IEEE, we can also obtain an appreciation of the capabilities and limitations of other protocol suites as well as the manner by which those suites support data flow from source to destination.

2.3.1 Layers of the OSI Reference Model

With the exception of layers 1 and 7, each layer in the ISO Reference Model is bounded by the layers above and below it. Layer 1, the physical layer, which is responsible for moving bits in electrical or optical form, can be considered to be bound below by the interconnecting medium over which transmission flows. In comparison, layer 7 is the upper layer and has no upper boundary. Within each layer is a group of functions that can be viewed as providing a set of defined services to the layer that bounds it from above, resulting in layer n using the services of layer n-1. To obtain an appreciation of the manner in which the ISO's Reference Model operates, let us turn our attention to each of the layers in the model.

Layer 1: the physical layer

At the lowest or most basic layer, the physical layer represents a set of rules that specifies the electrical, optical, and physical connection between devices and the transmission medium. Typically, the physical layer can include the coding method by which data is placed onto the medium as well as the cabling interface to include the operation of different pins on the cabling connection.

Layer 2: the data link layer

The data link layer defines how a device gains access to the medium specified by the physical layer as well as the data formats to include framing, error control procedures, and other link control activities. The data format specification includes procedures employed to correct transmission errors, thus, layer 2 becomes responsible for the reliable delivery of information.

At the data link layer information is grouped into entities referred to as frames. As a minimum, each frame contains control information that enables the receiver to synchronize itself to an incoming frame, addressing information that identifies the source and destination, a field containing the actual information being transmitted from source to destination, and a field used for verifying the integrity of the data.

One important characteristic of data link protocols is the fact that they do not have network addresses and as such are non-routable. As we will note later in this chapter, Ethernet, Token-Ring, and FDDI represent examples of data link protocols.

Because the development of OSI layers was originally targeted towards wide area networking, its applicability to local area networks required a degree of modification. Under IEEE 802 standards, the data link layer was subdivided into two sublayers: Logical Link Control (LLC) and Media Access Control (MAC). The LLC layer is responsible for generating and interpreting commands that control the flow of data and perform recover operations in the event of errors. In comparison, the MAC layer is responsible for providing access to the local area network, which enables a station on the network to transmit information. Later in this chapter we will discuss the subdivision in additional detail.

Layer 3: the network layer

The third layer in the ISO Reference Model is the network layer. As its name implies, this layer is responsible for arranging a logical connection through a network to include the selection and management of a route for the flow of information between source and destination based upon the available paths in a network. Services provided by this layer are associated with the movement of data packets through a network, including addressing, routing, switching, sequencing, and flow control procedures. In a complex network, the source and destination may not be directly connected by a single path, but instead require a path to be established that consists of many subpaths. Thus, routing of data through the network onto the correct paths is an important feature of this layer.

Several protocols represent commonly used layer 3 protocols. Those protocols include the X.25 packet protocol, which governs the flow of information within a packet network, Novell's Internet Packet Exchange (IPX), and the Internet Protocol (IP).

Layer 4: the transport layer

The fourth layer in the ISO's Reference Model is the transport layer. This layer is responsible for guaranteeing that the transfer of information occurs correctly after a route has been established by the network layer protocol. Thus, the primary function of this layer is to control the communications session between nodes once a path has been established by the network control layer. Error control, sequence checking, and other end-to-end data reliability factors are the primary concern of this layer. In addition, to support the transfer of different types of data between source and destination, this layer is also responsible for multiplexing and de-multiplexing data streams between upper layer application processes.

Although most transport layer protocols provide an end-to-end reliability mechanism, this is an optional feature associated with this layer. Similarly, although most transport layer protocols are connection-oriented, requiring the destination to acknowledge its ability to receive data prior to a transmission session being established, this is also an optional feature.

Instead of operating as a connection-oriented protocol, a transport layer protocol can operate on what is referred to as a best-effort basis. This means that the protocol will initiate transmission without knowing if the destination is ready to receive data or even if it is powered on and operational. Although this method of operation may appear awkward, the originator will set a timer that decrements in value. If no response is received to the initial packet flow by the time the timer expires, the originator will assume that the destination is not reachable and terminate the session. The use of a connectionless protocol avoids the relatively long handshaking process associated with some connection-oriented transport layer protocols. Examples of transport layer protocols include Novell's Sequenced Packet Exchange (SPX) as well as the Transmission Control Protocol (TCP) and the User Datagram Protocol (UDP). TCP is a connection-oriented, error-free delivery protocol. In comparison, UDP is a connectionless, best effort protocol.

Layer 5: the session layer

The fifth layer in the OSI Reference Model is the session layer. This layer provides a set of rules for establishing and terminating data streams between nodes in a network. The services that the session layer can provide include establishing and terminating node connections, flow control, dialogue control, and end-to-end data control.

Layer 6: the presentation layer

The sixth layer in the ISO's OSI Reference Model is the presentation layer. This layer is primarily responsible for formatting, data transformation, and syntax-related operations. One of the primary functions of this layer that is both visible and probably overlooked as we take it for granted is the conversion of transmitted data at the receiver into a display format for a receiving device. Concerning the receiving device, different presentation layers reside on different devices, since the manner in which data is displayed on a PC would more than likely differ from the manner in which data is displayed on a dumb terminal. Other functions that can be performed by the presentation layer include encryption/decryption and compression/decompression.

Layer 7: the application layer

The seventh and top layer of the OSI Reference Model is the application layer. This layer can be viewed as functioning as a window through which the application gains access to all of the services provided by the seven-layer model. Examples of functions that can be performed at the application layer include file transfer, electronic mail transmission, and remote terminal access.

While the first four layers in the Reference Model are fairly well defined, the functions associated with the upper three layers can vary considerably, based upon the application, the type of data transported, and the manner in which the attributes of the display of a device are used for the presentation of information. As we will note later in this chapter, such popular Internet protocols as the File Transfer Protocol (FTP), Telnet, and the HyperText Transport Protocol (HTTP) represent a blend of layer 5 through layer 7 functions.

2.3.2 Data flow

As data flows within an ISO network each layer appends appropriate heading information to frames of information flowing within the network while removing the heading information added by a lower layer. In this manner, layer *n* interacts with layer *n*-1 as data flows through an ISO network.

Figure 2.3 illustrates the appending and removal of frame header information as data flows through a network constructed according to the ISO Reference Model. Since each higher level removes the header appended by a lower level, the frame traversing the network arrives in its original form at its destination.

2.3.3 Layer subdivision

Prior to examining the major components of the TCP/IP protocol suite, a discussion of layer subdivision resulting from the efforts of the Institute of Electrical and Electronic Engineers (IEEE) is in order. The IEEE is responsible for developing LAN standards in the USA, and its efforts are commonly incorporated by the American National Standards Institute (ANSI) into US standards, either as is or with slight modification.

During the early development of LAN standards, the IEEE recognized that it would be desirable to subdivide the data link layer. The result of this subdivision was the creation of Logical Link Control (LLC) and Media Access Control (MAC) sublayers. The MAC sublayer, which resides at the bottom of the portion of the data link layer that was subdivided, defines the manner by which a station gains access to a LAN. Examples of MAC methods include Ethernet's Carrier Sense Multiple Access/Collision Detection (CSMA/CD) scheme and Token-Ring's free token acquisition method. Above the MAC layer, which differs for each type of LAN, is the LLC layer. The LLC layer, which is common for each IEEE network, is used for controlling the establishment, maintenance, and termination of logical connections between stations on a network.

Addressing

Access to an IEEE network is accomplished through the MAC layer. Frames placed on an IEEE network include two address-related fields: destination and source address. Each address normally represents a 6-byte address burnt into read-only memory (ROM) on the network adapter card of the frame

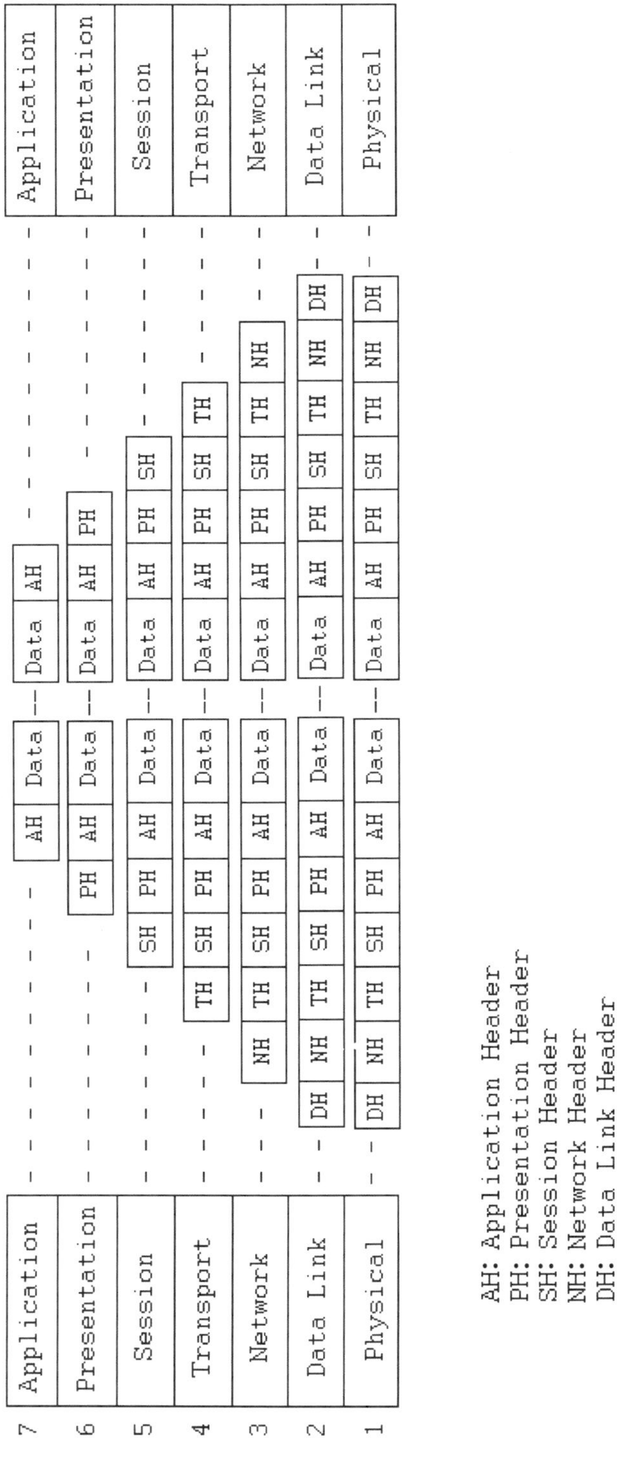

Layer		Sending	Receiving	
7	Application	AH Data	Data AH	Application
6	Presentation	PH AH Data	Data AH PH	Presentation
5	Session	SH PH AH Data	Data AH PH SH	Session
4	Transport	TH SH PH AH Data	Data AH PH SH TH	Transport
3	Network	NH TH SH PH AH Data	Data AH PH SH TH NH	Network
2	Data Link	DH NH TH SH PH AH Data	Data AH PH SH TH NH DH	Data Link
1	Physical	DH NH TH SH PH AH Data	Data AH PH SH TH NH DH	Physical

AH: Application Header
PH: Presentation Header
SH: Session Header
NH: Network Header
DH: Data Link Header

Figure 2.3 Data flow within an ISO Reference Model network

originator (source address) and the frame recipient (destination address). The first three bytes of the 6-byte network adapter card address are assigned by the IEEE to an adapter card manufacturer, and represent the manufacturer identification (ID) portion of the address. The next three bytes are used by the adapter card manufacturer to uniquely identify each adapter card that it manufactures. If the manufacturer is so successful that it runs out of its allocated 3-byte sequence of numbers, it will request another manufacturer ID from the IEEE and use that ID for producing a new series of network adapter cards.

Figure 2.4 illustrates the general format of an IEEE Mac address. When used as a source address, a bit composition of all binary 1s represents a broadcast address and results in each station copying the contents of the frame off the network. Depending upon the type of LAN, the setting of different bits within the 6-byte source MAC address can be used to identify different groups. Then, each workstation associated with the group identifier would copy the frame off the network. If the frame's destination address is neither a broadcast nor a group address, it will only be copied off the network by the station whose adapter address matches the destination address in the frame.

Universally vs. locally administered addresses

Two types of addresses can be associated with stations on an IEEE network: universally administered and locally administered. When a burnt-in ROM address is used, it is referred to as a universally administered address, as it is uniquely assigned by the IEEE. In comparison, a second type of address results from the effort of a LAN administrator or network manager to override the universally administered address. This second type of MAC address results from the creation of a batch file statement being used to set a locally generated address that overrides the burnt-in ROM address. Because this address is developed locally, it is referred to as a locally administered address. Note that, regardless of the type of MAC address, it is a layer 2 address that is 48 bits in length. Because TCP/IP addresses are 32 bits in length (IPv4) and represent both a network address and a host address on a network, a translation process is required to associate a layer 3 IP address to a layer 2 MAC address. Later in this book we will examine the address resolution process that performs the required translation.

2.4 THE TCP/IP PROTOCOL SUITE

In the previous section we have an overview of the functions of the seven layers in the ISO Reference Model to provide a frame of reference when examining the TCP/IP protocol suite. In actuality, TCP/IP represents one of the earliest developed layered protocol suites and preceded the development of the ISO's OSI Reference Model by approximately 20 years. Although it predates the OSI Reference Model, we can obtain an appreciation of the protocol suite by comparing it with that model.

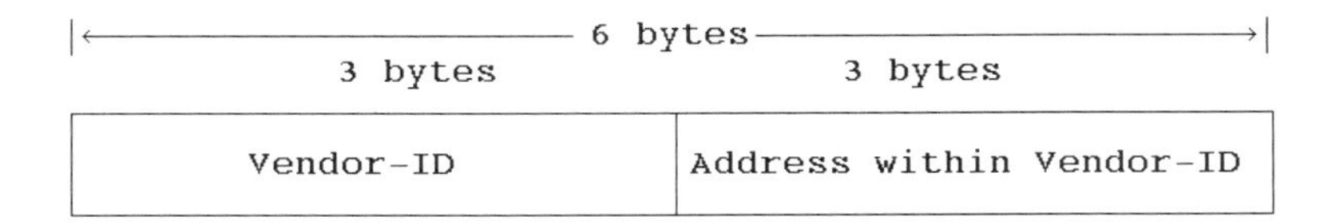

Figure 2.4 The IEEE MAC address format

2.4.1 Comparison with the ISO Reference Model

Similar to the ISO Reference Model, the TCP/IP protocol suite is subdivided into distinct layers, commencing at the network layer. Although the protocol suite does not include equivalents to the lower two layers of the ISO Reference Model, it does provide a mechanism to translate addressing from the network layer of the reference model to MAC addresses used by LANs at the lower portion of the data link layer. This enables the TCP/IP protocol suite to use the physical layer supported by different LANs.

A second key difference between the ISO Reference Model and the TCP/IP protocol suite occurs at the top of the suite. TCP/IP applications can be considered to represent the equivalent of layers 5 through 7 of the OSI Reference Model. Based upon the preceding, Figure 2.5 provides a general comparison of the TCP/IP protocol suite with the ISO Reference Model. Note that, as previously mentioned, the TCP/IP protocol suite commences at the equivalent of layer 3 of the ISO Reference Model. Thus, the dashed lines surrounding Ethernet, Token-Ring, and FDDI layer 2 protocols and their physical layers indicate that they are not actually part of the TCP/IP protocol suite. Instead, the Address Resolution Protocol (ARP), which can be viewed as a facility of the Internet Protocol (IP), provides the translation mechanism that enables IP addressed packets to be correctly delivered to workstations that use MAC addresses. In fact, the TCP/IP protocol suite can also run over ATM, with a special type of address resolution used to resolve IP to ATM addresses. Thus, address resolution enables the TCP/IP protocol suite to be transported by other protocols and use the physical layer specified by those protocols.

Now that we have an appreciation for the general relationship between the TCP/IP protocol stack and the ISO's Open System Interconnection Reference Model, let's turn our attention to the actual layers of the protocol suite.

The network layer

The Internet Protocol (IP) represents the network layer protocol employed by the TCP/IP protocol suite. IP packets are formed by the addition of an IP header to the layer 4 protocol data entity, which is either the Transport Control Protocol (TCP) or the User Datagram Protocol (UDP).

IP headers contain 32-bit source and destination addresses that are normally subdivided to denote a network address and host address on the

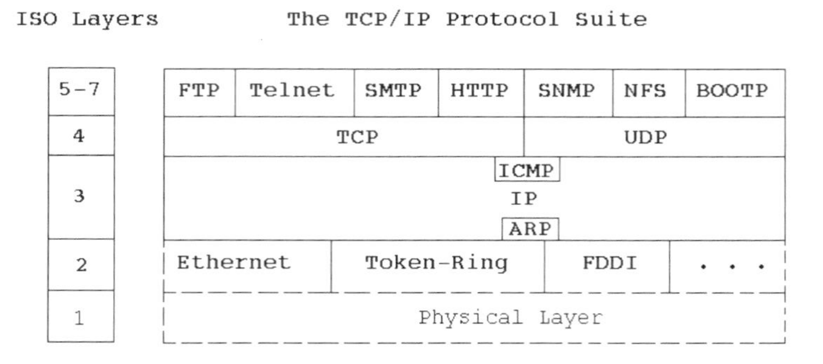

ARP: Address Resolution Protocol
BOOTP: Bootstrap Protocol
FTP: File Transfer Protocol
HTTP: HyperText Transmission Protocol
SNMP: Simple Network Management Protocol
ICMP: Internet Control Message Protocol

Figure 2.5 Comparing the TCP/IP Protocol Suite with the ISO Reference Model

network. In actuality, the host address is really an interface on the network, since a host can have multiple interfaces, with each having a distinct address. However, over the years the terms host address and interface address have been used synonymously—although this is not technically correct. In Chapter 3 we will examine the IP header in detail.

ICMP

The Internet Control Message Protocol (ICMP) represents a diagnostic testing and error reporting mechanism that enables devices to generate various types of status and error reporting messages. Two of the more popularly employed ICMP messages are the Echo Request and Echo Response packets generated by the Ping application.

Although Figure 2.5 indicates that ICMP is a layer 3 protocol, from a technical perspective an ICMP message is formed by the addition of an IP header to an ICMP message with the Type field within the IP header set to indicate it is transporting an ICMP message. When we examine IP in Chapter 3, we will also turn our attention to the Internet Message Protocol.

The transport layer

The designers of the TCP/IP protocol suite recognized that two different types of data delivery transport protocols would be required. This resulted in two transport protocols supported by the protocol suite.

TCP TCP is a reliable, connection-oriented protocol used to transport applications that require reliable delivery and for which actual data should not be exchanged until a session is established. From Figure 2.5 you will note that FTP, Telnet, SMTP, and HTTP are transported by TCP.

Because TCP is a connection-oriented protocol, this means that actual data will not be transferred until a connection is established. While this makes sense when you are transmitting a file or Web pages, it also delays actual data transfer.

UDP A second transport protocol supported by the TCP/IP protocol suite is UDP. UDP represents a connectionless protocol that operates on a best effort basis. This means that instead of waiting for confirmation that a destination is available, UDP will commence actual data transfer, leaving it to the application to determine if a response was received. Examples of applications that use UDP include SNMP, NFS, and BOOTP.

The use of UDP and TCP results in the prefix of an appropriate header to application data. When TCP is used as the transport layer protocol, the TCP header and application data are referred to as a TCP segment. When UDP is used as the transport layer protocol, the UDP header and application data transported by UDP is referred to as a UDP datagram.

Port numbers Because TCP and UDP were designed to transport multiple types of application data between a source and the same or different destinations, a mechanism was needed to distinguish one type of application from another. This mechanism is obtained by port number fields contained in TCP and UDP headers and explains how a Web server can also support FTP and other applications. In Chapter 4 we will turn our attention to the composition of TCP/IP transport protocol headers and the use of different port numbers.

2.4.3 Application data delivery

In concluding this chapter we will examine the use of TCP/IP and LAN headers to facilitate the delivery of application data from a host on one

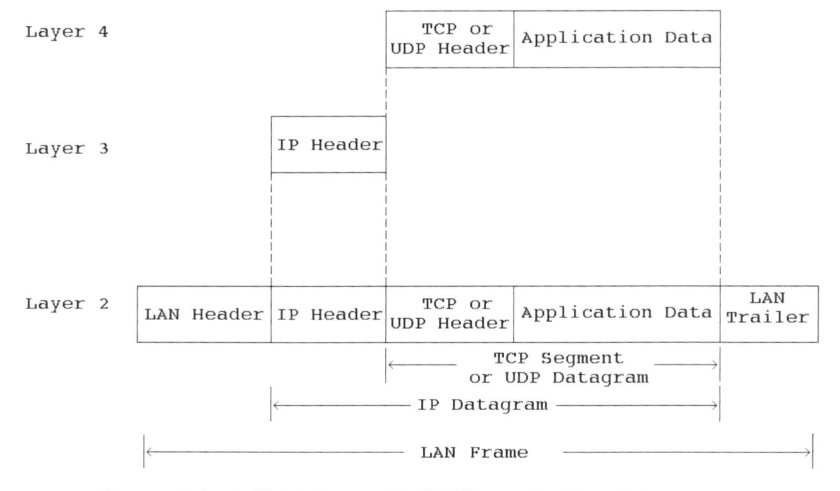

Figure 2.6 LAN delivery of TCP/IP application data

network to a host on another network. Figure 2.6 illustrates the manner by which a LAN frame containing TCP/IP application data is formed. The LAN frame header uses a MAC destination address to direct the frame to a router. The router removes the LAN header and trailer and uses a Wide Area Network (WAN) protocol to transport the IP datagram. At the destination network another router receives the inbound packet, removes the WAN header and trailer, and encapsulates the IP datagram into a LAN frame for delivery to the appropriate IP address. However, since LAN frames use MAC addresses while TCP/IP applications use IP addresses, the router will either check its memory to determine if it previously discovered the MAC address associated with the destination IP address or use the Address Resolution Protocol (ARP) to discover the MAC address. Once the destination MAC address is known, the router can complete the formation of the LAN frame and transmit it onto the network for delivery to the appropriate device.

3

THE INTERNET PROTOCOL

In this chapter we continue to acquire a foundation of knowledge concerning the TCP/IP protocol suite by focusing attention upon the network layer in the suite. The Internet Protocol (IP) represents both the network layer protocol in the TCP/IP protocol suite as well as the data delivery mechanism that enables packets to be routed from source to destination.

We will first examine the composition of the fields within the IP header. This will include a detailed examination of IP addressing, since many network-related problems can be traced to this area. Because Internet Control Message Protocol (ICMP) messages are transported via IP, we will also examine the ICMP in this chapter. Once this has been accomplished, we will conclude this chapter by turning our attention to the evolving replacement of the present version of the IP, IPv4. That replacement is IPv6, which is sometimes referred to as the next generation IP or IPng.

3.1 THE IPv4 HEADER

As noted above, the current version of the IP is version 4. Therefore, we will commence our examination of the network layer of the TCP/IP protocol suite by turning our attention to the IPv4 header.

The fields in the IPv4 header are illustrated in Figure 3.1. In examining that illustration note that the header contains a minimum of 20 octets of data. Also note that the width of each field is shown in Figure 3.1 with respect to a 32-bit word.

In this chapter and succeeding chapters we will use the term octet to reference the width of different header fields. The term octet was employed by standards organizations to explicitly reference 8 bits operated upon as an entity at a time when computers were manufactured with different numbers of bits per byte. To alleviate potential confusion when referencing a group of 8 bits, standards organizations turned to the term octet. Today essentially all computers use 8-bit bytes, and the terms byte and octet are commonly used synonymously. To obtain an appreciation for the functions performed by the

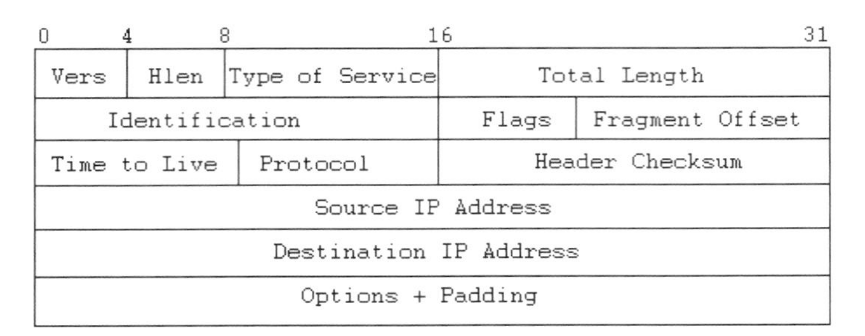

Figure 3.1 The IPv4 header

IPv4 header, let us turn our attention to reviewing the functions of each of the fields in the header.

3.1.1 Vers field

The Vers field consists of four bits that identify the version of the IP used to create the datagram. The current version of the IP is 4 and the next generation of the IP is assigned version number 6. As we will note later in this chapter, the Vers field retains its meaning in both the IPv4 and IPv6 headers.

3.1.2 Hlen and Total Length fields

The second and fourth fields in the IPv4 header indicate the length of the header and the total length of the datagram, respectively. The Hlen field indicates the length of the IPv4 header in 32-bit words. In comparison, the Total Length field indicates the total length of the datagram to include its header and higher layer information, such as a following TCP or UDP header and application data following either of those headers. Because the Total Length field consists of 16 bits, an IP datagram can be up to 2^{16}, or 65 535 octets in length.

3.1.3 Type of Service field

The Type of Service (TOS) field is 1 octet or 8 bits in length. The purpose of this field is to denote the importance of the datagram (precedence), delay, throughput, and reliability requested by the originator.

Figure 3.2 illustrates the assignment of bit positions within the TOS field. Because the TOS field provides a mechanism to define priorities for the routing of IP datagrams, it would appear that the TOS field could be used to provide a Quality of Service (QoS) for IP. Applications can set the appropriate values in the TOS field to indicate the type of routing path they would like. For example, a file transfer would probably request normal delay, high throughput, and normal reliability. In comparison, a real time video

0 1 2	3	4	5	6 7
PRECEDENCE	D	T	R	UNUSED

bit positions

```
Bits 0–2: Defines precedence
                              111: Network control
                              110: Internetwork control
                              101: Critic/ECP
                              100: Flash override
                              011: Flash
                              010: Immediate
                              001: Priority
                              000: Routine
Bit 3 Delay:                  0 = normal, 1 = low delay
Bit 4 Throughput:             0 = normal, 1 = high
Bit 5 Reliability:            0 = normal, 1 = high
Bits 6 and 7:                 Reserved for future use (set to 0)
```

Figure 3.2 The Type of Service field

application would probably select low delay, high throughput, and high reliability. While this concept appears to provide a QoS, this is not the case, as it does not provide a mechanism to reserve bandwidth. For example, 10 stations, each requiring 512 Kbps, could all define an immediate priority for flowing through a router connected on a T1 circuit operating at 1.544 Mbps. Another problem associated with the TOS field is the fact that many routers ignore its settings. This is due to the fact that, to support the TOS field, a router would have to construct and maintain multiple routing tables, which in the era of relatively slow processors when the Internet evolved was not an attractive option with router manufacturers. Thus, although this field provides a precedence definition capability, its use on a public network can be limited. Recognizing this limitation, plans were being developed to reuse the TOS field as a mechanism to differentiate services requested when a data stream enters a network. This action resulted in a proposal to rename the TOS byte as a Diff Service field, and an RFC was being developed to define its use when this book was written.

3.1.4 Identification field

The Identification field is two octets or 16 bits in length. This field is used to identify each fragmented datagram and is one of three fields that govern fragmentation. The other two fields that govern fragmentation are the Flags field and the Fragmentation Offset field.

IP fragmentation results when data flow between networks encounters different size maximum transmission units (MTUs). The MTU is commonly set when a device driver initializes an interface and represents the payload portion of a frame, i.e., the frame length less frame overhead. Most protocol stacks support MTUs up to 64K–1 octets (65 535). Another MTU is a per route MTU, which represents the MTU that can be used without causing fragmentation from source to destination. Per route MTUs are usually

Table 3.1 Flag field bit values

Bit 0: Reserved (set to 0)
Bit 1: 0 = may fragment, 1 = don't fragment
Bit 2: 0 = last fragment, 1 = more fragment(s) follow

maintained as a value in a host's routing table and are set either by manual configuration or via a discovery process. When a route has interfaces with different MTUs and a large datagram must be transferred via an interface with a smaller MTU, the routing entity will either fragment the packet or drop it. As we will note in the next section, if the DON'T_FRAGMENT bit is set in the flag field the router will drop the datagram. This will result in the router generating an ICMP 'Destination Unreachable–Fragmentation Needed' message to the originator, which will cause the MTU discovery algorithm to select a smaller MTU for the path and subsequent transmissions.

3.1.5 Flags field

This 3-bit field indicates how fragmentation will occur. Bit 0 is reserved and set to zero, while the values of bits 1 and 2 define whether or not fragmentation can occur and if the present fragment is the last fragment or if one or more fragments follow. Table 3.1 lists the values associated with the three bits in the Flags field.

3.1.6 Fragment Offset field

The third field in the IPv4 header that is involved with fragmentation is the Fragment Offset field. This field is 13 bits in length and indicates where the fragment belongs in the complete message. The actual value placed in this field is an integer that corresponds to a unit of 8 octets and provides an offset in 64-bit units.

IP fragmentation places the burden of effort upon the receiving station and the routing entity. When a station receives an IP fragment, it must fully reassemble the complete IP datagram prior to being able to extract the TCP segment, resulting in a requirement for additional buffer memory and CPU processing power at the receiver. In doing so it use the values in the Fragment Offset field in each datagram fragment to correctly reassemble the complete datagram. Because the dropping of any fragment in the original datagram requires the original datagram to be present, most vendor TCP/IP protocol stacks set the DON'T_FRAGMENT bit in the Flag field. As mentioned above, setting that bit causes oversized IP datagrams to be dropped and results in an ICMP 'Destination Unreachable–Fragmentation Needed' message transmitted to the originator. This action results in the MTU discovery algorithm selecting a smaller MTU for the path and using that MTU for subsequent transmissions.

3.1.7 Time-to-Live field

The Time-to-Live (TTL) field is one octet in length. This field contains a value that represents the maximum amount of time a datagram can live. The use of this field prevents a mis-addressed or mis-routed datagram from endlessly wandering the Internet or a private IP network.

The value placed in the TTL field can represent router hops or seconds, with a maximum value for either being 255. Because an exact time is difficult to measure and requires synchronized clocks, this field is primarily used as a hop count field. That is, routers decrement the value in the field each time a datagram flows between networks. When the value of this field reaches zero, the datagram is sent to the great bit bucket in the sky. The current recommended default time-to-live value for IP is 64.

3.1.8 Protocol field

The purpose of the Protocol field is to identify the higher layer protocol being transported within an IP datagram. By examining the value of this field, networking devices can determine if they have to look further into the datagram or should simply forward the datagram towards its destination. For example, a router that receives an IP datagram whose Protocol field value is 6 and which indicates the higher layer protocol is TCP would simply forward the datagram towards its destination.

The 8-bit positions in the Protocol field enables up to 256 protocols to be uniquely defined. Table 3.2 lists the current assignment of Internet protocol numbers. Although TCP and UDP by far represent the vast majority of upper layer protocol transmission, other protocols can also be transported that govern the operation of networks, such as the Exterior Gateway Protocol (EGP) and Interior Gateway Protocol (IGP) that govern the interconnection of autonomous networks. In examining the entries in Table 3.2 note that a large block of numbers are currently unassigned. Also note that the evolving IPv6 uses a Next Header field in place of the Protocol field but uses the values contained in the table.

3.1.9 Checksum field

The tenth field in the IPv4 header is the Checksum field. This 16-bit or 2-octet field protects the header and is also referred to as the Header Checksum field.

3.1.10 Source and Destination address fields

Both the Source and Destination address fields are 32 bits in length. Each field contains an address that normally represents both a network address and a host address on the network. Because it is extremely important to understand IP addressing, this topic will be covered in detail in Section 3.2 below.

Table 3.2 Assigned Internet Protocol numbers

Decimal	Keyword	Protocol
0	HOPOPT	IPv6 Hop-by-Hop Option
1	ICMP	Internet Control Message
2	IGMP	Internet Group Management
3	GGP	Gateway-to-Gateway
4	IP	IP in IP (encapsulation)
5	ST	Stream
6	TCP	Transmission Control Protocol
7	CBT	CBT
8	EGP	Exterior Gateway Protocol
9	IGP	any private interior gateway (used by Cisco for their IGRP)
10	BBN-RCC-MON	BBN RCC Monitoring
11	NVP-II	Network Voice Protocol Version 2
12	PUP	PUP
13	ARGUS	ARGUS
14	EMCON	EMCON
15	XNET	Cross Net Debugger
16	CHAOS	Chaos
17	UDP	User Datagram
18	MUX	Multiplexing
19	DCN-MEAS	DCN Measurement Subsystems
20	HMP	Host Monitoring
21	PRM	Packet Radio Measurement
22	XNS-IDP	XEROX NS IDP
23	TRUNK-1	Trunk-1
24	TRUNK-2	Trunk-2
25	LEAF-1	Leaf-1
26	LEAF-2	Leaf-2
27	RDP	Reliable Data Protocol
28	IRTP	Internet Reliable Transaction
29	ISO-TP4	ISO Transport Protocol class 4
30	NETBLT	Bulk Data Transfer Protocol
31	MFE-NSP	MFE Network Services Protocol
32	MERIT-INP	MERIT Internodal Protocol
33	SEP	Sequential Exchange Protocol
34	3PC	Third Party Connect Protocol
35	IDPR	Inter-Domain Policy Routing Protocol
36	XTP	XTP
37	DDP	Datagram Delivery Protocol
38	IDPR-CMTP	IDPR Control Message Transport Protocol
39	TP++	TP++ Transport Protocol
40	IL	IL Transport Protocol
41	IPv6	Ipv6
42	SDRP	Source Demand Routing Protocol
43	IPv6-Route	Routing Header for IPv6
44	IPv6-Frag	Fragment Header for IPv6
45	IDRP	Inter-Domain Routing Protocol
46	RSVP	Reservation Protocol

(*Continued*)

Table 3.2 Assigned Internet Protocol numbers (*Continued*)

Decimal	Keyword	Protocol
47	GRE	General Routing Encapsulation
48	MHRP	Mobile Host Routing Protocol
49	BNA	BNA
50	ESP	Encap Security Payload for IPv6
51	AH	Authentication Header for IPv6
52	I-NLSP	Integrated Net Layer Security
53	SWIPE	IP with Encryption
54	NARP	NBMA Address Resolution Protocol
55	MOBILE	IP Mobility
56	TLSP	Transport Layer Security Protocol (using Kryptonet key management)
57	SKIP	SKIP
58	IPv6-ICMP	ICMP for IPv6
59	IPv6-NoNxt	No Next Header for IPv6
60	IPv6-Opts	Destination Options for IPv6
61		any host internal protocol
62	CFTP	CFTP
63		any local network
64	SAT-EXPAK	SATNET and Backroom EXPAK
65	KRYPTOLAN	Kryptolan
66	RVD	MIT Remote Virtual Disk Protocol
67	IPPC	Internet Pluribus Packet Core
68		any distributed file system
69	SAT-MON	SATNET Monitoring
70	VISA	VISA Protocol
71	IPCV	Internet Packet Core Utility
72	CPNX	Computer Protocol Network Executive
73	CPHB	computer Protocol Heart Beat
74	WSN	Wang Span Network
75	PVP	Packet Video Protocol
76	BR-SAT-MON	Backroom SATNET Monitoring
77	SUN-ND	SUN ND PROTOCOL–Temporary
78	WB-MON	WIDEBAND Monitoring
79	WB-EXPAK	WIDEBAND EXPAK
80	ISO-IP	ISO Internet Protocol
81	VMTP	VMTP
82	SECURE-VMTP	SECURE-VMPT
83	VINES	VINES
84	TTP	TTP
85	NSFNET-IGP	NSFNET-IGP
86	DGP	Dissimilar Gateway Protocol
87	TCF	TCF
88	EIGRP	EIGRP
89	OSPFIGP	OSPFIGP
90	Sprite-RPC	Sprite RPC Protocol
91	LARP	Locus Address Resolution Protocol
92	MTP	Multicast Transport Protocol

(*Continued*)

Table 3.2 Assigned Internet Protocol numbers (*Continued*)

Decimal	Keyword	Protocol
93	AX.25	AX.25 Frames
94	IPIP	IP-within-IP Encapsulation Protocol
95	MICP	Mobile Internetworking Control Protocol
96	SCC-SP	Semaphore Communications Sec. Protocol
97	ETHERIP	Ethernet-within-IP Encapsulation
98	ENCAP	Encapsulation Header
99		any private encryption scheme
100	GMTP	GMTP
101	IFMP	Ipsilon Flow Management Protocol
102	PNNI	PNNI over IP
103	PIM	Protocol Independent Multicast
104	ARIS	ARIS
105	SCPS	SCPS
106	QNX	QNX
107	A/N	Active Networks
108	IPPCP	IP Payload Compression Protocol
109	SNP	Sitara Networks Protocol
110	Compaq-Peer	Compaq Peer Protocol
111	IPX-in-IP	IPX in IP
112	VRRP	Virtual Router Redundancy Protocol
113	PGM	PGM Reliable Transport protocol
114		any 0-hop protocol
115	L2TP	Layer 2 Tunneling Protocol
116	DDX	D-II Data Exchange (DDX)
117–254		Unassigned
255		Reserved

3.1.11 Options and Padding fields

The IP includes a provision for adding optional header fields. Such fields are identified by a value greater than zero in the field. Table 3.3 indicates IP Option field values based upon the manner by which the Option field is subdivided. That subdivision includes a 1-bit copy flag, a 2-bit class field, and a 5-bit option number. The value column in Table 3.3 indicates the value of the 8-bit field. IP options are commonly referred to by this value.

Options whose values are 0 and 1 are exactly 1 octet long, which is their Type field. All other options have their 1-octet Type field followed by a 1-octet length field followed by one or more octets of option data. The optional padding occurs when it becomes necessary to expand the header to fall on a 32-bit word boundary.

3.2 IP ADDRESSING

IP addressing provides the mechanism that enables packets to be routed between networks as well as to be delivered to an appropriate host on a

Table 3.3 IP Option field values

Copy	Class	Number	Value	Name	
0	0	0	0	EOOL:	End of Options List
0	0	1	1	NOP:	No Operation
1	0	2	130	SEC:	Security
1	0	3	131	LSR:	Loose Source Route
0	2	4	68	TS:	Time Stamp
1	0	5	133	E-SEC:	Extended Security
1	0	6	134	CIPSO:	Commercial Security
0	0	7	7	RR:	Record Route
1	0	8	136	SID:	Stream ID
1	0	9	137	SSR:	Strict Source Route
0	0	10	10	ZSU:	Experimental Measurement
0	0	11	11	MTUP:	MTU Probe
0	0	12	12	TRUR:	MTU Reply
1	2	13	205	FINN:	Experimental Flow Control
1	0	14	142	VISA:	Experimental Access Control
0	0	15	15	ENCODE	
1	0	16	144	IMITD:	IMI Traffic Descriptor
1	0	17	145	EIP:	
0	2	18	82	TR:	Traceroute
1	0	19	147	ADDEXT:	Address Extension
1	0	20	148	RTRALT:	Router Alert
1	0	21	149	SDB:	Selective Directed Broadcast
1	0	22	150	NSAPA:	NSAP Addresses

destination network. As noted earlier in this chapter, there are two versions of the IP. The current version in IPv4, while the next generation IP, which is currently operated on an experimental portion of the Internet, is IPv6. Because there are significant differences in the method of addressing used by each version of IP, we will cover both versions in this section. First, we will focus our attention upon the 32-bit addressing scheme employed under IPv4. Once we have obtained an appreciation for the manner by which IPv4 addresses are formed and used, we will then turn our attention to IPv6.

3.2.1 Overview

IPv4 uses 32-bit IP addresses to identify distinct device interfaces, such as interfaces that connect routers, workstations, and gateways to networks, as well as to route data to those devices. Each device interface in an IP network must be assigned a unique IP address to enable it to receive communications addressed to the interface. Normally workstations have a single interface in the form of a LAN connection, which would be assigned an IP address. However, routers typically have more than one interface, and some high performance servers may have two network connections. In such instances each device interface would have a separate IP address.

As previously noted, each IPv4 header contains a 32-bit source and 32-bit destination address. The use of a 32-bit binary number provides an address pace that is capable of supporting 2 294 967 296 distinct addressable devices. When the TCP/IP protocol suite was developed, the address space exceeded the world's population! However, the proliferation of workstations connected to LANs, the projected growth in the use of cable modems that require individual IP addresses, and the fact that every interface on an IP network must have a distinct IP address contributed to a rapid depletion of available IP addresses. Recognizing the possibility that hundreds of millions of to-be-developed communications appliances such as personal digital assistants may need access to the Internet, the Internet Activities Board (IAB) commenced work on a replacement for IPv4 during 1992. While the addressing limitations of IPv4 was of primary concern, the efforts of the IAB resulted in a complete revamp of the next generation version of IP known as IPv6. Improvements in IPv6 include the use of 128-bit addresses for source and destination devices, as well as a next header indicator that facilitates the daisy chaining of headers to perform different functions. The latter enables intermediary devices along a path to note whether or not they have to look further into the contents of a datagram or simply relay the datagram towards its destination. The ability to avoid looking further into a datagram considerably enhances the packet processing capability of routers, and can be expected to enhance the flow of data through the Internet. Because this section is focused upon IP addressing, we will examine the addressing schemes, address notation, host address restrictions, and special addresses associated with both IPv4 and IPv6 in this section.

3.2.2 IPv4

The IP was standardized in September, 1981. At that time the standard included a requirement for each host connected to an IP-based network to be assigned a unique, 32-bit address value for each network connection. This resulted in some networking devices that have multiple interfaces, such as routers, gateways, and servers, being assigned a unique IP address for each network interface.

To illustrate the assignment of IP addresses to interfaces, consider Figure 3.3, which shows two bus-based Ethernet LANs connected by a pair of routers. Note that each workstation and server on each LAN has a single interface in this example. Thus, each of those devices would be associated with a single IP address by the assignment of one IP address to each device interface. Each router has two interfaces, one being connected to a LAN and the second representing a serial port used to interconnect routers via a WAN. Thus, each router would be assigned two IP addresses, one being assigned to its LAN interface and the other to its serial interface. Through the assignment of addresses to each device interface, this method of addressing enables datagrams to be correctly routed when a device has two or more network connections.

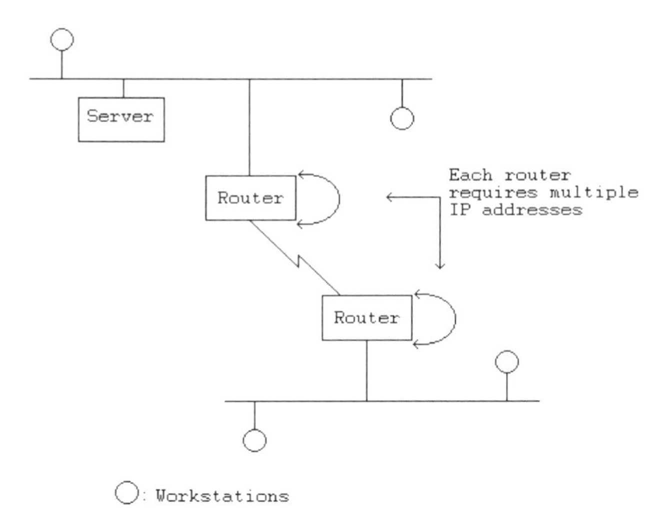

Figure 3.3 IP network addressing results in a unique 32-bit network address assigned to each device network interface

The basic addressing scheme

During the development of the TCP/IP protocol suite, it was recognized that internetworking two or more networks required distinct network addresses to differentiate one network from another. Because each host on a network also required a distinct address, a mechanism was required to identify a network as well as a host connected to a network. Although some protocols, such as NetWare's IPX, use separate network and host addresses, the designers of the TCP/IP protocol suite looked for a method to subdivide IP address space so that one address field in the IP header could identify the network and the host on the network, with the latter actually the interface since IP addresses are assigned to interfaces. The result was the development of a two-level addressing hierarchy, which is illustrated in Figure 3.4.

Under the two-level IP addressing scheme, all hosts on the same network must be assigned the same network prefix; however, each host must have a unique address to differentiate it from another host on the same network. Similarly, two hosts on different networks must be assigned different network prefixes; however, the hosts can have the same host address.

Under the two-level IP addressing hierarchy, the 32-bit IP address is subdivided into network and host portions. The composition of the first four bits of the 32-bit word specifies whether the network portion is 1, 2, or 3 bytes in length, resulting in the host portion being either 3, 2, or 1 bytes in length, respectively.

Figure 3.4 The two-level IP addressing hierarchy

Address classes

The two-level IP addressing scheme illustrated in Figure 3.4 represents the most common method of routing data from source to destination over an IP network. However, as we will soon note, IPv4 supports other addressing schemes. Those schemes, as well as the two-level IP addressing scheme, were developed in recognition of the fact that the use of a single method of subdivision of the IPv4 32-bit address space would be wasteful with respect to the assignment of addresses. For example, if the address space was split evenly into a 16-bit network and a 16-bit host number, the result would be a maximum of 65 535 ($2^{16}-1$) networks, with up to 65 535 hosts per network. In actuality there are certain host addresses that cannot be used, which slightly reduces the number of hosts that can reside on an IP network. Later in this section we will turn our attention to those addresses. However, returning to our address splitting example, the assignment of a network number to an organization that only had 200 computers would result in a waste of 65 334 host addresses that could not be assigned to another organization. Recognizing this problem as well as recognizing the need to obtain flexibility in assigning address space to different organizations resulted in the subdivision of the 32-bit address space into different address classes. Today, IPv4 address space consists of five address classes that are referred to as Classes A through E. Of the five address classes, Classes A through C are subdivided into a network identifier and host identifier. Classes D and E do not incorporate two-level addressing, as they represent special IP addressing. Class D addresses are used for IP multicasting, where a single message is distributed to a group of hosts dispersed across one or more networks that join a multicast group to receive the message. Through IP multicasting, a single voice or video data stream can be transmitted to multiple recipients on the same or different networks, significantly reducing the use of precious bandwidth. Class E addresses are reserved for experimental use. Although Class D and Class E addresses are single-level addresses, they are similar to Class A through C addresses in that they are 32 bits in length and are identified in the same manner.

Address formats

Figure 3.5 illustrates the five IPv4 address formats to include the bit allocation in the first byte of each format, which identifies the address class. Once an address class has been identified the subdivision of the remainder of the address into network and host address portions can be automatically noted.

Addresses within a specific address class are assigned by the Internet Network Information Center (InterNIC). To obtain an appreciation for the use of each address class, let us first turn our attention to the composition and notation of IP address. Once this has been accomplished, we will then examine special IP addresses and each address class.

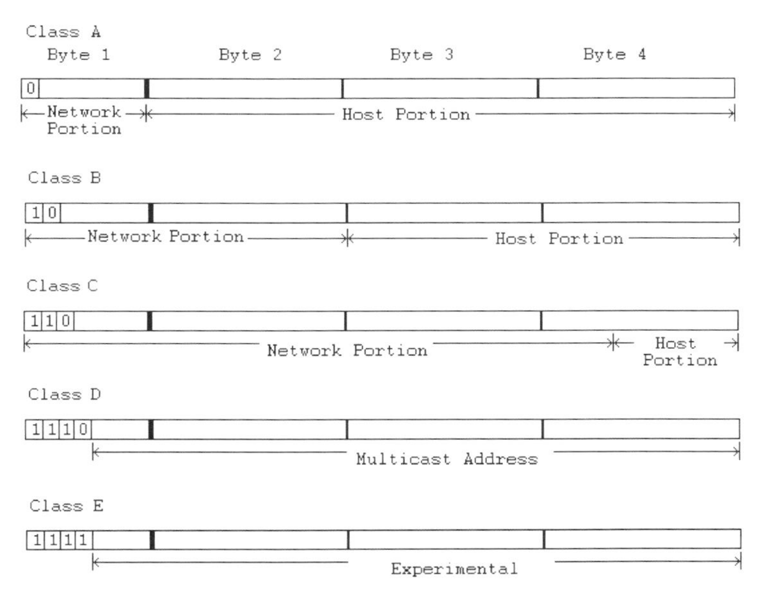

Figure 3.5 IPv4 address formats

Address composition and notation IPv4 addresses are specified in terms of 32 bits for source and destination. Because we would prefer not to work with 32-bit strings, dotted decimal notation was developed as a method to facilitate specifying IP address.

Under dotted decimal notation, the binary value of an 8-bit byte is expressed as a decimal number between 0 and 255. Because a 32-bit IPv4 address is equivalent to four bytes, the use of dotted decimal notation permits four decimal numbers to be used in place of a 32-bit binary string. Since periods are used to separate each decimal number, the term dotted decimal notation is used to reference the use of a string of four decimal numbers separated from one another by periods to represent a 32-bit binary IP address.

To illustrate the use of dotted decimal notation, consider the following IP address expressed as a 32-bit binary number:

11001111 00000010 10000000 10101010

If you remember, the bit values of a byte are as follows:

128 64 32 16 8 4 2 1

Then, the dotted decimal address for the above 32-bit binary IP address becomes

207.2.128.170

Special IP addresses There are several special IP addresses that have predefined functions. Those addresses and their meanings are listed in Table 3.4.

Table 3.4 Special IP addresses

Address	Description
Network = 0, Host = 0	This host on this network
Network = all 1s, Host = all 1s	Direct broadcast to network
Subnet = all 1s, Host = all 1s	Direct broadcast to all subnets on the network
Host = all 1s on any subnet	Direct broadcast to all hosts on the specified subnet
Network = 127, Host = any	Internal host loopback address

In examining the entries in Table 3.4, note that a subnet represents a subdivision of a network obtained by the expansion of the network portion of the address to the detriment of the host portion of the address. The expanded portion of the network address is used to denote a subnet that is only applicable internally within an organization's network structure. Later in this chapter we will examine subnets in detail.

Returning to Table 3.4, note that an IP address of either all 0s or all 1s has a special meaning. Because host values of all 0s and all 1s are part of the first two special addresses in Table 3.4, they cannot be used for a host address. This means that when we compute the number of possible hosts on a Class A, B, or C network, we must reduce the total by two. Similarly, we must reduce the number of hosts on a subnet by two to take into account a host value of zero and a host value of all 1s that has special subnet meanings.

In examining Table 3.4, note that any address with all 0s in the network portion of the address is used to represent 'this' network. Also note that an old form of broadcasting known as the all 0s takes the form of 0.0.0.0. You should not use this old broadcasting address, as that address is now used in a routing table to indicate a default route.

Finally, note that an address prefix of 127 represents an internal host loopback address and cannot be assigned to any host as a unique address. This address can be used to determine if a host's TCP/IP protocol stack is operational. For example, in a Windows NT/Windows 2000 environment you can Ping yourself by using 127.0.0.1. In fact, because 127.*anything* represents a loopback, you could also use 127.0.0.2 or even 127.255.255.255. Now that we have an appreciation for special IP addresses, let us turn our attention to the five IPv4 classes.

Class A

A Class A IPv4 address is defined by a 0-bit value in the high order bit position of the 32-bit address. This setting indicates that the address has the following 4-byte format:

<network number.host.host.host>

Because the network portion of the Class A address uses 1 bit for identification, only 7 bits in the first byte are available for network addressing. Out of the 128 combinations, all 0s and all 1s cannot be used. This is because all 0s represent 'this host on this network' and all 1s provide an internal loopback address. Thus, a maximum of 126 network addresses are permitted. Those addresses range from 1 to 126.

The 3-byte host field of a Class A address cannot have all 0s or all 1s. Thus, it supports $2^{24}-2$ or 16 277 214 hosts per network. Due to the relatively small number of Class A networks that can be defined and the large number of hosts that can be supported per network, Class A addresses are primarily assigned to large organizations and countries that have national IP-based networks. Another use of Class A addresses is for use by Internet Service Providers as a mechanism to issue Classless InterDomain Routing Protocol (CIDR) addresses. CIDR addressing is covered later in this section.

Class B

The setting of the two high order bits in an IPv4 address to a value of '10' indicates a Class B address. This address takes the following form:

<network number.network number.host.host>

The network portion of a Class B address cannot use the first two bit positions, as they identify the network class. Thus, the number of distinct networks becomes 2^{14} or 16 384 network numbers. Each network number is capable of supporting $2^{16}-2$ or 65 354 hosts.

Because the first two bits in the Class B address are set to a value of '10,' network numbers are restricted to the decimal range of 128 to 191 in the first portion of the dotted decimal notation for Class B addresses. Since a Class B address supports a large but not extravagant population of hosts, such addresses are normally assigned to relatively large organizations with tens of thousands of employees. Today, just about all Class B addresses are allocated, and only when a previously allocated Class B address is returned is it possible to obtain the use of this type of IPv4 address.

Class C

A Class C address is identified by the setting of the three high-order bits in a 32-bit IPv4 address to a value of '110.' This results in the form of the Class C address being noted as follows:

<network number.network number.network number.host>

The use of three bits to identify the network address as a Class C address reduces the number of bits that can be used to identify a particular network address from 24 to 21. This enables 2^{21} or 2 097 152 possible network addresses to be supported. Because a Class C address uses 8 bits for the portion of the host address, this means that each Class C address can support up to 2^8-2 or 254 hosts.

Because the first byte of a Class C address will always have the composition 110*xxxxx*, where *x* represents any binary value, the allowable network range is decimal 192–233 in the first field used for dotted decimal representation of a Class C address. Class C addresses are primarily assigned for use by relatively small networks, such as an organizational LAN requiring a connection to the Internet. Because it is common for many organizations to have multiple LANs, it is also common for multiple Class C addresses to be assigned to organizations that require more than 254 host addresses.

Class D

The assignment of a value of '1110' to the first four bit positions in a 32-bit IPv4 address defines a Class D address. The remaining 28 bits in the address are used to define 2^{28} or approximately 268 million possible multicast addresses.

Multicasting is an addressing technique that allows a source to send a single copy of a datagram to a specific group of recipients through the use of a multicast address. Each recipient dynamically registers to join the multicast group. As multicast traffic flows through a network, only recipients registered to receive an appropriate multicast session read the traffic denoted by the lower order 28 bits within a 32-bit Class D address. Other stations that are not members of a multicast group only have to read the first four bits of the address to note it is a Class D address and can then ignore the remainder of the address.

To obtain an appreciation for the manner by which a Class D address conserves bandwidth, consider a digitized audio or video presentation routed from the Internet onto a private network for which a dozen employees on the network wish to receive the presentation. Without a multicast transmission capability, a dozen separate audio or video data streams would be transmitted onto the private network, with each stream containing packets with a dozen distinct host addresses. In comparison, the use of a multicast address allows one data stream to be routed to the private network on which each registered station reads appropriate traffic. Because audio or video data streams can require a relatively large amount of bandwidth, the ability to eliminate multiple data streams via multicast transmission can prevent networks from becoming saturated and can also considerably reduce traffic on the Internet. Since the first four bits in a Class D address are set to a value of '1110,' the range of Class D addresses lies between 224 and 239 for the first decimal position when the address is expressed as a dotted decimal number.

Class E

The assignment of the binary value '1111' to the first four bits in a 32-bit IPv4 address denotes a Class E address. This address is reserved for experimentation.

Class E addresses range between 240 and 254 in their first decimal position when the IPv4 address is expressed as a dotted decimal number. Table 3.5 summarizes IPv4 address classes based upon values permissible in the network or first byte portion of the address when expressed in dotted decimal notation.

Reserved addresses

No discussion of IPv4 address classes would be complete without focusing attention upon three blocks of reserved addresses. Such addresses were originally reserved for networks that would not be connected to the Internet, and are defined in RFC 1918, 'Address Allocation for Private Internets.'

Table 3.6 lists the three address blocks defined in RFC 1918. The use of addresses in one or more address blocks defined by RFC 1918 is primarily based upon security considerations as well as the difficulty organizations can face in attempting to obtain relatively scarce Class B or Class A IPv4 addresses.

Because the use of any private RFC 1918 Internet address by two or more organizations connected to the Internet would result in addressing conflicts and the unreliable delivery of information, those addresses are not directly used. Instead, organizations either use a router with a Network Address Translation (NAT) capability or a proxy firewall to provide an address translation capability between a large number of private Internet addresses used on the internal private network and a lessor number of assigned IP addresses. For example, an organization with a thousand workstations could assign one RFC Class B address internally and translate those addresses to one Class C issued address, permitting up to 254 IP sessions at a time to be supported. In addition to enabling organizations to connect large internal networks to the Internet without having to obtain relatively scarce Class A or Class B addresses, NAT hides internal addresses from the Internet community. This action results in a degree of security, as any hacker who attempts to attack a host cannot directly do so. Instead, they must attack an organization's router or proxy firewall, which hopefully is hardened by the manufacturer to resist such attacks.

Table 3.5 Permissible IPv4 first byte values

Class	Length of Network Address (Bits)	Decimal Values
Class A	8	0–127
Class B	16	128–191
Class C	24	192–223
Class D	N/A	224–239
Class E	N/A	240–254

Table 3.6 Reserved IPv4 addresses for private internet use (RFC 1918)

Address blocks
10.0.0.0–10.255.255.255
172.16.0.0–172.31.255.255
192.168.0.0–192.168.255.255

Subnetting and the subnet mask

The use of IP addresses represents a precious resource. Recognizing the limited number of network addresses available for use as well as the need of organizations to create more manageable networks, the IETF approved subnetting in RFC 950 as a mechanism to share a single network address among two or more networks. To better understand the need for subnetting, consider a Class B address. That address permits up to 65 535 hosts. However, it would be both a performance and an administrative nightmare to have one network with that number of hosts. Thus, subnetting provides users with the ability to subdivide a Class B network as well as Class A and Class C networks into more manageable entities.

Subnetting represents an extension of the network portion of a Class A, B, or C address internally to an organization. Through the process of subnetting, the two-level IPv4 address hierarchies of Class A, B, and C addresses are turned into three-level hierarchies.

Figure 3.6 illustrates the creation of a subnet by the extension of the network address to the detriment of the host portion of an address. Note that the resulting action produces a subnet field and a reduced length host field, which reduces the number of hosts that can reside on each subnet.

Through subnetting a Class A, B, or C network address can be extended, with the extension divided into different subnet numbers. Each subnet number can be used to identify a different network internal to an organization. However, because the network portion of the address does not change, all subnets appear externally to be located on the same network. This means that routing tables on devices that form the backbone of the

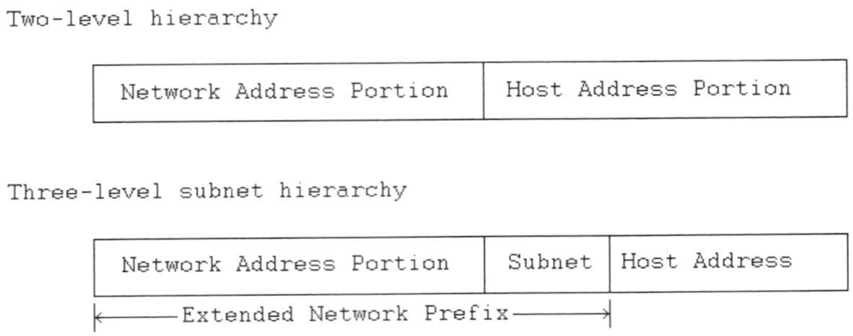

Two-level hierarchy

Network Address Portion	Host Address Portion

Three-level subnet hierarchy

Network Address Portion	Subnet	Host Address

|←Extended Network Prefix→|

Figure 3.6 Creating a subnet by extending the network prefix into the host address portion of a Class A, B or C address

Internet need to recognize a lesser number of network addresses, which simplifies routing. This also means that routers within an organization must be able to differentiate between different subnets.

To illustrate the subnet process, consider an organization that within a building operates five Ethernet networks, with between 20 and 30 stations on each network. Although the organization could apply for five Class C addresses and assign one address to each network, doing so would waste precious Class C address space since each Class C address supports a maximum of 254 devices. In addition, the assignment of five Class C addresses would result in configuring numerous routers on the Internet to note those addresses. This in turn would adversely effect bandwidth utilization on the Internet, as five router table entries would be transmitted each time routers broadcast the contents of their routing tables.

Because we need to support five networks at one location, we would extend the network portion of one Class C address by 3 bit positions. This is because 2^3 provides 8 subnets, while 2^2 provides 4, which is insufficient. Because a Class C address uses one 8-bit byte for host identification, this also means that a maximum of five bit positions (8−3) can be used for the host number. This reduces the number of hosts that can reside on each subnet to 2^5-2 or 30, which is sufficient for our example.

Let us assume that we obtained the Class C network address 205.131.175.0 for our subnetting effort. To use that network, we would extend its network prefix by three bit positions as illustrated in Figure 3.7.

In examining Figure 3.7, note that the top entry labeled 'Base Network' represents the Class C network address with a host address byte field set to all 0s. Because we previously decided to use three bits from the host portion of the Class C IPv4 address, the entries below the base network indicate the use of three bits from the host position in the address to create extended prefixes to identify all possible distinct subnets.

For each subnet there are several addressing restrictions that reduce the number of hosts, or more correctly, interfaces that can be supported. First, you cannot use a base subnet of all 0s or all 1s. Thus, for subnet 0 in Figure 3.7 valid host addresses would range from 1 to 30, while for subnet 1 valid host addresses would range from 33 to 61, and so on.

A second limitation on subnetting for Class A, B, and C addresses concerns the subdivision of the last byte of an IP address. Because a subnet must be able to have some hosts residing on it, you can only use up to 6 bits in the last byte when you create a subnet mask. Thus, the maximum class C subnet would be 6 bits, while the maximum Class A and B subnets would be 22 and 14 bits, respectively.

Although the use of a 3-bit subnet mask permits eight subnets to be defined, our requirement was to assign subnets to five LANs. Thus, we will use subnets 0 through 4 shown in Figure 3.7, although we could select any five of the eight subnets. To the router connecting the organization's network to the Internet, all five subnets that we will use would appear as the network address 205.131.175.0, with the router of our organization being responsible for directing traffic to the appropriate subnet. It is important to note that external to the organization there is no knowledge of the dotted decimal

Base Network: 11001101.10000011.10101111.00000000 = 205.131.175.0

Subnet #0: 11001101.10000011.10101111.00000000 = 205.131.175.0

Subnet #1: 11001101.10000011.10101111.00100000 = 205.131.175.32

Subnet #2: 11001101.10000011.10101111.01000000 = 205.131.175.64

Subnet #3: 11001101.10000011.10101111.01100000 = 205.131.175.96

Subnet #4: 11001101.10000011.10101111.10000000 = 205.131.175.128

Subnet #5: 11001101.10000011.10101111.10100000 = 205.131.175.160

Subnet #6: 11001101.10000011.10101111.11000000 = 205.131.175.192

Subnet #7: 11001101.10000011.10101111.11100000 = 205.131.175.224

Figure 3.7 Creating extended network prefixes via subnetting

numbers shown in the right column of Figure 3.7, which represent distinct subnets. This results from the fact that routers external to the organization view the binary value of the first byte of each dotted decimal number and note that the first two bits are set. This informs each router that the address is a Class C address and that the first three bytes represent the network portion of the IPv4 address, while the last byte represents the host address. Thus, to the outside world the 205.131.175.32 address would not be recognized as subnet 1 on network 205.131.175.0. Instead, routers external to the organization would interpret the address as network 205.131.175.0 with a host address of 32. Similarly, subnet four would be recognized as a Class C network address of 205.131.175.0 with a host address of 128. However, within the organization internally each of the IPv4 addresses listed in the right column in Figure 3.7 would be interpreted and recognized as a subnet. Figure 3.8 illustrates the difference between viewing the network internally and externally.

Host addresses on subnets Although we have briefly discussed some subnet addressing rules, we have yet to denote how we assign host addresses to devices connected to different subnets, nor how routers can examine an IPv4 address so it can correctly route traffic to an appropriate subnet. Therefore, let us turn our attention to these topics.

In Figure 3.7 we subdivided the host portion of the Class C address into a 3-bit subnet field and a 5-bit host field. Because we cannot use a host field address of all 0s or all 1s, this means each subnet can support a maximum of 2^5-2 or 30 addresses. Thus, we could use host addresses 1 through 30 on subnet 0, 33 through 62 on subnet 1, and so on. Remembering the restriction that we cannot use all 0s or all 1s in the host portion of a subnet governs our ability to assign host addresses by subnet. This is illustrated in Figure 3.9, which indicates how we could assign host addresses to subnet 2, whose

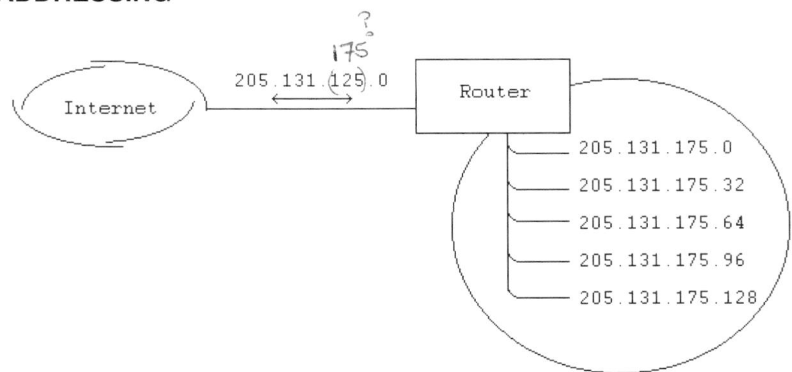

Figure 3.8 Internet vs. internal network view of subnets

creation was previously indicated in Figure 3.7. In examining Figure 3.9, note that we commence our addressing operation with the subnet address 205.131.175.64, for which the first three bits in the fourth byte are used to indicate the subnet. Then, we use the remaining five bits to define the host address on each subnet. Thus, the address 205.131.175.64 represents the second subnet on the 205.131.175.0 network, while addresses 205.131.175.65 through 205.131.175.94 represent hosts 1 through 30 that can reside on subnet 2.

While the previously presented information explained how we can create subnets and host addresses on subnets, an unanswered question is how devices on a private network recognize subnet addressing. For example, assume that an IP datagram with the destination address of 205.131.175.65 arrives at your organization's router. How does that router know to route the datagram onto subnet 2? The answer to this question is the use of a subnet mask, so let us turn our attention to this topic.

The subnet mask The subnet mask is a sequence of binary 1s that indicates the length of the network address to include any subnetting that has occurred. Thus, the subnet mask provides a mechanism that enables communications devices on a network to determine the separation of an IP address into its network, subnet, and host portions.

To illustrate the use of the subnet mask, let us assume that our network address is 205.131.175.0 and that we need to develop a subnet mask that can be used to identify the extended network as well as the subnet and host on the subnet. Because we previously extended the network by 3 bits, the subnet would become

11111111.11111111.11111111.11100000

Note that the above mask can be expressed in dotted decimal notation as 255.255.255.224.

The subnet mask tells a communications device which bits in an IP address should be treated as an extended network address consisting of network and subnet addresses. Then, the remaining bits that are not set indicate the host on the extended network address. Because the first or first few bits in an IP

Subnet 2: 11001101.10000011.10101111.01000000 = 205.131.175.64

Host 1: 11001101.10000011.10101111.01000001 = 205.131.176.65

Host 2: 11001101.10000011.10101111.01000010 = 205.131.176.66

. . . . . .

. . . . . .

. . . . . .

Host 30: 11001101.10000011.10101111.01011110 = 205.131.176.94

Figure 3.9 Assigning host addresses to subnet 2

address denote the address class, a communications device that examines those bits determines the number of bits in the network portion of the address. Subtracting that value from the number of bits in the subnet mask indicates the subnet field, allowing the device to note the subnet. For example, consider the IPv4 address 205.131.175.66 and the subnet mask 255.255.255.244. The address 205.131.175.66 has the first two bits in the address set, representing a Class C address that uses 3 bytes or 24 bits for the network address. Because the subnet mask represents 27 set bits, this indicates that the subnet is 27–24 or 3 bits in length and occurs in bit positions 25 through 27 in the IP address. Because bits 25 through 27 have the bit composition 010, this indicates the subnet is subnet 2. Because the last five bits have the value 00010, this indicates host 2 on subnet 2. Figure 3.10 illustrates the relationship above described.

To facilitate the ability to work with subnets, Table 3.7 contains a listing of the number of subnets that can be created for IPv4 Class B and C addresses, their subnet masks, the number of hosts that can reside on a network, and the total number of hosts capable of being supported by a particular subnet mask. Thus, this table can be used as a guide for considering the extension of a network address internally to form a subnet. In examining the entries in Table 3.7, note that the total number of hosts can vary considerably based upon the use of different subnet masks, and should be carefully considered prior to performing the subdivision of a network.

In Table 3.8 you will find an additional subnet mask reference. This table indicates the dotted decimal value associated with the use of different numbers of subnet bits in the last byte of a subnet mask.

Configuration examples There is an old adage that states 'the proof of the pudding is in its eating.' We can apply this adage to the information previously discussed about IP addressing by turning our attention to the manner by which workstations and servers are configured to operate on a TCP/IP network.

When configuring a workstation or server to operate on a TCP/IP network, most operating systems will require you to enter a minimum of three IP addresses and an optional subnet mask. The three IP addresses you will normally configure include the IP address assigned to the interface of a workstation or server, the IP address of the gateway responsible for relaying packets with a destination address different from the local network off the local network, and the address of a name resolver. The gateway represents an old term for the modern router. The resolver is a computer responsible for translating host addresses entered in a browser, FTP or Telnet client applications, or another application into its IP address, as routing in a TCP/IP environment is based upon IP addresses. The name resolver is also referred to as a Domain Name Server or DNS.

Figure 3.11 illustrates the first configuration screen in a series of screens displayed by the Windows 95 TCP/IP properties dialog box. In the example illustrated in Figure 3.11 note that we entered an IP address of 205.131.175.66 and a subnet mask of 255.255.255.224. This informs the protocol stack that the workstation is a Class C address, since a 205 value in the first byte indicates this fact. Because the subnet mask has 27 set bits (224 decimal is 11100000 in binary) and the network portion of a Class C address is 24 bits, the protocol stack knows that the first three bits in the fourth byte of the IP address represents the subnet. Because decimal 66 has a binary value of 01000010, this indicates the workstation resides on subnet 2. Similarly, because the last five bits in the fourth byte of the IP address represent the host on the subnet, a decimal value of 66 indicates host 2 on subnet 2. Because we previously specified in this section that we would use a 3-bit subnet, a subnet mask of 255.255.255.224 was entered in Figure 3.11. The last decimal digit corresponds to the 3-bit subnet entry in Table 3.8 and indicates how you can use that table if you have an aversion to working with binary numbers.

Returning to the TCP/IP properties display shown in Figure 3.11, note that by clicking on different tabs you can display different configuration screens. For example, clicking on the tab labeled 'Gateway' results in the display shown in Figure 3.12. In this screen you would enter the IP address of one or more gateways so the TCP/IP protocol stack on the workstation will know where to send IP datagrams whose destination addresses are not on the local network.

```
IP address:       205.131.175.66        11001101.10000011.10101111.01000010
Subnet mask:      255.255.255.224       11111111.11111111.11111111.11100000

Extended Network Address                |<------ 27 bit positions ------>|

First 2 bits of IP address              = 11
  indicate Class C address

Network portion of Class C address     |<-24 bit positions ------>|

Subnet field = Extended Address - Network Portion of Address  |<3>|
  Value of subnet = 010 = 2                                    bits
Host portion of subnet is 8 - 3 bits                               |<-5->|
  Value of host on subnet 2 is 00010 or 2                           bits
```

Figure 3.10 The relationship between an IP address, its subnet mask, subnet value, and host value on the subnet

Table 3.7 Subnet Mask reference

Number of Subnet bits	Subnet Mask	Number of Subnetworks	Hosts/ Subnet	Total Number of Hosts
Class B				
1	—	—	—	—
2	255.255.192.0	2	16 382	32 764
3	255.255.224.0	6	8 190	49 140
4	255.255.240.0	14	4 094	57 316
5	255.255.248.0	30	2 046	61 380
6	255.255.252.0	62	1 022	63 364
7	255.255.254.0	126	510	64 260
8	255.255.255.0	254	254	64 516
9	255.255.255.128	510	126	64 260
10	255.255.255.192	1 022	62	63 364
11	255.255.255.224	2 046	30	61 380
12	255.255.255.240	4 094	14	57316
13	255.255.255.248	8190	6	49 140
14	255.255.255.252	16 382	2	32 764
15	—	—	—	—
16	—	—	—	—
Class C				
1	—	—	—	—
2	255.255.255.192	2	62	124
3	255.255.255.224	6	30	180
4	255.255.255.240	14	14	196
5	255.255.255.248	30	6	170
6	255.255.255.252	62	2	124
7	—	—	—	—
8	—	—	—	—

A third configuration screen that is used to specify the address of the name resolver is shown in Figure 3.13. At this level clicking on the tab labeled DNS Configuration provides you with the ability to specify your computer's host and domain name as well as the IP address of one or more DNS servers. In Figure 3.13 the host name is shown entered as 'gil' while the DNS domain name was entered as 'feds.gov.' This informs the domain server at the

Table 3.8 Subnet masks

Subnet Bits	Host Bits	Decimal Mask
0	8	0
1	7	128
2	6	192
3	5	224
4	4	240
5	3	248
6	2	252
7	1	254
8	0	255

indicated address that requests to access the host address gil.feds.gov should be routed to the IP address previously entered into the IP configuration screen. If no network users will be addressing a computer by its host name, you can leave those entries blank. Chapter 5 contains detailed information concerning the DNS.

Classless networking Although the term classless normally refers to a person without taste, when applied to IPv4 addressing it represents a technique to more efficiently assign addresses to organizations. Classless networking obtains its name as it represents a technique that does away with network classes, enabling the inefficiencies associated with allocating Class A, B, and C addresses to organizations that have a limited number of devices to be overcome.

Under classless networking an organization is assigned a number of bits for use as the local part of its address that best corresponds to the number of addresses it requires. For example, assume that your organization requires 4000 IP addresses. Because 2^{12} provides 4096 distinct addresses that best correspond to your organization's requirements, you would be assigned 12 bits for use as the local portion of your organization's address. The remaining 20 bits in the 32-bit IPv4 address space are then used as a prefix to denote what is referred to as a supernetwork. Thus, the format of a classless address is as follows:

Supernetwork address: network address

The forward slash is used to denote the network portion of a classless network. That character is then followed by the number of bits in the network address. Thus, the previously described classless network would be denoted as /20.

Address allocations employed for classless networking are taken from available Class C addresses. This means that obtaining a 20-bit classless network prefix to support up to 4096 devices is equivalent to obtaining 16 continuous Class C addresses. Table 3.9 provides a list of classless address blocks that can be assigned from available Class C address space.

Another key advantage of classless networking concerns the use of router tables and router performance. Under classless addressing a router becomes able to forward traffic to an organization using a single routing entry. This permits a reduction in router table entries, which in turn allows the router to perform lookup operations faster. Thus, classless addressing both provides a mechanism to extend the availability of IP addresses in a more efficient manner and enables routers to operate more efficiently until IPv6 is deployed.

3.3 THE IPv6 HEADER

Although IPv4 addresses were rapidly being depleted, this was only one of the several issues that was the driving force for the development of a replacement

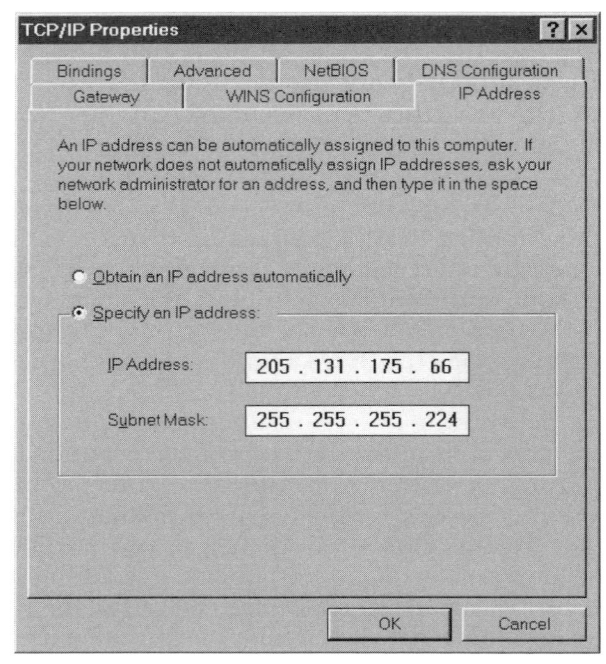

Figure 3.11 The Windows 95 TCP/IP Properties screen provides users with the ability to set the IP address of the host operating a TCP/IP protocolstack

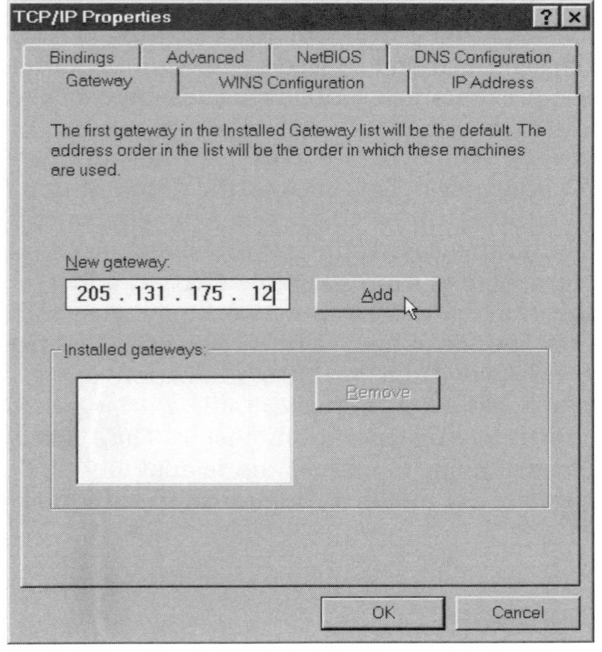

Figure 3.12 Through the use of the tab labeled Gateway in the TCP/IP Properties dialog box, you can configure the location of one or more routers to service packets addressed to different networks

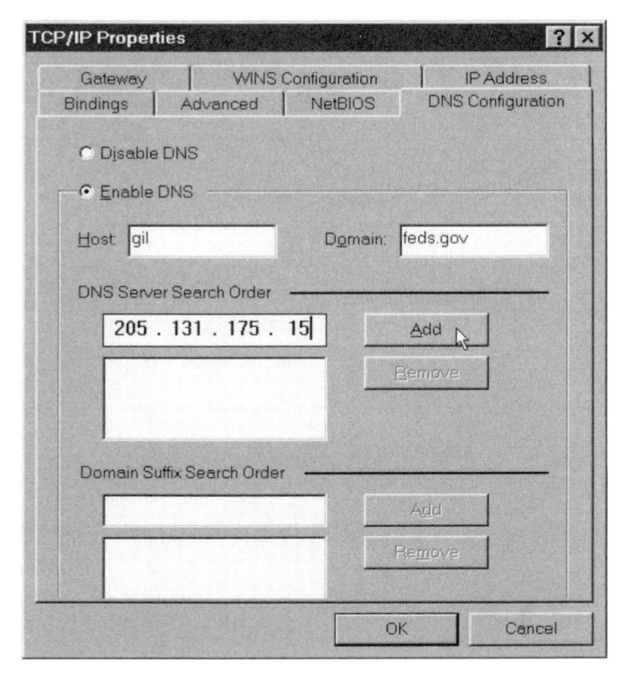

Figure 3.13 The Windows 95 DNS Configuration tab in the TCP/IP Properties dialog box enables a host to be configured so it can be identified by its host name as well as by its IP address

for that version of the IP. Other issues include the need to improve support for extending the IP header to accommodate features and functions yet to be developed, simplify the header to reduce packet processing time, and enable the labeling of packets belonging to a flow as a mechanism to provide a Quality of Service capability. These issues resulted in a complete redesign of the IP header.

Figure 3.14 illustrates the format of an IPv6 header. Note that the resulting header is significantly simpler than the IPv4 header, consisting of only seven fields. Although the header is simpler than an IPv4 header, we will soon note that through the use of the Next Header field we can daisy-chain IPv6 headers, one after another, which considerably expands the capability of an IPv6 header, enabling such security options as authentication to be added. To obtain an appreciation of the IPv6 header, let us examine each field in the header.

3.3.1 Ver field

The Ver field is four bits in length and functions in the same manner as its IPv4 cousin. That is, it identifies the version of the IP. For IPv6 a binary value of 0110 is placed in this field. Because the Ver field in an IPv6 header, like its IPv4 counterpart, is located at the beginning of the header, it enables a

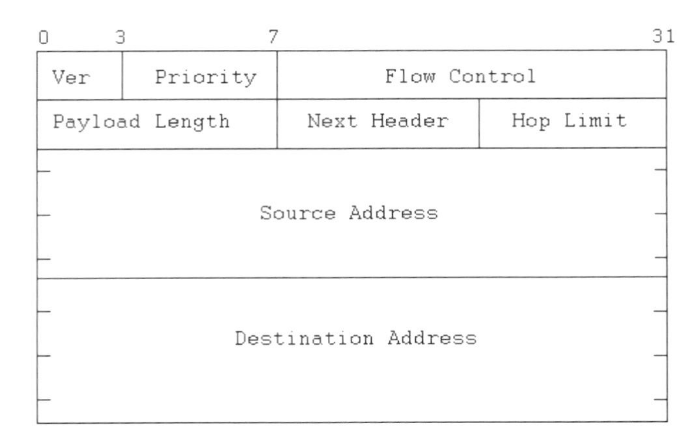

Figure 3.14 The IPv6 header

communications device to rapidly distinguish an IPv4 packet from an IPv6 packet.

3.3.2 Priority field

The priority field enables a data originator to identify the desired delivery priority of a packet. Because this field is 4 bits, up to 16 priorities (0 to 15) can be specified.

Table 3.10 lists presently defined Priority field values. In examining the recommended priority values listed in Table 3.10, Netnews and email can be considered to represent applications that could be assigned priorities of filler and unattended data transfer, respectively. FTP could represent an example of an attended bulk transfer priority, while Telnet and SNMP could represent interactive traffic and internet control traffic. Priority values of 8 through 15 are reserved for non-congestion controlled traffic or traffic consisting of real-time packets that does not back off in response to network congestion. The lowest priority value (8) should be used for packets that the originator is most

Table 3.9 Classless network address assignments

Network Part	Local Bits	Equivalent Number of Class C Addresses	Distinct Addresses
124	8	1	256
123	9	2	512
122	10	4	1024
121	11	8	2048
120	12	16	4096
119	13	32	8192
118	14	64	16 284
117	15	128	32 768

willing to have discarded when congestion effects data flow, while the highest value (15) should be assigned to packets that the originator is least willing to have discarded.

3.3.3 Flow Label field

The Flow Label field is 24 bits in length. This field enables a source to identify a sequence of related packets that require special handling by intermediate routers when the packets travel from source to destination. An example of a flow could be a real time video conference. Through the identification of a traffic flow it becomes possible for a Quality of Service (QoS) protocol, such as the Resource Reservation Setup Protocol (RSVP), to use the Priority and Flow Label fields to provide special packet handling based upon the allocation of bandwidth.

3.3.4 Payload Length field

The Payload Length field is 16 bits in length, and its value indicates the payload carried by the packet after the header. Because an IPv6 header's length is fixed at 40 octets, the total length of the packet is the value of the Payload Length field plus 40.

3.3.5 Next Header field

The Next Header field identifies the type of header that immediately follows the IPv6 header. Although the values used for this 8-bit field are the same values as used in the IPv4 Protocol field (refer to Table 3.2), the use of this field considerably differs from that of the IPv4 Protocol field. The latter simply indicates the protocol that follows the IPv4 header. In comparison, the Next Header field can indicate an optional header, a higher layer protocol, or no protocol above IP. This significantly expands the capability of IPv6, since it enables headers to be daisy-chained to support a particular function or series of functions.

Figure 3.15 illustrates an example of the extension of an IPv6 packet through the daisy-chaining of headers. In this example a value of 43 in the IPv6 header indicates that the next header is a routing header. The routing header in turn would have a value of 51 in its Next Header field to indicate that Authentication follows, resulting in an Authentication header following the routing header. The authentication header would then have a value in its Next Header field to indicate that a TCP header follows.

3.3.6 Hop Limit field

This 8-bit field indicates the maximum number of routers that a packet can traverse prior to being discarded. This field acknowledges that the use of a time entry in the IPv4 TTL field was impractical, and the earlier protocol's more practical use of a hop count is officially used by the newer protocol.

Table 3.10 Priority field value assignments

Priority Field Value	Recommended Assignment
0	uncharacterized traffic
1	filler traffic
2	unattended traffic
3	reserved
4	attended bulk transfer
5	reserved
6	interactive traffic
7	Internet control traffic

3.3.7 Source and Destination address fields

Both Source and Destination address fields were extended to 128 bits from IPv4's 32-bit addresses. Because IPv6 is based upon the same architecture used in IPv4, each network interface still requires a distinct IP address. However, under IPv6 an interface can be identified by several types of addresses, each of which contains 128 bits, or 96 more bits than an IPv4 address. In the remainder of this section we will turn our attention to some of the address types supported under IPv6, their notation or representation, and the manner by which IPv6 address space is allocated.

3.3.8 Address types

IPv6 defines three types of addresses—unicast, multicast, and anycast—with the latter representing a new address category that was not included in IPv4. An anycast address identifies a group of devices similar to a multicast address. However, a packet with a multicast address is delivered to only one host, usually the closest one to the source, with 'close' defined by the routing protocol. The use of anycast addressing can be expected to facilitate network restructuring while minimizing the number of configuration changes required to support a new network topology. This can be accomplished since an anycast address could be used to reference a group of routers, and the alteration of a network when IPv6 is used with anycast addressing would allow the stations to continue to access the nearest router without having to change the station's address configuration in its TCP/IP protocol stack.

3.3.9 Address notation

IPv6 addresses are written in dotted decimal notation similar to the manner in which they are used to specify an IPv4 address. However, because an IPv6 address consists of 128 bits or 32 decimals, the format by which dotted decimal numbers are used changed. Instead of a period or dot to separate individual numbers, under IPv6 a colon (:) is used to separate 16-bit or 4-hex-character address entities. Because IPv6 addresses will often have a long

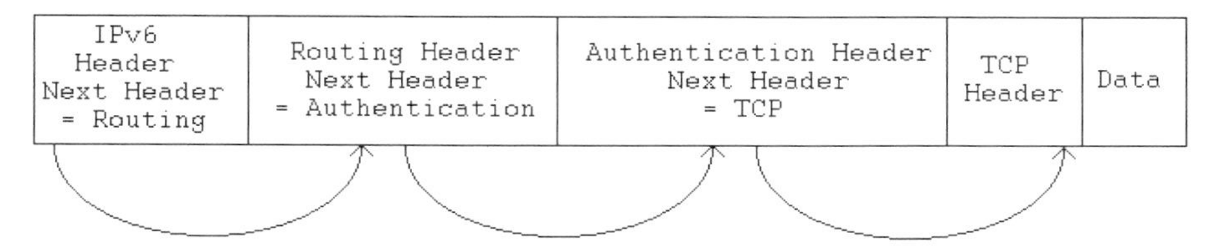

Figure 3.15 An example of an IPv6 packet with daisy-chained extension headers

string of zeros such as a 32-bit IPv4 address represented in a 128-bit IPv6 address field, a double-colon (::) is used to represent contiguous multiple 16-bit (4-hex-character) blocks of zeros. To illustrate IPv6 address notation, let us examine the use of colons and double-colons. For example, consider the following IPv6 address:

1ACD:4ABD:0003:0000:0000:0001:31BC:010A

To facilitate the entry of IPv6 addresses, you can ignore or skip the leading zeros in each hexadecimal 4-byte component. Thus, this allows you to rewrite the previous IPv6 network address as

1ACD:4ABD:3:0:0:1:31BC:10A

We can simplify the address further by replacing consecutive null 16-bit numbers by double-colons. Doing so we obtain

1ACD:4ABD:3::1:31BC:10A

Note that the double-colon can only be used once inside an IPv6 address. This is because the reconstruction of the address requires the number of integer fields in the address to be subtracted from 8 to determine the number of consecutive fields of zero value that the double-colon represents. Otherwise, the use of two or more double-colons would result in an ambiguity that would prevent the address from being correctly reconstructed.

3.3.10 Address allocation

The use of a 128-bit address space for IPv6 results in a much higher degree of address assignment flexibility than is obtainable when using IPv4 addresses. For example, under IPv6 addressing Internet Service Providers can be identified. In addition, IPv6 addressing provides the ability to identify local and global multicast addresses, private site addresses for use within an organization, hierarchical global unicast addresses, and other types of addresses. In Table 3.11 you will note a list of the initial allocation of address space under IPv6.

Table 3.11 IPv6 address space allocation

Allocation	Prefix (Binary)	Fraction of Address Space
Reserved	0000 0000	1/256
Unassigned	0000 0001	1/256
Reserved for NSAP allocation	0000 001	1/128
Reserved for IPX Allocation	0000 010	1/128
Unassigned	0000 011	1/128
Unassigned	0000 1	1/32
Unassigned	0001	1/16
Unassigned	001	1/8
Provider-Based Unicast Address	010	1/8
Unassigned	011	1/8
Reserved for Geographic-Based Unicast Address	100	1/8
Unassigned	101	1/8
Unassigned	110	1/8
Unassigned	1110	1/16
Unassigned	1111 0	1/32
Unassigned	1111 10	1/64
Unassigned	1111 110	1/128
Unassigned	1111 1110 0	1/512
Link-Local Use Addresses	1111 1110 10	1/1024
Site-Local Use Addresses	1111 1110 11	1/1024
Multicast Addresses	1111 1111	1/256

IPv6 address space allocation was assigned to the Internet Assigned Numbers Authority (IANA). IANA distributes portions of IPv6 address space to regional registries, such as the InterNIC in North America, RIPE in Europe, APNIC in Asia, and similar organizations. To obtain an appreciation for the potential use of IPv6 addresses, let us examine several types of IPv6 addresses.

Provider-Based Unicast addresses

A Provider-Based Unicast address is assigned by an Internet Service Provider to a customer. Based upon the initial allocation of IPv6 addresses shown in Table 3.11, each Provider-Based Unicast address has the three-bit prefix '010.' That prefix is followed by fields that identify the Internet address registry from which the ISP obtained the address (Registry Identifier), the address block assigned to the ISP by the registry authority, which identifies the ISP (Provider Identifier), the ISP's subscriber (Subscriber Identifier), and the subscriber address in the form of a 16-bit subnet identifier (Subnet) and a 48-bit interface identifier (Station). The latter can represent the MAC address of a station and when used also represents a unique address. Figure 3.16 illustrates the format of a Provider-Based Unicast address.

Prefix	Registry ID	Provider ID	Subscriber ID	Subnet ID	Station ID

Prefix:	three bits set to 010
Registry:	5 bits identifies organization that allocated the address
Provider:	24 bits, with 16 used to identify ISP and 8 used for future extensions
Subscriber:	32 bits, with 24 used to identify the subscriber and 8 for extension
Subnet:	16 bits to identify the subnetwork
Station:	48 bits to identify the station

Figure 3.16 IPv6 Provider-Based Unicast address structure

Multicast addresses

As indicated in Table 3.11, a multicast address begins with eight 1s set (hex FF). This address is used in IPv6 to identify a group of nodes, with the term node used to indicate a device that supports the new version of the IP and that replaces the term interface used under IPv4. A node can belong to any number of multicast groups, which allows the device to participate in multiple multicast sessions.

Figure 3.17 illustrates the IPv6 Multicast address format. Note that after the first eight set bits the next four bits represent a set of flag bits. The first three flag bits are currently set to zero, while the fourth bit is used to indicate a permanently assigned (0) or transient (1) multicast address. The next field consists of four bits that indicate the scope of the address. Here the Scope field denotes the portion of the network for which the multicast is relevant, in effect providing a mechanism to focus a multicast to a specific area. Scope values include node-local (0001), link-local (0010), site-local (0101), organization-local (1000), and global (1111). The remaining 112 bits in the 128-bit address represent a Group Identifier field that identifies the multicast group for a permanent or transient application within a given scope.

3.3.11 Transporting IPv4 addresses

In a mixed IPv4 and IPv6 environment communications devices that do not support IPv6 will have their addresses mapped using a special form of IPv6. That form is as follows:

0:0:0:0:0:FFF:w.x.y.z

In the above format w.x.y.z represents the original 32-bit IPv4 address. Thus, an organization with a considerable investment in time and effort in configuring hundreds or thousands of workstations and servers can migrate to IPv6, yet the organization's router only needs to be upgraded to support

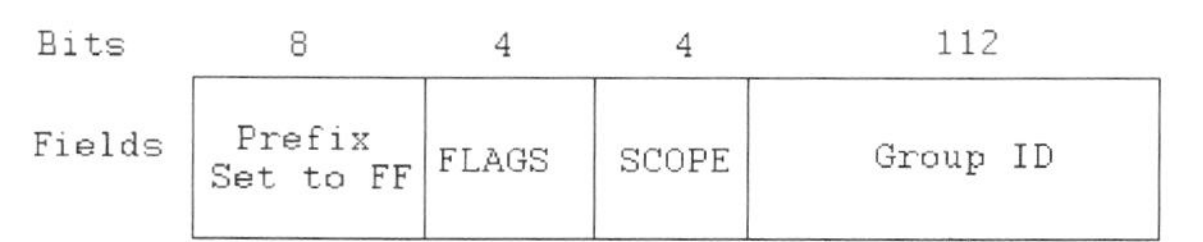

Figure 3.17 IPv6 Multicast address format

IPv6 addressing when they deploy the new version of IP. Then, if they wish to do so, they can gradually convert existing IPv4 addresses to IPv6 addresses.

3.4 ICMP AND ARP

In concluding this chapter on the IP, we will turn our attention to two special protocols that facilitate the operation of the IP: the Internet Control Message Protocol (ICMP) and the Address Resolution Protocol (ARP).

3.4.1 ICMP

In this section we will focus our attention upon the Internet Control Message Protocol (ICMP). ICMP represents an error reporting mechanism that is transported via IP datagrams.

The format of an ICMP message and its relationship to an IP datagram is illustrated in Figure 3.18. Note that although each ICMP message has its own format, they all begin with the same three fields: an 8-bit Type field, an 8-bit Code field, and a 16-bit Checksum field.

Similar to IP, there are two versions of ICMP you should obtain familiarity with: ICMPv4 and ICMPv6. In this section we will first examine the values of the Type and Code fields associated with ICMPv4. Once this has been accomplished, we will turn our attention to the values of those fields when ICMPv6 is employed.

ICMPv4

When ICMPv4 is used to transport control messages, an IPv4 header precedes the ICMP header. This results in the use of 32-bit IPv4 addressing for the routing of datagrams.

Type field The ICMP Type field defines the meaning of the message as well as its format. Perhaps the two most familiar ICMP messages are type 0 and type 8.

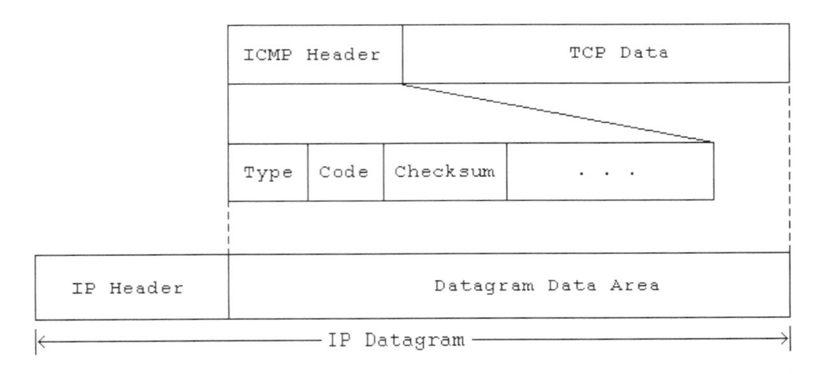

Figure 3.18 ICMP messages are transported via encapsulation within an IP datagram

A Type field value of 8 represents an Echo Request, while a Type 0 ICMP message denotes an Echo Reply. Although their official names are Echo Reply and Echo Request, most persons are more familiar with the term Ping, which is used to reference both an ICMP Echo Request and ICMP Echo Reply. Table 3.12 lists the values of the ICMP Type fields that identify specific types of ICMP messages.

Code field The second field common to each ICMP header is the Code field. The Code field provides additional information about the message and may not be meaningful for certain messages. For example, both Type field values of 0 and 8 always have a Code field value of 0. In comparison, a Type field value of 3 (Destination Unreachable) can have one of 16 possible code field values that further defines the reason why the destination was unreachable.

Table 3.12 ICMP Type Field values

Type	Name
0	Echo Reply
1	Unassigned
2	Unassigned
3	Destination Unreachable
4	Source Quench
5	Redirect
6	Alternate Host Address
7	Unassigned
8	Echo Request
9	Router Advertisement
10	Router Selection
11	Time Exceeded
12	Parameter Problem
13	Timestamp
14	Timestamp Reply
15	Information Request
16	Information Reply
17	Address Mask Request
18	Address Mask Reply
19	Reserved (for Security)
20–29	Reserved (for Robustness Experiment)
30	Traceroute
31	Datagram Conversion Error
32	Mobile Host Redirect
33	IPv6 Where-Are-You
34	IPv6 I-Am-Here
35	Mobile Registration Request
36	Mobile Registration Reply
37	Domain Name Request
38	Domain Name Reply
39	SKIP
40	Photuris
41–255	Reserved

Table 3.13 lists the Code field values presently assigned to ICMPv4 based upon their Type field values. As we will note when we discuss security issues later in this book, many firewalls and routers provide administrators with the ability to filter all or selected ICMP messages. You could use this feature to prevent malicious Pinging either as an attempt by a hacker to learn the devices on your organization's internal network or as a mechanism to deny service to a host by flooding it with Pings.

ICMPv6

ICMPv6 is functionally very similar to ICMPv4, using a similar message format that results in the message being transported within an IP datagram. However, ICMPv6 is transported within an IPv6 datagram while the IPv6 header uses a Next Header field value of 58 to indicate that an ICMPv6 header follows.

Type field In comparison with ICMPv4, the universe of ICMPv6 type field values is considerably reduced. Table 3.14 lists presently defined ICMPv6 Type field values. In examining the entries in Table 3.14, note that only four types of error messages are defined. A Destination Unreachable error message indicates that a packet cannot be delivered to its destination address other than due to congestion causing the packet to be discarded. A Packet Too Big error message is transmitted from a router to the packet originator when the packet exceeds the MTU of the outgoing link and cannot be forwarded.

The third message, Time Exceeded, is sent by a router when a packet's Hop Limit value reaches zero or if all fragments of a datagram are not received within the fragment reassembly time. The fourth error message, Parameter Problem, is generated by a device that notes a problem in a field in the packet header that makes it impossible to process the header. Thus, by observing the type of ICMPv6 error messages, you can obtain an appreciation for the cause of several common types of network-related problems.

In addition to four error messages, the ICMPv6 Type field currently defines 11 informational messages. The first two listed in Table 3.14, Echo Request and Echo Reply, are used for diagnostic purposes, while the three 'Group' messages are used to convey information about multicast group membership from IPv6 nodes to their neighboring routers.

Code field Similar to ICMPv4, certain ICMPv6 messages can have more than one meaning, and depend upon a Code field value to clarify the message. Table 3.15 lists presently defined ICMPv6 Code field values based upon their association with a specific Type field value. Note that if a Type field value only has a zero code field value, that entry was not placed in Table 3.15.

3.4.2 ARP

The Address Resolution Protocol (ARP) was developed as a mechanism to translate layer 3 addresses such as the Internet Protocol used by the TCP/IP protocol suite into layer 2 addresses used primarily by LAN delivery systems,

Table 3.13 ICMP Code field values based on message type

Message Type	Code Field Values	
3	Destination Unreachable	
	Codes	
	0	Net Unreachable
	1	Host Unreachable
	2	Protocol Unreachable
	3	Port Unreachable
	4	Fragmentation Needed and Don't Fragment was Set
	5	Source Route Failed
	6	Destination Network Unknown
	7	Destination Host Unknown
	8	Source Host Isolated
	9	Communication with Destination Network is Administratively Prohibited
	10	Communication with Destination Host is Administratively Prohibited
	11	Destination Network Unreachable for Type of Service
	12	Destination Host Unreachable for Type of Service
	13	Destination Host Unreachable for Type of Service
	14	Communication Administratively Prohibited
	15	Precedence cutoff in effect
5	Redirect	
	Codes	
	0	Redirect Datagram for the Network (or subnet)
	1	Redirect Datagram for the Host
	2	Redirect Datagram for the Type of Service and Network
	3	Redirect Datagram for the Type of Service and Host
6	Alternate Host Address	
	Codes	
	0	Alternate Address for Host
11	Time Exceeded	
	Codes	
	0	Time to Live exceeded in Transit
	1	Fragment Reassembly Time Exceeded
12	Parameter Problem	
	Codes	
	0	Point Indicates the Error.
	1	Missing a Required Option
	2	Bad Length
40	Photuris	
	Codes	
	0	Reserved
	1	Unknown security parameters index
	2	Valid security parameters, but authentication failed
	3	Valid security parameters, but decryption failed

Table 3.14 ICMPv6 Type field values

Type	Name
1	Destination Unreachable
2	Packet Too Big
3	Time Exceeded
4	Parameter Problem
128	Echo Request
129	Echo Reply
130	Group Membership Query
131	Group Membership Report
132	Group Membership Reduction
133	Router Solicitation
134	Router Advertisement
135	Neighbor Solicitation
136	Neighbor Advertisement
137	Redirect Message
138	Router Renumbering

Table 3.15 ICMPv6 Code field values based upon message type

Type	Name	
1	Destination Unreachable	
	Code	
	0	no route to destination
	1	communication with destination administratively prohibited
	2	not a neighbor
	3	address unreachable
	4	port unreachable
2	Packet Too Big	
	Code	
	0	
3	Time Exceeded	
	Code	
	0	hop limit exceeded in transit
	1	fragment reassembly time exceeded
4	Parameter Problem	
	Code	
	0	erroneous header field encountered
	1	unrecognized Next Header type encountered
	2	unrecognized IPv6 option encountered
138	Router Renumbering	
	Code	
	0	Router Renumbering Command
	1	Router Renumbering Result

such as Ethernet, Token-Ring, and FDDI. Technically ARP is not a layer 3 protocol, since it does not run on top of IP and its format, as we will note, is conspicuous by the absence of an IP header. Although ARP actually operates on top of the datalink layer, its translation of layer 2 to layer 3 address results in most authors discussing it after IP. This also explains why a diagram of the TCP/IP protocol suite when compared with the ISO OSI Reference Model typically shows ARP above layer 2, usually as a small block within the network layer portion of the protocol suite.

As noted earlier in this book, popular LANs use 6-byte source and destination addresses that represent universally or locally administrated addresses. A 3-byte universally administrated address prefix is assigned by the IEEE to vendors who burn a 6-byte address into network adapter cards they manufacture. The first three bytes have a fixed value that identifies the manufacturer of the network adapter card and is obtained from the IEEE. The values for the last three bytes are altered by the adapter manufacturer to uniquely identify each adapter produced. When software is used to override the burnt-in address on the adapter with a locally generated address, the addressing technique is referred to as a locally administered address.

Need for address resolution

To understand the need for address resolution, consider a station on a network that operates a TCP/IP protocol stack and needs to transmit a packet to the Internet via a router connected to the Internet and the local network. The workstation's TCP/IP stack is configured to set an IP address to the interface on the network. Similarly, the router is configured so that its interface on the local network is assigned another IP address. When the workstation needs to transfer a packet to an IP address that is not located on the local network where it resides, it must forward the packet to the router. While this appears to be a simple process, in actuality it is not. This is because data flows on the LAN in frames that use layer 2 addresses. Thus, while the workstation knows the router's IP address, it needs to determine the router's MAC address so it can transmit one or more LAN frames containing the layer 3 packet to the router. A similar but opposite problem occurs when a communications device knows an IP address but needs to find the associated IP address. Both of these translation problems are handled by protocols developed to provide an address resolution capability. One protocol, known as the Address Resolution Protocol (ARP), translates an IP address into a hardware (layer 2) address. The Reverse Address Resolution Protocol (RARP), as its name implies, performs a reverse translation or mapping, converting a hardware layer 2 address into an IP address.

Operation

Figure 3.19 illustrates the format of an ARP packet. Note that the numbers contained in some fields represent the byte position when the field spans a

0	8	16	31
Hardware Type		Protocol Type	
Hardware Length	Protocol Length	Operation	
SENDER HARDWARE ADDRESS (0–3)			
SENDER HARDWARE ADDRESS (4–5)		SENDER IP ADDRESS (0–1)	
SENDER IP ADDRESS (2–3)		TARGET HARDWARE ADDRESS (0–1)	
TARGET HARDWARE ADDRESS (2–5)			
TARGET IP ADDRESS			

Figure 3.19 The Address Resolution Protocol (ARP) packet format

4-byte boundary. When a station knows a destination IP address but needs to learn the destination layer 2 address associated with the destination IP address, it will transmit an ARP packet as a broadcast frame.

By setting the target IP address in the last field in the packet, the sender provides a mechanism for each station on the network to determine if it has the desired IP address. If so, the station will return the packet to the originator after it fills in its hardware address, enabling the sender to create a frame using an appropriate MAC address to transport data to the desired IP address. Now that we have an overview of the manner by which ARP operates, let us turn our attention to each field in the header.

Hardware Type field The first field in the ARP packet is a 16-bit Hardware Type field. A value is placed in this field to indicate the type of hardware generating the ARP. Table 3.16 lists presently defined Hardware Type field values.

Protocol Type field The Protocol Type field indicates the protocol for which an address resolution process is being performed. Thus, as you might surmise, ARP can be used to resolve addresses between many types of network protocols and hardware addresses. For IP the Protocol Type field has a value of hex 0800.

Hardware Length field The Hardware Length field denotes the number of bytes in the hardware address. Similar to the Protocol Type field permitting many types of network addresses to be resolved, the Hardware Length field permits many types of hardware products to include shared media LANs and ATM switch-based networks to be identified. The value of this field is 6 for both Ethernet and Token-Ring network adapter cards.

Protocol Length field The Protocol Length field is similar to the Hardware Length field, indicating the length of the address in bytes for the protocol to

Table 3.16 ARP Hardware Type field values

Field Value	Hardware Type (hrd)
1	Ethernet (10 Mb)
2	Experimental Ethernet (3 Mb)
3	Amateur Radio AX.25
4	Proteon ProNET
5	Chaos
6	IEEE 802 Networks
7	ARCNET
8	Hyperchannel
9	Lanstar
10	Autonet Short Address
11	LocalTalk
12	LocalNet (IBM PCNet or SYTEK LocalNET)
13	Ultra link
14	SMDS
15	Frame Relay
16	Asynchronous Transmission Mode (ATM)
17	HDLC
18	Fibre Channel
19	Asynchronous Transmission Mode (ATM)
20	Serial Line
21	Asynchronous Transmission Mode (ATM)
22	MIL-STD-188-220
23	Metricom
24	IEEE 1394.1995
25	MAPOS
26	Twinaxial
27	EUS-64

be resolved. For IPv4 the value of this field is 4, while for IPv6 the value of this field is 16.

Operation field The Operation field has a value of 1 for an ARP Request. When a target station responds, the value of this field is changed to 2 to denote an ARP reply. This explains how the same ARP packet can be transmitted as a broadcast to all stations on a LAN and returned with the resolved layer 2 address to the originator.

Sender Hardware Address field The Sender Hardware Address field is 6 bytes in length. The entry in this field indicates the address of the station generating the ARP Request or ARP Reply.

Sender IP Address field Because ARP is used to resolve an IP address the station generating the ARP request must have an operational TCP/IP stack. Thus, the sender's IP address is placed in this field.

Target Hardware Address field Because a station is attempting to determine the target hardware address, this field is set to a value of zero in an ARP request. When a destination station recognizes its IP address, it enters its hardware address in this field.

Target IP Address field The last field in the ARP packet is the Target IP address field. A station requiring the hardware address associated with a particular IP address enters the IP address in this field. When this field is completed and returned in a packet with an Operation field value of 2, the address resolution process is completed.

Because the need for address resolution occurs quite frequently, the broadcasting of frames transporting ARP packets and the appropriate ARP reply can adversely effect the utilization of the local network. To lower the level of network utilization as well as facilitate the prompt resolution of previously learned addresses, the originator of ARP requests maintains a table of resolved addresses referred to as an ARP cache. The use of the ARP cache allows subsequent datagrams with previously learned correspondences between IP addresses and MAC addresses to be quickly noted.

Under the ARP standard, devices on a network also update their ARP tables with MAC and IP address pairs found in ARP replies that flow on the network. Most modern operating systems include an ARP command that provides you with the ability to view the contents of the ARP cache as well as to modify entries in the cache. In a Windows NT environment you can use its ARP command to both view and alter ARP cache entries. Figure 3.20 illustrates the format of the ARP command while Figure 3.21 illustrates the use of the command with its –a option to view the contents of the ARP cache.

ARP notes

In concluding our examination of ARP there are two additional items that we should note. These items are what are referred to as gratuitous ARP and proxy ARP.

A gratuitous ARP occurs when a communications device initializes its network connection. This type of ARP results in a device issuing an ARP request for its own IP address by entering its address in the Target IP Address field. While this action may appear silly, after all, the protocol stack needs to be configured with an IP address, assuming a response occurs from a different hardware address contained in the Target Hardware Address field in the ARP Reply. This would indicate that another device on the network is using the IP address configured on the device, and explains how the IP address conflict error message occurs.

The second ARP variation is a proxy ARP. A proxy ARP permits a device to andwer an ARP Request on behalf of another device. To understand the rationale for a proxy ARP, consider a Class C network subdivided into two or more subnets via the use of a router. If the originating device erronously had a standard Class C subnet mask of 255.255.255.0, it would assume that any address within its Class C network resides on its own physical network and if

```
C:\>arp

  Displays and modifies the IP-to-Physical address translation tables used by
  address resolution protocol (ARP).

  ARP -s inet_addr eth_addr [if_addr]
  ARP -d inet_addr [if_addr]
  ARP -a [inet_addr] [-N if_addr]

  -a            Displays current ARP entries by interrogating the current
                protocol data. If inet_addr is specified, the IP and Physical
                addresses for only the specified computer are displayed. If
                more than one network interface uses ARP, entries for each ARP
                table are displayed.
  -g            Same as -a.
                inet_addr Specifies an internet address.
  -N if_addr    Displays the ARP entries for the network interface specified by if_addr.
  -d            Deletes the host specified by inet_addr.
  -s            Adds the host and associates the Internet address inet_addr
                with the Physical address eth_addr. The Physical address is
                given as 6 hexadecimal bytes separated by hyphens. The entry is permanent.
  eth_addr      Specifies a physical address.
  if_addr       If present, this specifies the Internet address of the
                interface whose address translation table should be modified.
                If not present, the first applicable interface will be used.

  C:\>
```

Figure 3.20 Windows NT ARP command options

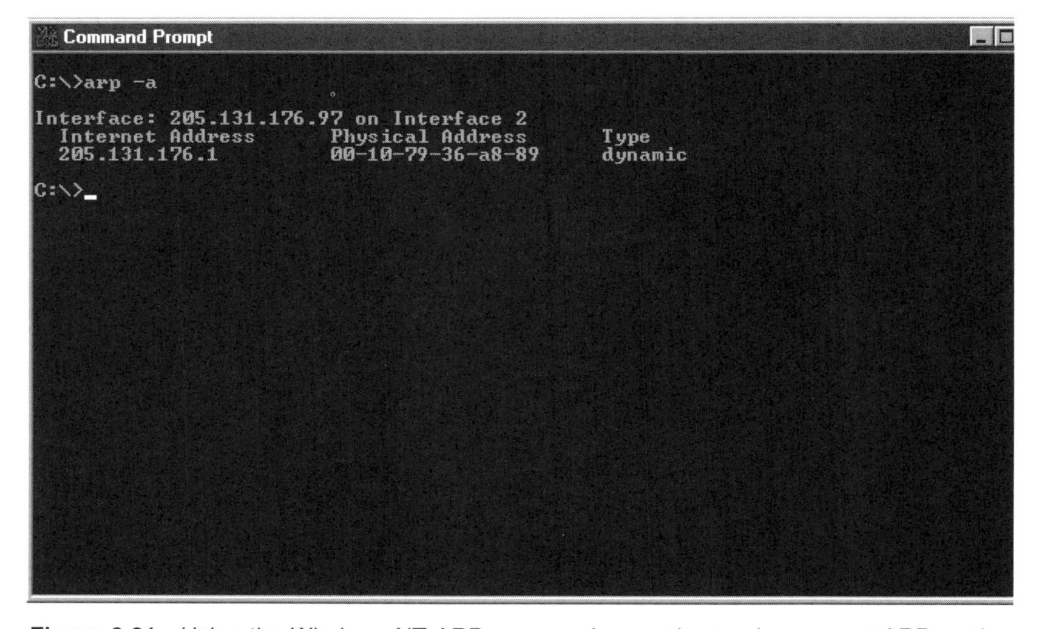

Figure 3.21 Using the Windows NT ARP command –a option to view current ARP entries

powered on should respond to an ARP Request. However, if the destination was on the other side of the router, it would not receive the ARP. However, because the router must be configured to note each subnet, it can answer an ARP Request on behalf of other devices on the destination subnet by entering its own layer 2 address in an ARP Reply. This results in the originating device placing the router's MAC address in its ARP cache and sending packets to the router, which forwards them onto the correct subnet. However, it should be noted that not all routers support a proxy ARP capability, and those that do may require you to enable this feature.

4

THE TRANSPORT LAYER

In the TCP/IP protocol suite the transport layer permits multiple applications to flow to a common destination, either from the same source or from different data originators. The transport layer resides between the network layer and the application layer, receiving application data, encapsulating the data with a transport header that identifies the type of application, and providing the encapsulated data and header to the network layer for transmission onto the network.

Included in the TCP/IP protocol suite are two transport layer protocols—the Transmission Control Protocol (TCP) and the User Datagram Protocol (UDP)—both of which are the focus of this chapter. TCP represents a connection-oriented, reliable transport protocol that creates a virtual circuit for the transfer of information. In comparison, UDP represents a connection-less, best-effort transport layer protocol. As we turn our attention to each protocol in this chapter we will note their suitability for different applications. For example, a connection-oriented protocol requires a significant amount of setup time when used to transport small packets of information. Recognizing the fact that different applications will have different requirements concerning both the need to establish a connection prior to transmitting data and the need to acknowledge the fact that data was delivered correctly and without error resulted in the support of two transport protocols by the TCP/IP protocol suite. As we investigate each protocol it will become clearer that there are valid reasons for some applications using one transport protocol while other applications are designed to use a different transport protocol.

4.1 TCP

As previously noted, TCP represents a connection-oriented, reliable transport protocol. The need for this type of protocol was recognized when the TCP/IP protocol suite was being developed during the 1970s. At that time local area networks were in the conceptual phase of development and TCP/IP hosts communicated via analog serial lines that were prone to transmission errors. This meant that a transport protocol was required that could detect missing packets as well as transmission errors that occurred in the contents of a packet, resulting in TCP using sequence numbers to order packets and to note when one or more became missing, as well as using an error detection algorithm and acknowledgments to detect errors occurring within a packet and then attempting to correct errors by requesting the retransmission of erroneous packets. Because an understanding of the manner by which the TCP protocol operates is facilitated by an understanding of the fields in the protocol header, we will turn our attention to the header prior to examining TCP's data flow.

4.1.1 The TCP header

In Chapter 3 we noted that IP does not provide a guaranteed delivery mechanism to enable datagrams to be correctly received with respect to both their content and their sequence. Instead, reliable delivery is provided at the transport layer through the use of the TCP protocol. As we will shortly note, the TCP header includes fields that are used to ensure datagrams are received correctly with respect to both their content and sequence. Another important function of the TCP header is to denote the type of application data carried by each datagram. This function is accomplished by the use of port fields that identify the process or application transported in the datagram. In actuality, the TCP header plus data is referred to as a segment, resulting in the port number identifying the type of data in the segment. When the IP header is prefixed to the TCP segment, the resulting datagram contains the source and destination IP address and enables the segment to be delivered via the network. Concerning the TCP segment, at the transport layer TCP accepts application data in chunks up to 64 Kbytes in length. Those chunks are fragmented into a series of smaller pieces representing segments, which become datagrams through the addition of IP headers. TCP segments are commonly 512 or 1024 bytes in length, with the actual length used dependent upon the length supported by source and destination networks as well as any intermediate networks. As we will note later in this chapter, the selection of an appropriate datagram length requires the TCP layer at each end to denote the maximum length they support and select the smallest mutually supported length. Figure 4.1 illustrates the fields of the TCP header. By examining the contents and utilization of each field in the header, we can obtain an appreciation for various protocol functions that will be discussed and described once we complete our examination of the header fields.

Source and Destination Port Fields

The Source and Destination Port fields are 16 bits in length. The source port contains a port number that theoretically denotes the application associated with the data generated by the originating station. The reason the term

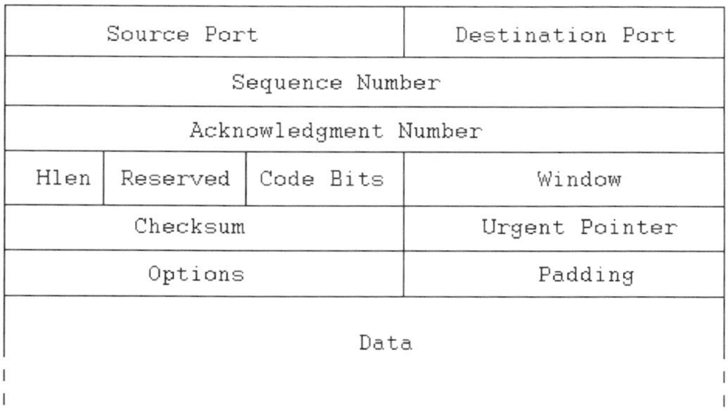

<table>
<tr><td colspan="3">Source Port</td><td>Destination Port</td></tr>
<tr><td colspan="4">Sequence Number</td></tr>
<tr><td colspan="4">Acknowledgment Number</td></tr>
<tr><td>Hlen</td><td>Reserved</td><td>Code Bits</td><td>Window</td></tr>
<tr><td colspan="3">Checksum</td><td>Urgent Pointer</td></tr>
<tr><td colspan="3">Options</td><td>Padding</td></tr>
<tr><td colspan="4">Data</td></tr>
</table>

Figure 4.1 The TCP header

'theoretically' was used is that in most transmissions the source port number is randomly generated by the originator. If the source port is not used, its value is set to 0. In comparison, the Destination Port field contains a port number that identifies a user process or application for the receiving station, enabling it to distinguish different applications transported from a common location. For example, when a station initiates a file transfer, it might open FTP to transfer data using port number 1234 as the source port while later in the day a second file transfer might occur with the station using source port 2345. However, for all FTP transfers the destination port would be fixed at 21, which is the standard port number for which FTP incoming data is received. When the destination station receives the incoming data, it responds by creating a segment and placing the source port number in the destination port field. This action enables the file originator to correctly identify the response to its datagram. As we will note later in this section, each TCP segment contained in an IP datagram has a segment number that enables the opposite end of the transmission to ensure that datagrams are received in their correct order and none are lost. Because there are three types of port numbers that can be used in the Port fields, and both TCP and UDP headers have the same Source and Destination Port fields, we will first turn our attention to the port number universe prior to continuing our examination of the fields in the TCP header.

Port numbers Both TCP and UDP headers contain 16-bit Source and Destination Port fields, permitting port numbers in the range of 0 through 65 535. This results in 65 536 distinct port numbers being available for utilization. This 'universe' of port numbers is subdivided into three ranges, referred to as well-known ports, registered ports, and dynamic or private ports.

Well-known ports Well-known ports are also referred to as assigned ports, as their assignment is controlled by the Internet Assigned Numbers Authority (IANA).

Well-known or assigned ports are in the range of 0 through 1023 and are used to indicate the transportation of standardized processes. Where possible, the same well-known port number assignments are used with TCP and UDP. Ports used with TCP commonly provide connections that transport relatively long-term conversations, such as file transfers and remote access. In dated literature references to well-known port numbers are specified as being in the range of values from 0 through 255. While that range was correct many years ago, the range for assigned ports managed and controlled by the IANA is now from 0 through 1023.

Table 4.1 provides a summary of some of the more popular well-known ports to include the service supported by a particular port and the type of port (TCP or UDP) for which the port number is primarily used. Note that such common applications as Telnet, File Transfer Protocol (FTP), and World Wide Web (HTTP) traffic are transported via the use of well-known ports.

Registered port numbers Assigned port numbers range from 0 through 1023 out of the universe of 65 536 available numbers. Port numbers that exceed 1023 can be used by any process or application; however, doing so in a haphazard manner could create incompatibilities between vendor products. Recognizing this potential problem, the IANA permits vendors to register the use of port numbers. The result is the use of the range of port number values from 1024 through 49 151 for registered ports. Although an application or process may be registered, its registration does not hold legal implications, and is primarily to enable other vendors to develop compatible products as well as to enable end-users to set up equipment appropriately. For example, if a new application uses a registered port number, it is relatively easy to adjust a router access list or firewall to enable many compatible products behind the router or firewall to interact with the new application. Although developers can use any port number beyond 1023, many respect registered port numbers.

Dynamic port numbers Dynamic port numbers are in the range from 49 152 through 65 535. Port numbers in this range are typically used by vendors implementing proprietary network applications, such as a method to transmit digitized voice. Another common use of dynamic port numbers includes the random selection of a port number by certain applications.

Sequence Number field

The third field in the TCP header is a 32-bit Sequence Number field. Unless a bit in the Code Bit field known as the SYN bit is set, each TCP segment is assigned a number. That number reflects the number of bytes in a TCP packet, as TCP represents a byte-oriented sequencing protocol. That is, a byte-oriented protocol results in every byte in each packet being assigned a sequence number. However, it is important to note that this does not mean that TCP transmits a packet containing a single byte. What it means is that TCP will transmit a group of bytes and assign the packet a sequence number based upon the number of bytes in the packet's data field. For example, assume that a station transmits three packets to the same destination with the first two packets each containing 512 bytes of data, while the third packet

Table 4.1 Well-known TCP and UDP services and port utilization

Keyword	Service	Port Type	Port Number
TCPMUX	TCP Port Service Multiplexer	TCP	1
RJE	Remote Job Entry	TCP	5
ECHO	Echo	TCP and UDP	7
DAYTIME	Daytime	TCP and UDP	13
QOTD	Quote of the Day	TCP	17
CHARGEN	Character Generator	TCP	19
FTD-DATA	File Transfer (Default Data)	TCP	20
FTP	File Transfer (Control)	TCP	21
TELNET	Telnet	TCP	23
SMTP	Simple Mail Transfer Protocol	TCP	25
MSG-AUTH	Message Authentication	TCP	31
TIME	Time	TCP	37
NAMESERVER	Host Name Server	TCP and UDP	42
NICNAME	Who Is	TCP	43
DOMAIN	Domain Name Server	TCP and UDP	53
BOOTPS	Bootstrap Protocol Server	TCP	67
BOOTPC	Bootstrap Protocol Client	TCP	68
TFTP	Trivial File Transfer Protocol	UDP	69
FINGER	Finger	TCP	79
HTTP	World Wide Web	TCP	80
KERBEROS	Kerberos	TCP	88
RTELNET	Remote Telenet Service	TCP	107
POP2	Post Office Protocol Version 2	TCP	109
POP3	Post Office Protocol Version 3	TCP	110
NNTP	Network News Transfer Protocol	TCP	119
NTP	Network Time Protocol	TCP and UDP	123
NETBIOS-NS	NetBIOS Name Server	UDP	137
NETBIOS-DGM	NetBIOS Datagram Service	UDP	138
NETBIOS-SSN	NetBIOS Session Service	UDP	139
NEWS	News	TCP	144
SNMP	Simple Network Management Protocol	UDP	161
SNMTTRAP	Simple Network Management Protocol Traps	UDP	162
BGP	Border Gateway Protocol	TCP	179
HTTPS	Secure HTTP	TCP	413
RLOGIN	Remote Login	TCP	513
TALK	Talk	TCP and UDP	517

contains 1024 bytes of data. The first packet would have a sequence number of 512 while the second packet would have a sequence number of 1024. Then, the third packet would have a sequence number of 2048. At the receiving station the fact that the first packet was received with a sequence number of 512 and contains 512 bytes of data tells the receiver to expect the next sequence number to be 1024. Similarly, upon receipt of the second packet with a sequence number of 1024, the counting of 1024 data bytes results in the receiver expecting the next sequence number to be 2048.

Acknowledgment Number field

The fourth field in the TCP header is the Acknowledgment Number field. This 32-bit field contains the number that indicates the next sequence number the destination expects to receive. Similar to the sequence number field, the Acknowledgment Number field value is based upon a data byte count, and it indicates the next byte the receiver expects to receive. Because it would be inefficient to acknowledge each datagram when transmission occurs over relatively error-free circuits, the TCP protocol supports a variable window. For example, returning an Acknowledgment Number field value of 2049 would indicate the receipt of all data that was transmitted by the prior three datagram examples consisting of 512, 512, and 1024 bytes of data. To ensure that lost datagrams or lost acknowledgments do not place this transport protocol in an infinite waiting period, the originator will retransmit data if it does not receive a response within a predefined period of time. This period is controlled by one of several TCP timers, whose operation will be described later in this chapter.

The use of the Acknowledgment Number field is referred to as Positive Acknowledgment or Retransmission (PAR). PAR requires that each unit of data must be explicitly acknowledged. If a unit of data is not acknowledged by the time the originator's time-out period is reached, the previous transmission is retransmitted. When the Acknowledgment field is in use, the second bit position in the Code Bit field, which is the ACK flag, is set.

Hlen field

The Hlen field is 4 bits in length and contains a value that indicates the length of the header in 32-bit words. Thus, the Hlen field denotes the location where the TCP header ends and data being carried by the protocol begins.

The rationale for the use of the Hlen field results from the fact that the potential inclusion of options results in a variable length header. Without a field that indicates where the header ends, it would not be possible to support a variable length header. Because the minimum length of the TCP header is 20 bytes, the minimum value of the Hlen field is five 32-bit words. It should be noted that in many references to TCP the Hlen field is referred to as the Offset field.

Reserved field

Within the TCP header are 6 bits that are reserved for future use. The values of those bits are presently set to all zeros.

Code Bit fields

There are 6 bits in the TCP header whose individual settings control different functions. These bits are collectively referred to as code bits. They are also upon occasion referred to as control flags.

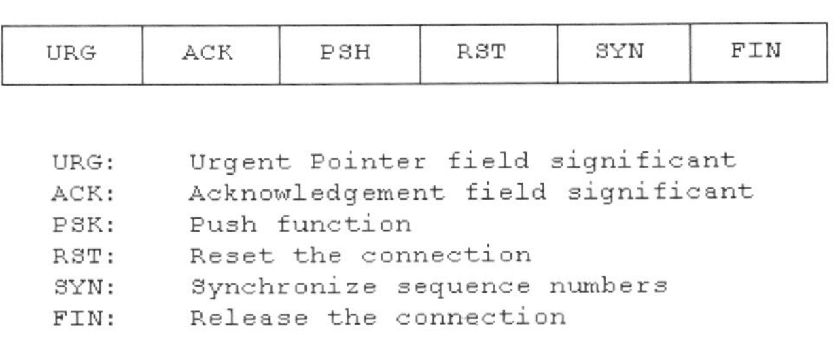

Figure 4.2 The flags in the Code Bit field

Figure 4.2 illustrates the position of the individual bits in the Code Bit field. Those bits include an urgent pointer (URG), acknowledgment (ACK), push function (PSH), connection reset (RST), sequence number synchronization (SYN), and connection release (FIN).

URG bit The urgent pointer bit or flag is set to denote an urgent or priority activity, such as when a user presses the CTRL–BREAK key combination. When a predefined urgent or priority activity occurs, the application sets the URG bit, which results in TCP immediately transmitting everything it has for the connection. If a destination station previously closed its Receive window, the setting of the URG bit tells the receiver to accept the packet. When the URG bit is set, it also indicates that the urgent pointer field is in use. That field contains the offset in bytes from the current sequence number where the urgent data is located.

ACK bit The second bit in the Code Bit field is the ACK bit. When set, this bit indicates that the segment contains an acknowledgment to a previously received datagram or series of datagrams.

PSH bit The third bit in the Code Bit field is the PSH (push) bit. Setting this bit tells the receiver to immediately deliver data to the application and forgo any buffering.

RST bit The fourth bit in the Code Bit field is the RST (reset) bit. The setting of this bit resets the current connection.

SYN bit The fifth bit in the Code Bit field is SYN (synchronization) bit. This bit is set at TCP startup as well as to inform a receiver that a sequence number is being established to synchronize sender and receiver.

FIN bit The sixth and last bit in the Code Bit field is the FIN (finish) bit. This bit is set to indicate the sender has no additional data to send and the connection should be released.

Window field

The eighth field in the TCP header is the Window field. This field is 2 octets in length and contains the number of data octets beginning with the one indicated in the Acknowledgment field that the sender of this segment is willing to accept. Thus, the use of the Window field provides a method of flow control between the source and destination. A large value can significantly improve the TCP performance, as it permits the originator to transmit more data without having to wait for an acknowledgment, while permitting the receiver to acknowledge the receipt of data carried in multiple segments with one acknowledgment.

The use of the Window field permits a variable window to be created that governs the amount of data that can be transmitted prior to requiring an acknowledgment. A window can be established in each direction between sender and receiver, since TCP is a full-duplex transmission protocol. This means that both ends of a transmission session can use the Window field to provide a bi-directional flow control mechanism. For example, by reducing the value of the Window field, one end of the conversation informs the other end to transmit less data.

Checksum field

The 16-bit Checksum field provides an error detection and correction capability for TCP. Instead of computing the checksum over the entire TCP header, the checksum calculation occurs over what is referred to as a 12-octet pseudo header. This header extends into the IP header and consists of such key fields as the 32-bit Source and Destination address fields in the IP header, the 8-bit Protocol field, and a Length field that denotes the length of the TCP header and data transported within the TCP segment. Thus, the checksum can be used to ensure that data arrives at its correct destination and the receiver has no doubt about the address of the originator or the length of the header and types of application data transported by TCP.

Urgent Pointer field

As previously noted, the setting of the URG bit in the Code Bit field indicates that the Urgent Pointer field is in use. This field contains the offset in octets from the current sequence number where the urgent data is located.

Options field

The Options field is variable in length and enables a host to specify options required by the TCP protocol. For example, through the use of values in this field a host can specify the use of an alternative checksum, a maximum segment size, or the use of authentication.

Table 4.2 lists currently defined option numbers used to specify the use of optional header fields. In examining the entries in Table 4.2 note that options 0 and 1 are exactly one octet in length. All other options have their 1 octet number field followed by a 1 octet length field, which in turn is followed by actual option data.

Padding field

Because the use of an option results in a variable length header, a padding field is included to ensure that the header ends on a 32-bit boundary. If the Options field is omitted or the use of an option results in the header following on a 32-bit boundary, the Padding field is not used.

4.1.2 Operation

As briefly noted earlier in this chapter, TCP is a connection-oriented protocol. This means that a connection between source and destination must be established prior to the transmission of data between the two. Within a TCP/IP protocol stack, applications such as FTP and Telnet initiate a connection request through the use of function calls. For example, an OPEN call will be requested by an application when a connection is required, while a CLOSE call is requested to terminate a previously established connection.

Table 4.2 TCP options numbers

Number	Length	Meaning
0	—	End of Option List
1	—	No-Operation
2	4	Maximum Segment Size
3	3	WSOPT—Window Scale
4	2	SACK Permitted
5	N	SAC
6	6	Echo (Obsoleted by option 8)
7	6	Echo Reply (Obsoleted by option 8)
8	10	TSOPT—Time Stamp Option
9	2	Partial Order Connection Permitted
10	3	Partial Order Service Profile
11		CC
12		CC.NEW
13		CC.ECHO
14	3	TCP Alternate Checksum Request
15	N	TCP Alternate Checksum Data
16		Skeeter
17		Bubba
18	3	Trailer Checksum Option
19	18	MD5 Signature Option

TCP alternate checksum numbers

Number	Description
0	TCP Checksum
1	8-bit Fletchers' algorithm
2	16-bit Fletchers' algorithm
3	Redundant Checksum Avoidance

Connection types

There are two types of connections many applications can establish: active and passive. An active connection is initiated by the issuance of an OPEN call when a connection to a remote station is required. In comparison, a passive connection occurs when an OPEN call is issued to enable connections to be received from a remote station. Because there are limits on the number of passive OPENs that can be issued by an application, one hacker attack commonly used is to flood an application on a target host with requests from randomly generated IP addresses.

Figure 4.3 illustrates an example of the TCP connection establishment process. In this example the host at 205.131.175.10 is initiating an FTP session with the server at IP address 205.131.175.20. Here the station at the first IP address issues an active OPEN call to the server at the second address. To enable a connection to be made, the server must have previously issued a passive OPEN request to enable incoming connections to be established. If a passive OPEN request was previously issued, the server's operating system will then generate a separate process to maintain the connection. Concerning this connection, the incoming datagram will contain the destination port number, which tells the server the application the originator wishes to use on the server. Thus, a server that supports multiple applications can conceivable accept certain applications while rejecting others.

The three-way handshake

A TCP session begins with a 'three-way handshake.' Only after the source and destination have exchanged datagrams via the three-way handshake will a pending connection become active.

The three-way handshake results in the exchange of three messages: SYN, SYN/ACK, and ACK. When a client application wants to establish a connection, the application places an active OPEN call to TCP, requesting a connection to an application on a remote station. The client TCP creates a header, setting the SYN bit in the Code Bit field and assigning an initial sequence number that can start at any value, including 0. After setting values in other header fields, the client sets a timer and passes the TCP segment to IP. IP adds its header, generating a datagram that is transmitted onto the network to its destination.

At the destination, assuming an appropriate passive OPEN was issued, the receiver creates a segment in response to the inbound SYN segment. The responding segment includes the setting of the SYN and ACK bits in the code field in the TCP header and the setting of appropriate values in the Sequence Number and Acknowledgment Number fields. Concerning the values in those fields, the Sequence Number field can be set to any value while the Acknowledgment Number field must contain a value of the sequence number in the inbound segment plus one. The TCP header is then passed to IP, which creates what is referred to as a SYN/ACK packet that must be received by the originator prior to its timer expiring. At the client, the SYN/ACK packet is noted as an acknowledgment to its connection request. The client then creates

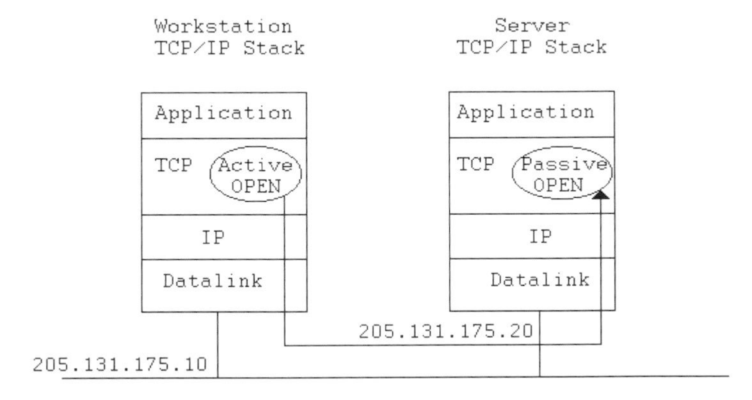

Figure 4.3 The TCP connection establishment process

a new TCP segment, setting the ACK bit in the Code Bit field, filling in the acknowledgment number to an appropriate value, and sending the segment to IP for the creation of a datagram. This datagram, which represents a TCP ACK, completes the three-way handshake. While this action completes the three-way handshake, the client also sets a retransmission timer based upon the round-trip delay occurring between the SYN and SYN/ACK datagrams. If succeeding data packets are not ACKed when the timer expires, this informs the client to retransmit a previously transmitted datagram.

Figure 4.4 illustrates an example of a three-way handshake between a client and server via the use of a time chart. In this example the client is shown setting the SYN bit in the Code Bit field and a value of 0 being placed in the Sequence Number field. The server is shown responding with an initial sequence number of 100 and sets the value of its Acknowledgment Number field to the received sequence number plus 1, or 1. The setting of the SYN and ACK bits in the TCP header's Code Bit field results in the response being referred to as a SYN/ACK response. In response to the SYN/ACK, the client constructs a response, setting the ACK bit in the Code Bit field, setting the acknowledgment number to one more than the received sequence number (101), and incrementing its sequence number from 0 to 1. Once the client has responded to the SYN/ACK with an ACK, it can then transmit one or more data segments.

Segment size support

When TCP transmits data, the number of segments that can be transmitted and the maximum length or size of each segment are governed by two different control mechanisms. The number of segments that can be transmitted depends upon the Window Field value, which we will cover in the next section. In comparison, the Maximum Segment Size (MSS) is set during the establishment of each TCP connection and is determined independently by each station via the use of one of the following algorithms.

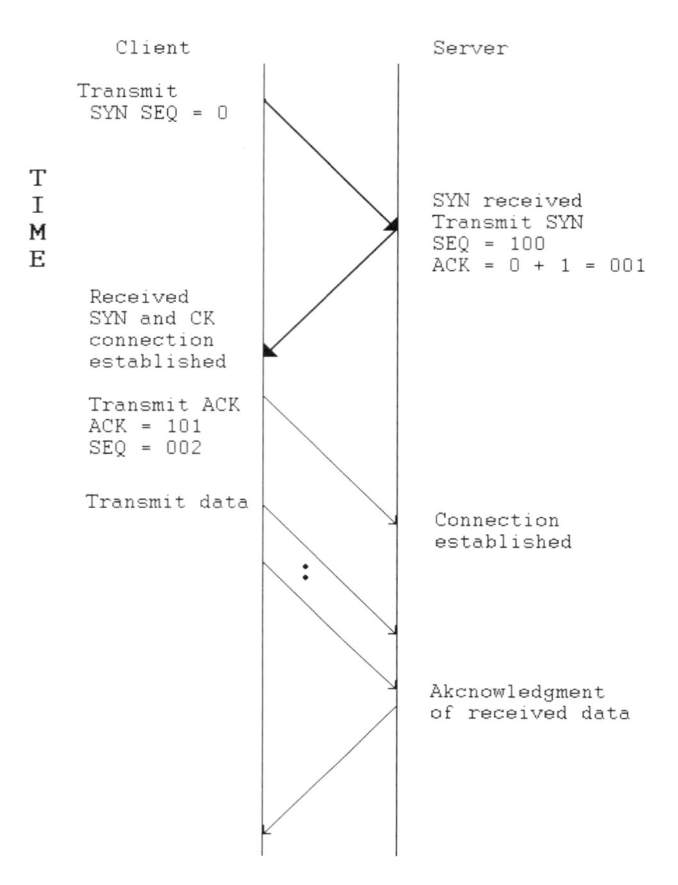

Figure 4.4 An example of the three-way handshake

During the exchange of SYN and SYN/ACK segments, each end station transmits their desired MSS to the remote endpoint. The desired MSS is the Maximum Transmission Unit (MTU) of the interface being used minus the IP and TCP header overhead, typically 40 bytes. Each end-station derives the MSS that will be used as the minimum of the MSS received in the SYN packet and the MTU of the local interface minus the IP and TCP header overhead. When two networks with differing MTUs are connected, the resulting MSS used is the smaller of the two, permitting communications without requiring IP fragmentation. Now that we have an appreciation for the manner by which a segment size is selected, let us turn our attention to the use of the Window field and how it can be used to control the flow of data on and between networks.

The Window field and flow control

We briefly noted earlier in this section that the Window field in the TCP header provides the mechanism that governs the number of segments that can be outstanding at any one point in time. Stations adjust the value in the Window field based upon their ability to process a series of segments, in effect providing a mechanism to regulate the flow of data between source and destination.

As TCP processes data in memory, it places a window over the data as it is structured into segments. The movement of the window over data results in a sliding window as more and more data is processed into segments.

Figure 4.5 illustrates an example of a TCP sliding window. In this example segments used with sequence numbers 200 through 205 were both transmitted to their destination and acknowledged. Segments with sequence numbers 206 through 210 were transmitted but have yet to be acknowledged. Segments with sequence numbers 211 through 215 are still in the originating station and are waiting to be transmitted, while segments with sequence numbers 216 through 218 represent data to be sent that is not presently in the window.

It is important to note that the length of the sliding window is variable and is controlled through the use of the Window field in the TCP header. If the receiving station is running out of buffer space, it can tell the originator to slow its transmission by reducing the value in the Window field returned in an acknowledgment. By setting the value in the Window field to zero, a receiving station informs the originator that it cannot accept additional data. The receiving device will continue to send datagrams with a Window field value of 0 until it cleans its buffer area to the point where it can again accept data. At that time it will transmit a datagram with a Window field value other than zero, indicating it is now capable of receiving data.

Timers

TCP retransmissions as well as the termination of an existing connection are governed by different types of timers. Four key timers used by TCP are a delayed ACK, FIN-WAIT-2, Persist, and Keep Alive timers.

Delayed ACK When TCP receives data that does not require an immediate acknowledgment, it can set a delayed ACK timer. This timer can be set to a value up to 20 ms, and enablesTCP to form a segment of data that can be transmitted with the ACK, conserving bandwidth.

FIN-WAIT-2 timer To gracefully terminate an existing connection,TCP sets the FIN bit in the Code Bit field in theTCP header. Upon receipt of the datagram, the destination station enters a FIN-WAIT-1 state. Because TCP supports a full-

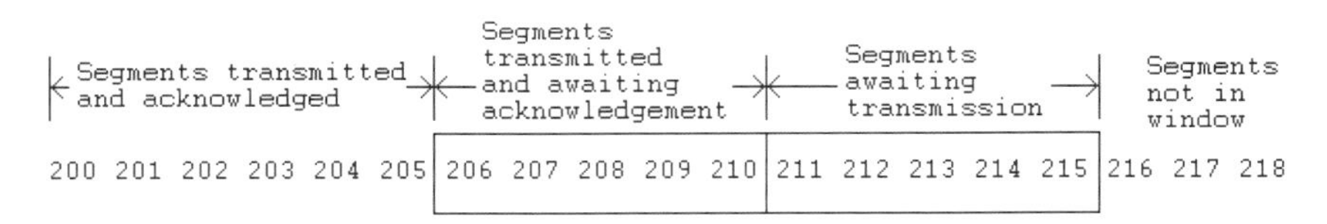

Figure 4.5 The TCP sliding window in operation

duplex connection, the originator expects an acknowledgment. Upon receipt of the acknowledgment, the originator enters a FIN-WAIT-2 state. At this time a FIN-WAIT-2 timer is set. The purpose of the timer is to avoid leaving the connection in a FIN-WAIT-2 state in the event that a final FIN is not received from the destination.

The FIN-WAIT-2 timers can be set to a value up to 10 minutes. Upon expiration, it is reset to 75 seconds, after which the connection is dropped.

Persist The Persist timer is set when the other side of a connection advertises a window size of zero, stopping TCP from transmitting additional data. Because it is possible that a datagram containing a window advertisement could be lost, TCP will transmit one byte of data after its Persist timer expires to see if the window opened up.

Keep Alive Once a connection has been established, there can be periods of time when data is not exchanged. During such periods, the Keep Alive timer is used to detect if the connection is up. That is, if the connection is idle for a fixed period of time, the Keep Alive timer expires, resulting in a probe segment being transmitted to the other side of the connection in an attempt to illicit a response. If no response is received, further attempts are optional and the connection will be closed.

Slow start and congestion avoidance

Early versions of the TCP/IP protocol stack enabled TCP to transmit multiple segments up to the window size generated by the receiver's acknowledgment during the initial phase of communications. While the ability to transmit a large amount of data when sender and receiver were on the same network rarely caused problems, if they were on different networks then intermediate devices and transmission facilities such as routers and serial communications links could easily become saturated, causing router buffers to overflow, packets to be dropped, and retransmissions to occur that would further compound an already bad situation. This potential situation resulted in the addition of a slow start mechanism to TCP as a method to avoid the generation of network congestion during and after startup.

Under slow start, TCP initially sets the value in its Window field to 1. Each time an ACK is received, the value in the returned Window field is increased exponentially. That is, the sender first transmits one segment and upon receipt of an ACK transmits two more. The next ACK results in the transmission of 4 segments, while the following ACK permits the sender to transmit 8 segments. The sender will continue to increase the number of segments transmitted at one time until the value reaches that provided during the initialization period. The value established during the initialization period is often referred to as the TCP advertised window value, while the value in the Window field during the slow start process is often referred to as the congestion window value.

Once the sender obtains the ability to transmit a group of segments that corresponds to the number advertised by the receiver, it will continue to do so until a time-out or duplicate ACKs are received. When either situation occurs, the sender compares the size of the congestion window and the TCP advertised window, selects and halves the smaller of the two, and stores it as a slow start threshold. The value stored must be a minimum of 2 segments; however, if a time-out occurred, the value will be stored as 1, resulting in a slow start process commencing anew. If a value exceeding 1 is stored, a congestion avoidance algorithm is used that results in the potential linear growth in the number of segments that can be transmitted in response to each ACK. This algorithm first multiplies the segment size by 2. Next, it divides the resulting value by the value of the congestion window, and then increases the resulting segment size using this algorithm each time an ACK is received. In comparison with slow start, the congestion avoidance algorithm results in a more controlled rate of growth in the number of segments that can be transmitted each time an ACK is received. Both slow start and congestion avoidance algorithms are now incorporated into modern TCP/IP protocol stacks. Through the use of these algorithms, a significant amount of network overloading is prevented; however, this is not without cost. The key cost associated with slow start is the inability of the protocol to recognize that a higher transfer capability could be used due to available network bandwidth. This means that TCP transfers are always initially slow, and are only enhanced as ACKs are received.

4.2 UDP

The User Datagram Protocol (UDP) represents the second network layer protocol in the TCP/IP protocol suite. UDP is based upon the datagram method of transport for applications for which an occasional lost packet is not considered serious. Thus, UDP represents a connectionless, unreliable, best-effort transport service. This means that UDP does not issue acknowledgments to the originator upon receipt of data nor provide order to incoming datagrams. Thus, UDP does not provide error detection or the capability to recover from the situation where packets become lost. Instead, it is up to the application to detect lost or missing data, typically by noting the absence of a response within a predefined period of time and then, if appropriate, retransmitting the data that was presumed to be lost.

4.2.1 The UDP header

Figure 4.6 illustrates the fields in the UDP header. In examining the fields in the header, you will note that it has a relatively low overhead in comparison with TCP, which enables a higher level of data transfer. Also note the absence of a Window field, which results in UDP being unable to adjust its rate of data

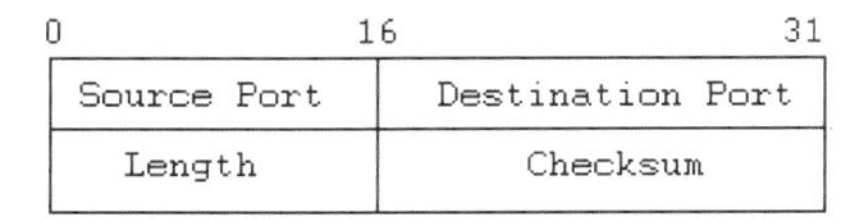

Figure 4.6 The UDP header

flow based on network conditions. Although UDP is similar to TCP in that it can transmit datagrams containing up to 65 535 bytes, unlike TCP there is no mechanism to communicate an MTU to a UDP stack. If a UDP datagram exceeds the MTU, IP fragmentation becomes necessary.

The ability of UDP to initiate transmission without first establishing a connection between source and destination results in a considerably faster data transfer capability. Examples of UDP applications include electronic mail, network management, and voice over IP transmission. To obtain a better understanding of UDP, let us examine the use of each field in the header.

Source and Destination Port fields

The Source and Destination Port fields in the UDP header are each 2 octets in length. Each field functions in a manner similar to their counterparts in the TCP header. That is, the Source Port field value is optional and can be either randomly selected or set to all 0s when not used. The Destination Port field contains a numeric that identifies the destination application or process.

Length field

The 2-octet Length field indicates the length of the UDP datagram, including its header and user data. This field has a minimum value of 8, which represents a UDP header without data.

Checksum field

The last field in the UDP header is the Checksum field. This field is 2 octets in length and is used to validate the UDP header and key data. When used in this manner, the checksum is computed on a pseudo header that covers the source and destination addresses and the protocol field from the IP header. By providing a mechanism to verify the two address fields in the IP header, the pseudo header checksum assures that the UDP datagram is delivered to the correct destination network and host. However, this checksum does not verify the contents of the datagram.

The use of the Checksum field is optional. If not used, the field is filled with 0s if the application does not require a checksum.

4.2.2 Operation

The simplified structure of the UDP header results in a simplified operating procedure, with error detection and correction left to the application layer. As a transport protocol, UDP breaks application data into pieces for transport, appending a header that identifies the application and the length of the datagram. UDP passes the resulting datagram to IP for the addition of an IP header that identifies sending and receiving stations. Because there is no error detection and correction mechanism, delivery occurs on a best-effort basis; however, the avoidance of a slow start procedure enables UDP to transfer data significantly faster than TCP. Because of this, UDP is well suited for digitized voice applications.

5

THE DOMAIN NAME SYSTEM

The Domain Name System (DNS) represents one of the most important areas of the TCP/IP protocol suite for both public and private networks. Without a DNS capability, we would be forced to reference hosts by their IP addresses instead of easy to remember host names. Thus, we can consider the DNS as a translation mechanism that enables host names to be mapped into IP addresses. However, as we will shortly learn in this chapter, that mapping or translation capability is one of several features performed by the DNS.

In this chapter we will first turn our attention to the reason why a DNS is required and its evolution from a centralized file to a hierarchical structure of DNS servers that support the use of resolver programs imbedded into TCP/IP protocol suites operating on workstations. Once the preceding has been accomplished, we will turn our attention to DNS parameters, including the structure and composition of a DNS database. Because many TCP/IP problems can be resolved by noting entries in the DNS database, we will conclude this chapter by examining the use of applications that can provide information about the contents of DNS databases and contact information that may be helpful in resolving a variety of DNS-related problems.

5.1 EVOLUTION

As ARPAnet expanded into the Internet, its initial incarnation linked together a relatively small number of computers. This enabled Stanford University, which was assigned the responsibiiity for maintaining an IP address to host name database prior to the development of the DNS, to maintain the database on a file named HOSTS.TXT.

5.1.1 The HOSTS.TXT file

The HOSTS.TXT file is essentially simplistic, with the first column containing IP addresses while the second contains the host name associated with each IP address followed by any aliases assigned to each address. As the Internet

```
# Copyright (C) 1993–1995 Microsoft Corp.
#
# This is a sample HOSTS file used by Microsoft TCP/IP for
Windows NT.
#
# This file contains the mappings of IP addresses to host names.
Each
# entry should be kept on an individual line. The IP address
should
# be placed in the first column followed by the corresponding
host name.
# The IP address and the host name should be separated by at
least one
# space.
#
# Additionally, comments (such as these) may be inserted on
individual
# lines or following the machine name denoted by a '#' symbol.
#
# For example:
#
#      102.54.94.97      rhino.acme.com      # source server
#      38.25.63.10       x.acme.com          # x client host

127.0.0.1      localhost
```

Figure 5.1 The Windows NT default HOSTS file

evolved, network administrators would periodically download the HOSTS.TXT file from Stanford University and store its contents on their networks. The file provided a mechanism to resolve host names to IP addresses, since the IP protocol mandates the use of IP addresses. In the Unix world this file is stored as */etc/hosts*.

As the Internet continued to expand, the HOSTS.TXT file grew in size until it became too bulky to download on a periodic basis. In addition, this file is a flat file, which in the era of relatively slow disk transfers and processing power made sequential searching a relatively long process. In spite of such problems, the hosts file is still used, as it remains a simple and efficient mechanism for resolving host names for small, non-Internet-connected TCP/IP networks. In a Windows NT environment the HOSTS.TXT file is located in *\winnt\system32\drivers\ect\hosts*. Figure 5.1 illustrates the default contents of that file. Note that the pound sign (#) is used to denote comments.

Based upon the previously described limitations associated with the use of the HOSTS.TXT file, a more eloquent solution was needed to provide an address resolution process. The resulting solution occurred through the development of the Domain Name System, which eliminated the performance problems associated with the centralized single HOSTS.TXT file.

5.2 DNS OVERVIEW

DNS can be viewed as a suite of protocols, software, servers, and databases that work together to provide a distributed host name to IP address translation process. In addition, DNS also provides a mechanism for reversing the translation process, as well as a number of additional functions to include facilitating the mail delivery process (which we will examine later in this chapter).

5.2.1 The domain structure

To facilitate the address resolution process, the address space of the Internet—and, for that matter, any public or private TCP/IP network—was structured in a hierarchical manner in the form of an inverted tree, similar to the familiar Unix and DOS file systems. At the top of the tree is the root, with branches in the form of domains at the next level. Under each domain a virtually unlimited number of subdomains can reside, providing an unlimited number of paths within a domain system.

Each top-level domain, such as *.edu* or *.gov*, contains all host names that have the suffix of that domain. Within each domain an infinite number of subdomains can reside, with a different prefix used to distinguish one subdomain from another. For example, *american.edu* and *harvard.edu* would represent two subdomains within the *.edu* domain. Within each subdomain different hosts are distinguished from one another by a prefix added to the subdomain name. Thus, *ftp.harvard.edu* and *ftp.american.edu* could represent FTP servers at each university. Note that a fully qualified domain name that identifies the host, subdomain, and domain consists of a sequence of labels separated by dots. Both top-level domain names and IP network addresses are assigned and maintained by the Internet Assigned Numbers Authority (IANA). Lower level domain names in the form of subdomains are registered for use for a nominal fee. At the time this book was written the firm Network Solutions had the contract for registering domain names in the USA.

Figure 5.2 illustrates the hierarchical structure of the Internet's domain name space. Unlike a file system where each branch represents a directory name in DNS, each branch is referred to as a label. Labels can be up to 63 characters in length, while the fully qualified domain name, such as *gil.goofy.com*, is limited to 255 characters.

In examining the DNS hierarchical domain name space shown in Figure 5.2, it should be noted that the top level domains may have significantly

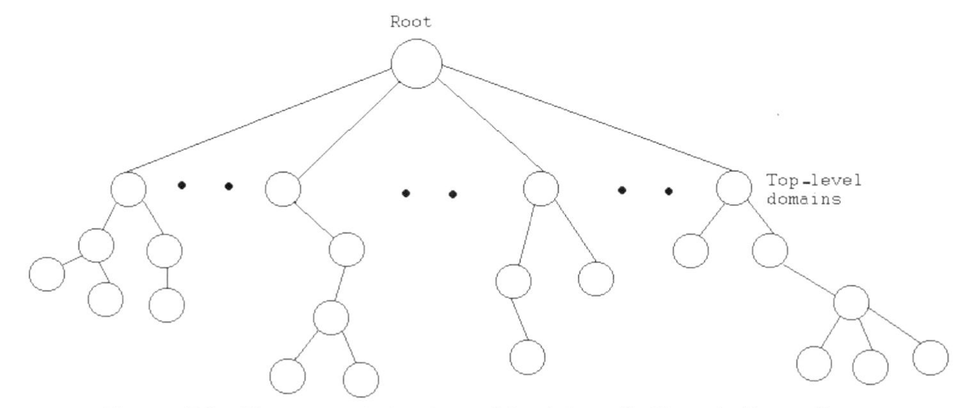

Figure 5.2 The general structure of the Internet's Domain Name Space

Table 5.1 Current and proposed assigned top-level domains

Current	
.com	commercial
.edu	educational
.gov	government
.mil	military
.org	non-profit organization
.net	network operator
.xx	a two-letter code to identify a country top-level domain such as .au (Australia), .ca (Canada), .il (Israel), .jp (Japan), and .us (USA)
Proposed	
.firms	business
.store	businesses offering goods to purchase
.web	organizations offering Web-related activities
.arts	cultural and entertainment activities
.rec	recreation/entertainment activities
.info	information services
.nom	individuals

increased in number by the time you read this book. Initially, top-level domains were restricted to the entries shown at the top of Table 5.1. However, during 1997, it was proposed to extend top-level domains by the addition of seven domains listed in the lower portion of Table 5.1. The proposed extension was still being considered during 1999.

Within each top-level domain the assignment of names is performed by an appropriate registry organization, such as Network Solutions in North America. Each registered domain name represents a top-level qualifier for a company, government agency, educational institution, etc., with the actual naming of hosts in the domain left up to individual organizations. However, since a path is known from the top-level domain to each organizational domain, the placement of domain servers at lower levels in the name space provides a mechanism to locate hosts within different domains. In fact, name servers represent one of several key DNS components. Thus, prior to examining in detail the name resolution process, let us first turn our attention to the three components of the modern DNS that enable the address resolution process to occur.

5.2.2 DNS components

The three key components of the modern DNS include resource records, name servers, and resolvers.

Resource records

Resource records represent a database of grouped names and addresses formatted to correspond to a tree-structured name space. Each domain maintains a set of records about local hosts, which collectively is referred to as a zone.

Name servers

A second key component of the DNS comprises computers that store a database of resource records and operate software that appropriately responds to queries. This type of computer is referred to as a name server.

Because a name server is responsible for storing information about a zone, it may not be able to resolve many resolution requests. Thus, the name server will contain records that function as pointers to other name servers that can be accessed in an attempt to locate required information. For example, a name server on an organizational LAN connected to the Internet may have a pointer to an ISP's name server, and the ISP's name server could have a pointer to a Network Service Provider's (NSP's) name server, and so on.

Resolvers

A third key component of the DNS comprises programs that reside on workstations and generate address resolution requests via the network to name servers. Such programs are referred to as resolvers, as they resolve host name to IP address translations or the reverse.

Resolvers are built into the TCP/IP protocol stack. When you configure the protocol stack, you must indicate the address of at least one name server, which is then used by the resolver as the first location to perform its address resolution effort. Figure 5.3 illustrates the DNS configuration process on a Windows NT platform. Note that the IP addresses for two servers were specified in this example and are used in the order in which they appear. You should list the nearest DNS server first, as this will facilitate the address resolution process.

The Domain Suffix Search Order box shown in Figure 5.3 represents an interesting capability that can be used by resolvers. If you enter one or more domain suffixes, such as *microsoft.com* or *ibm.com* and then request the resolution of the host portion of a name, the resolver will tack on each of the domain suffixes previously configured. For example, if you entered FTP, the resolver would attempt to translate *ftp.microsoft.com* and *ftp.ibm.com*, with the first match terminating the resolution process.

The resolution process If you supply a fully qualified domain name in the form of a host followed by a domain, such as *host.mydomain.ext*, where *ext* corresponds to one of the top-level qualifiers previously listed in Table 5.1, the resolver will send a request to the first DNS server address listed in Figure 5.3. If the DNS server has a resource record for the host, the server will return the IP address for the host. If the host is in the zone of authority for the server but a resource record does not exist, the server will return a 'host not found' message. A third possibility is that the host is not in the local server's zone of authority. For example, if your organization is in the domain *mouse.com* and the resolver requested resolution of *gil.goofy.edu* then, unless the results of a prior request were placed in cache memory in the local name server, that server will not be able to resolve the request. In this situation the local DNS name server will use a pointer record to pass the query to the first root server that is an authoritative DNS server for the

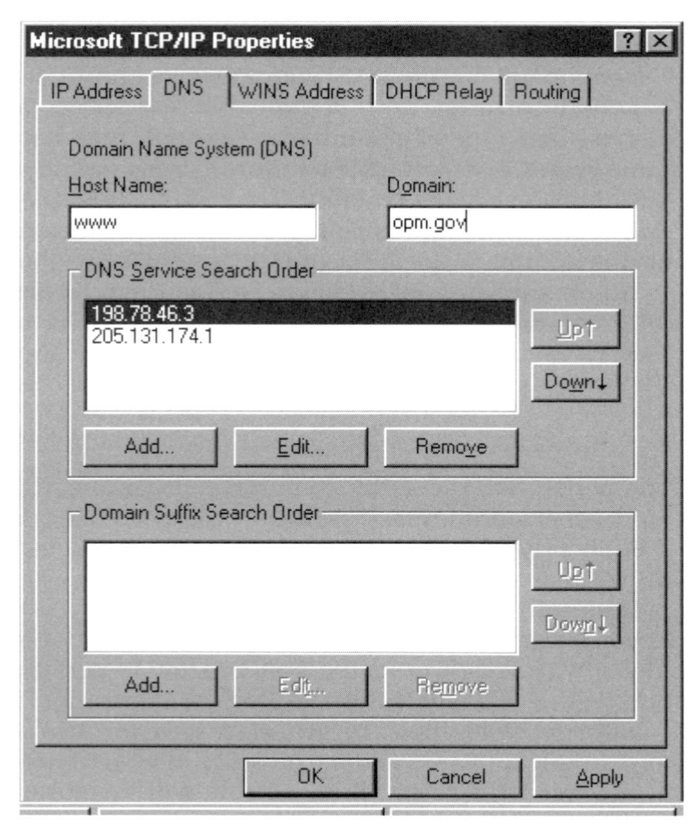

Figure 5.3 Configuring the DNS IP address on a Windows NT computer

domain *goofy.edu*. Here the term root server represents a special type of server, which contains the IP addresses of the name servers for specific domains. Currently there are nine root servers on the Internet, each of which functions as a traffic cop, directing queries to the appropriate name server for a specific domain. The resolver will then contact the indicated name server, which will either return the IP address of the host or report that the host was not found. The latter situation will be used by the application program requiring an address resolution to display a message similar to the one shown in Figure 5.4.

In examining the message displayed in Figure 5.4 by the Netscape Web browser, it should be noted that the lack of a resource record at the authoritative name server for the host address requiring resolution is but one of several reasons that can result in the display of the preceding message. Other reasons can include the specification of an incorrect domain, such as *.cov* instead of *.com*, or even the lack of a response due to the resolver request being dropped by a router because of a heavy traffic load occurring on the one or more of the tens of thousands of transmission facilities that make up the Internet. Because UDP is used to transport resolver requests, there is no guaranteed delivery mechanism and a timer is used internally to generate a message similar to the one illustrated in Figure 5.4 when a response to the

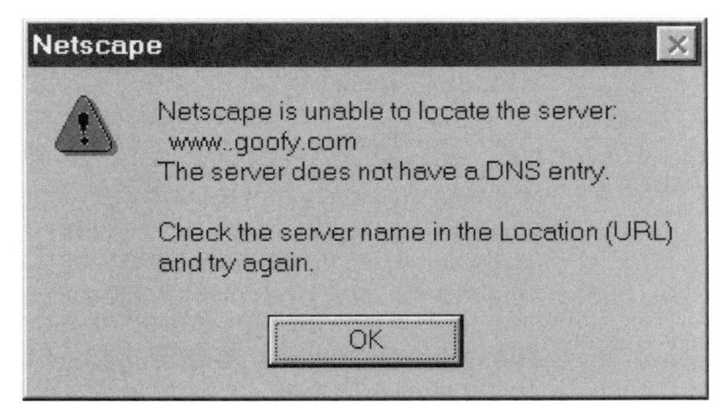

Figure 5.4 When a host name cannot be resolved, the resolver passes a message to the application requiring the resolution, which in turn displays a message to the user

resolver request is not received within a predefined period of time. This explains why a second or third entry of the same URL can often be successful, even after first receiving a message similar to the one shown in Figure 5.4. This also explains why a help desk should tell users complaining that they cannot access a particular Web site to check the URL and try again. Now that we have an appreciation of the address resolution process, let us turn our attention to the composition of records in the DNS database. This information will provide us with a base of knowledge that will make the use of application tools for examining the database more meaningful.

5.3 THE DNS DATABASE

The DNS database is similar to other types of databases in that it represents a collection of records. What sets the DNS database apart from other databases is the composition and use of resource records contained in its database.

5.3.1 Overview

When the Internet was in its early period of evolution, there were several methods referred to as domain name classes used for the Domain Naming System, including Chaos (CH), Hesiod (HS), and the Internet (IN). For the Internet class a total of 44 source records were defined. Table 5.2 provides a detailed list of Resource Record (RR) types defined for the Internet domain naming system class over the past 25 years. Note that some of the RRs are now obsolete while others are reserved for experimental purposes. Information for a majority of the RRs listed in Table 5.2 can be located in RFCs 1035, 1183, 2065, and 2168.

DNS represents a client–server querying system that uses operation code (OpCode) values to specify different types of operations or queries and responses. Those codes are used in UDP datagrams along with code values to indicate requests for specific types of resource records. Table 5.3 lists presently defined DNS OpCodes. Additional information concerning DNS OpCodes can be found in RFCs 1035, 1996, and 2136.

Table 5.2 Internet (IN) Class Resource Record types

Type	Value and Meaning
A	1 a host address
NS	2 an authoritative name server
MD	3 a mail destination (Obsolete - use MX)
MF	4 a mail forwarder (Obsolete - use MX)
CNAME	5 the canonical name for an alias
SOA	6 marks the start of a zone of authority
MB	7 a mailbox domain name (EXPERIMENTAL)
MG	8 a mail group number (EXPERIMENTAL)
MR	9 a mail rename domain name (EXPERIMENTAL)
NULL	10 a null RR (EXPERIMENTAL)
WKS	11 a well-known service description
PTR	12 a domain name pointer
HINFO	13 host information
MINFO	14 mailbox or mail list information
MX	15 mail exchange
TXT	16 text strings
RP	17 for Responsible Person
AFSDB	18 for AFS Data Base location
X25	19 for X.25 PSDN address
ISDN	20 for ISDN address
RT	21 for Route Through
NSAP	22 for NSAP address, NSAP style A record
NSAP-PTR	23
SIG	24 for security signature
KEY	25 for security key
PX	26 X.400 mail mapping information
GPOS	27 Geographical Position
AAAA	28 IP6 Address
LOC	29 Location Information
NXT	30 Next Domain
EID	31 Endpoint Identifier
NIMLOC	32 Nimrod Locator
SRV	33 Server Selection
ATMA	34 ATM Address
NAPTR	35 Naming Authority Pointer
KX	36 Key Exchanger
CERT	37 CERT
TKEY	249 TKEY
TSIG	250 Transaction Signature
IXFR	251 incremental transfer
AXFR	252 transfer of an entire zone
MAILB	253 mailbox-related RRs (MB, MG or MR)
MAILA	254 mail agent RRs (Obsolete - see MX)
*	255 a request for all records.

5.3.2 Resource records

Table 5.4 lists some of the more common DNS resource records and a description of each record type. Other less popular record types include a Generic Text record (TXT), Service Location record (SRV), and Responsible

Table 5.3 Domain Name System OpCodes

Domain Name System Operation Codes:

OpCode	Name
0	Query
1	IQuery
2	Status
3	Reserved
4	Notify
5	Update

Domain Name System Response Codes:

OpCode	Name	
0	NoError	No Error
1	FormERR	Format Error
2	ServFail	Server Failure
3	NXDomain	Non-Existent Domain
4	NotImp	Not Implemented
5	Refused	Query Refused
6	YXDomain	Name Exists when it should not
7	YXRRSet	RR Set Exists when it should not
8	NXRRSet	Set that should exist does not
9	NotAuth	Server Not Authoritative for zone
10	NotZone	Name not contained in zone

Table 5.4 Common resource records in a DNS database

Record Type	Representation	Description
SOA	Start of Authority	Indicates the authoritative name server for a given domain and such administrative information as the administrator's email address and the duration of data in its cache
A	Address record	Maps the host name in the current domain to a corresponding IP address; multiple names can be assigned to one IP address via a CNAME record
CNAME	Canonical Name record	Permits alias host names to be defined for an address in an A record
PTR	Pointer record	Associates a host name with a given IP address
MX	Mail Exchange record	Defines the mail system(s) for a given domain
NS	Name Server record	Defines the name server(s) for a given domain

Person (RP) record, with the latter indicating the text name of the person responsible for the domain DNS.

In examining the record types listed in Table 5.4, a few items warrant attention. First, although it is permissible to define multiple A records for a given IP address, it is usually easier to configure one A record and use CNAME records to define alias host names for the IP address in the A record. A second item that warrants attention is the PTR record. Since this record provides the reverse of information in an A type record, as you will probably surmise, PTR records are used for reverse name lookups.

5.3.3 Using a sample network

Now that we have an appreciation of the types of resource records in the DNS database, we can turn our attention to the fields in each record. However, since a DNS database reflects the composition of hosts in a domain, we can facilitate our examination of DNS resource records by first developing a sample network structure for illustrative purposes.

Figure 5.5 illustrates the structure, host, and IP address for the network domain *goofy.gov*. Note that the network address 198.78.46.0 is purely hypothetical.

5.3.4 DNS software configuration

As mentioned earlier in this chapter, the evolution of the TCP/IP protocol suite has a close association with the development of Unix. In the Unix world the DNS is referred to as BIND (Berkeley Internet Name Daemon), while under Windows NT the mnemonic DNS is used.

The BOOT file

When the DNS process is initialized, the name of a boot file must be specified. This file contains the names of database files containing DNS information as well as records that define certain DNS functions.

Figure 5.6 illustrates the contents of a boot file for the DNS for the domain *goofy.gov*. Note that we named the file *dnsinfo.boot* and the semicolons (;) is used as a prefix for placing comments in the file.

```
Domain:  goofy.gov
IP network address:  198.78.46.0
```

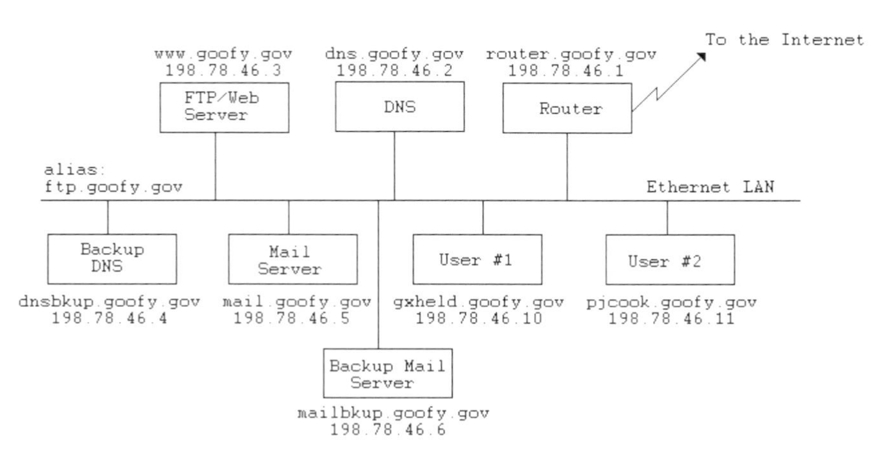

Figure 5.5 Network structure for the hosts in the domain *goofy.gov*

The first record in the boot file provides the directory path to the location where the DNS program files reside. In this example the directory */usr/local/name* represents the location where the DNS program files reside. The remaining records in the boot file contain three fields. One field defines the type of DNS information. The second field indicates the domain defined in the file, while the third field indicates the filename where the information is located. Note that the six records with the type of 'primary' reference the DNS databases for which the DNS server has primary responsibility.

The first primary record indicates that the resource records for the domain *goofy.gov* are in the file *goofy.gov.zone*. The second primary record indicates that the domain is associated with the Class C IP address 198.78.46.0. Note that the IP address for the domain is in reverse dotted decimal notation order. That address is followed by the term 'in-addr.arpa,' which denotes the use of IP addresses within the ARPA addresses assignment domain. Although ARPAnet has long ceased to exist, this nomenclature represents a holdover from the time when more than one type of address could be found and there were several addressing authorities. Also note that the file containing records for reverse lookup is *goofy.gov.rev*.

The fourth primary record, 'localhost,' represents a special IP name used to reference the local host. By convention it has the IP address 127.0.0.1. In our example this IP information is in the file named *localhost.zone*. In our examination of IP addresses we noted that the IP address 127.0.0.0 is reserved in IP for a local loopback operation. This address is not contained in or maintained by any global DNS registry. In our example the reverse lookup for the loopback address is stored in a file labeled *localhost.rev*.

The last two primary records define the files for all zeros (0) and all ones (253) addresses. As previously noted in Chapter 3, all ones denotes a broadcast while all zeros represents this network.

The last record listed in Figure 5.6 defines where DNS cache information for the root domain (.) can be located. Note that the information stored in the file *named.cache* represents a set of fixed pointers to higher-level DNS databases, such as root servers, and not cache memory. Now that we have an appreciation of the composition of the boot file, let us turn our attention to the records in the key DNS file that provides the resolution capability for the domain. In our example that file is the *goofy.gov.zone* file. Figure 5.7 illustrates an example of the possible contents of that file, with the resource records corresponding to the network structure shown in Figure 5.5.

```
;FILE: dnsinfo.boot
directory /usr/local/names

;TYPE     Domain                    Data File
primary   goofy.gov                 goofy.gov.zone
primary   46.78.198.in-add.arpa     goofy.gov.rev
primary   localhost                 localhost.zone
primary   0.0.127.in-add.arpa       localhost.rev
primary   0.in-addr.arpa            all-zero.zone
primary   255.in-adr.arpa           all-one.zone
cache     .                         named.cache
```

Figure 5.6 A sample BIND boot file

5.3.5 Using resource records

In examining Figure 5.7, note that all DNS resource records have a similar format. The first field normally contains a host name or an IP address. If omitted, then the host name or IP address from the previous record is implied. Each name and IP address is terminated with a trailing dot (.) if it is an absolute name or address. Absolute addresses provide a full path relative to the root; hence, they are also known as fully qualified domain names. In comparison, relative addresses are not terminated with a dot (.), which indicates they are relative to a default domain that may or may not be the root. Optionally, the first field can be followed by a Time to Live (TTL) value. This value, when included, indicates the length of time the information in the field is valid.

The second field in the DNS resource record indicates the record type. Here the string 'IN' (which is used as a prefix to each record type) indicates a class of data that represents an Internet address. In actuality, IN represents a carryover from early DNS databases when different classes existed in the DNS, and was originally used for compatibility with older systems. It is now included in RRs for historical purposes.

The third field indicates the address associated with the specific resource record. Now that we have an overview of the composition of the fields in the resource record, let us examine each of the records in the DNS database file shown in Figure 5.7.

```
;Start of Authority (SOA) record
goofy.gov.            IN SOA     dns.goofy.gov.     gil.goofy.gov(
                                  24601             ;serial number
                                  28800             ;refresh after 8 hours
                                   3600             ;retry after 1 hour
                                 604800             ;expire after 1 week
                                  86400)            ;TTL of 1 day

;Name Server (NS) records
goofy.gov.            IN NS      adns.goofy.gov
                      IN NS      dnsbkup.goofy.gov.

;Mail Exchange (MX) records
goofy.gov             IN MX 10   mail.goofy.gov
                      IN MX 20   mailbkup.goofy.gov.

;Address (A) records
localhost.goofy.gov   IN A       127.0.0.1
router.goofy.gov.     IN A       198.78.46.1
dns.goofy.gov.        IN A       198.78.46.2
www.goofy.gov.        IN A       198.78.46.3
dnsbkup.goofy.gov.    IN A       198.78.46.4
mail.goofy.gov.       IN A       198.78.46.5
mailbkup.goofy.gov.   IN A       198.78.46.6
gxheld.goofy.gov.     IN A       198.78.46.10
pjcook.goofy.gov.     IN A       198.78.46.11

;Aliases in Canonical Name (CNAME) records
ftp.goofy.gov. IN CNAME www.goofy.gov
```

Figure 5.7 An example of the *goofy.gov.zone* file

SOA record

The Start of Authority record represents the first entry in a DNS database file. This resource record indicates the name server for the domain, the email address of the person responsible for the name server, and certain administrative data. Note that instead of a conventional email address of the form *user@domain*, the format *user.domain* is used. In our example the DNS address for the domain *goofy.gov* is *dns.goofy.gov.*, with a terminating dot (.), while the email address of the person responsible for the name server is *gil.goofy.gov.*, also with a terminating dot (.).

The five lines following the SOA record's left parenthesis (() represent administrative information that will be used by a secondary DNS server. Here the serial number identifies the version of the DNS database. This number is incremented whenever the file's content is changed, providing the mechanism by which a secondary server will know that the database was modified. Some systems begin counting at 1 and increment each time a change occurs, while other systems may use the year, month, and day, with the latter sufficient only if the DNS database is limited to one change per day.

The refresh value tells the secondary server(s) how often in seconds to check for updated information. In our example 28 800 seconds represents 8 hours. If the secondary name server is unable to contact the primary name server, the retry value tells the secondary when to retry. In our example the retry is configured for 3600 seconds or 1 hour.

As a safety mechanism, the expire time value tells the secondary server when to stop answering queries about the domain if it is unable to contact the primary for a period of 'expire' seconds. Here it is assumed that after 'expire' seconds the data on the secondary may have aged to the point where it does more harm than good. In our example 'expire' is shown set to a value of 604 800 seconds, which represents a period of one week.

The Time to Live (TTL) value represents the value returned with all responses to database queries. This value tells requestors operating on clients as well as other DNS servers how long to cache information. In our example the TTL value is 86 400 seconds, or one day; however, this TTL value can be overridden by a TTL value included with specific resource records.

NS records

The SOA resource record is commonly followed by one or more Name Server (NS) records, with one entry for each name server in your domain. In the example shown in Figure 5.7 there are two name servers in the *goofy.gov* domain; hence, there are two NS records. Although the host name is indicated as *dnsbkup*, many organizations simply assign sequential numbers when they have multiple name servers or mail servers, such as ns1, ns2, and so on.

MX records

Although the DNS does not send or receive mail, it provides information on the mail server(s) for a given address, which facilitates the delivery of mail. In the DNS database this is accomplished through the use of the MX record, where MX is a mnemonic for mail exchange.

The MX record specifies a mail exchanger for a domain. In the DNS database listed in Figure 5.7 there are two mail exchangers specified for the domain *goofy.gov*. Note that each MX type value in the resource record is followed by a precedence value that can range between 0 and 65 535. When a remote user transmits email to a user at *goofy.gov*, the remote mail system will look up the MX records for the *goofy.gov* domain. The remote mailer will attempt to establish a Simple Mail Transfer Protocol (SMTP) connection with the mail system in the *goofy.gov* domain that has the lowest preference value. In the example shown in Figure 5.7 this would be the mail server *mail.goofy.gov*. If that server is not available, the remote mailer will try the server with the next higher precedence value, which in our example is *mailbkup.goofy.gov*.

A records

A records are address records and are used to indicate the name-to-IP address mappings for hosts in the domain. Note that A records are for mapping IPv4 32-bit addresses. Another type of record worth noting is the AAAA address record. This record derives its name from the fact that an IPv6 address is 128 bits in length, or four times the length of an IPv4 address. In examining the entries in Figure 5.7, note that there is an A record for each host in the domain illustrated in Figure 5.5.

CNAME records

As previously noted, a Canonical Name (CNAME) record permits an alias host name to be associated with a previously defined host name contained in an A record. In the example illustrated in Figure 5.7 the alias *ftp.goofy.gov* is associated with the host *www.goofy.gov*.

PTR records

As previously noted, Pointer (PTR) records are used to map IP addresses to host names. This provides a reverse lookup capability, which is commonly used by Web sites to gather statistical information concerning users accessing the site. Because routing occurs via IP addressing, hits on a server are identified by IP addresses. To obtain more meaningful data, Web server statistical analysis programs perform a reverse lookup to obtain domain and host information about users accessing the system.

The use of PTR records occurs in the file *goofy.gov.rev* for our hypothetical domain. Figure 5.8 illustrates an example of the contents of the file *goofy.gov.rev* for the hosts in our hypothetical network.

In examining Figure 5.8, note that the reverse zone's address is constructed by first reversing the IP address qualifiers after the host portion of the address is removed and then adding *in-addr.arpa*. Also note that the IP

```
;Start of Authority  (SOA)  record
46.78.198.in-addr.arpa.          IN SOA     dns.goofy.gov
gil.goofy.gov
                                 (24601     ;serial number
                                  28800     ;refresh after 8 hours
                                   3600     ;retry after 1 hour
                                 604800     ;expire after 1 week
                                  86400)    ;TTL of 1 day

;Name Server (NS) records
46.78.198.in-addr.arpa.          IN NS      dns.goofy.gov.
                                 IN NS      dnsbkup.goofy.gov.

;Pointer (PTR) records for reverse lookups
1.46.78.198.in-addr.arpa.          IN PTR   router.goofy.gov.
2.46.78.198.in-addr.arpa.          IN PTR   dns.goofy.gov.
3.46.78.198.in-addr.arpa.          IN PTR   www.goofy.gov.
4.46.78.198.in-addr.arpa.          IN PTR   dnsbkup.goofy.gov.
5.46.78.198.in-addr.arpa.          IN PTR   mail.goofy.gov.
6.46.78.198.in-addr.arpa.          IN PTR   mailbkup.goofy.gov.
10.46.78.198.in-addr.arpa.         IN PTR   gxheld.goofy.gov.
11.46.78.198.in-addr.arpa.         IN PTR   pjcook.goofy.gov.
```

Figure 5.8 an example of the file *goofy.gov.rev* used for reverse lookups

address associated with each PTR record represents the full reverse of the IP address associated with the host name included in the resource record.

Loopback files

Facilitating forward and reverse address lookups for the loopback address requires two additional files. Those files are named *localhost.zone* and *localhost.rev*, respectively. Figure 5.9 illustrates an example of the contents of the file *localhost.zone* for the domain *goofy.gov*, while Figure 5.10 illustrates an example of the contents of the file *localhost.rev* for the same domain.

All-zero/all-ones files

Two additional files required are the *all-zero.zone* and *all-one.zone* files. Examples of the contents of those files for the domain *goofy.gov* are shown in Figures 5.11 and 5.12.

```
localhost.         IN SOA         dns.goofy.gov    gil.goofy.com.(
                    23579         ;serial number
                    28800         ;refresh after 8 hours
                     3600         ;retry after 1 hour
                   604800         ;expire after 1 week
                    86400)        ;TTL of 1 day

localhost.         IN NS          dns.goofy.gov.
                   IN NS          dnsbkup.goofy.gov.

localhost.         IN A           127.0.0.1
```

Figure 5.9 An example of the contents of the file *localhost.zone*

```
;Start of Authority (SOA) record

0.0.127.in-addr.arpa    IN SOA      dns.goofy.gov.  gil.goofy.gov(
                         23579      ;serial number
                         28800      ;refresh after 8 hours
                          3600      ;retry after 1 hour
                        604800      ;expire after 1 week
                         86400)     ;TTL of 1 day

;Name Server (NS) records
0.0.127.in-addr.arpa    IN NS       dns.goofy.gov.
                        IN NS       dnsbkup.goofy.gov.

;Only need one PTR record
1.0.0.127.in-addr.arpa  IN PTR      localhost.
```

Figure 5.10 An example of the contents of the file *localhost.rev.*

```
0.in-addr.arpa          IN SOA      dns.goofy.gov. gil.goofy.gov(
                         24601      ;serial number
                         28800      ;refresh after 8 hours
                          3600      ;retry after 1 hour
                        604800      ;expire after 1 week
                         86400)     ;TTL of 1 day

0.in-addr.arpa          IN NS       dns.goofy.gov.
                        IN NS       dnsbkup.goofy.gov.
```

Figure 5.11 Example of an *all-zero.zone* file

```
255.in-addr.arpa        IN SOA      dns.goofy.gov. gil.goofy.gov(
                         24601      ;serial number
                         28800      ;refresh after 8 hours
                          3600      ;retry after 1 hour
                        604800      ;expire after 1 week
                         86400)     ;TTL of 1 day

255.in-addr.arpa        IN NS       dns.goofy.gov.
                        IN NS       dnsbkup.goofy.gov.
```

Figure 5.12 Example of the contents of the *all-one.zone* file

For further resolution

When a name cannot be resolved by the local name server, it needs to know where to forward the query. This information is included in the file *named.cache*, which contains the addresses of the top-level DNS servers. You can download this file from the Internet Network Information Center (InterNIC) at *ftp://rs.internic.net/domain/named.cache*. It should be noted that this file is also labeled *named.root* and *named.ca.*

The file *named.cache* primarily consists of a series of NS and A resource records with a significant number of imbedded comments. Figure 5.13 illustrates an example of a portion of the contents of the *named.cache* file.

```
;           last update:    month day, year
;           related version of root zone: r23456
.                           3600000   IN NS   A.ROOT-SERVERS.NET.
A.ROOT-SERVERS.NET          3600000      A    w.x.y.z
.                           3600000      NS   B.ROOT-SERVERS.NET.
B.ROOT-SERVERS.NET          3600000      A    w.x.y.z

.
.
.                           3600000      NS   N.ROOT-SERVERS.NET.
NROOT-SERVERS.NET           3600000      A    w.x.y.z
```

Figure 5.13 An example of a *named.cache* file

Note that currently there are nine root servers, labeled A through N. Also note that for brevity of illustration a majority of the comments in the *named.cache* file were removed, and that the sequence w.x.y.z represents the IP address of each root server.

5.3.6 Accessing a DNS database

Within many TCP/IP protocol suites is an application program called nslookup. This program can be used to query a DNS server and obtain a formatted list of the contents of its database. Most versions of Unix, Windows NT, and Windows 2000 include nslookup. However, Windows 95 does not. Thus, you should access an appropriate computer to use this program.

nslookup

nslookup has two modes of operation: interactive and non-interactive. If you only require the ability to look up a single piece of data, you should use the program's non-interactive mode of operation. Otherwise you should use its interactive mode of operation, as this will allow you to issue repeated queries without having to restart the program for each query. In this section we will illustrate the use of nslookup on a Windows NT 4.0 system. While implementations of nslookup can differ slightly from one system to another, the information in this section will provide you with examples of how you can query different DNS databases in an attempt to isolate many TCP/IP address resolution and mail delivery problems.

The general format of the nslookup command is as follows:

nslookup [-options] [computer-to-locate|-[server]]

To use the program's interactive mode, you would either enter the command by itself without further data or you would type a hyphen (-) for the first argument followed by the name or IP address of a DNS server. If you do not specify a DNS server in the command line, the program will use the DNS server specified in your TCP/IP configuration as a default.

Figure 5.14 illustrates the use of the nslookup command in its non-interactive mode of operation. In this example the program was used to look up the domain opm.gov. The first two lines of the response indicate the

```
C:\>nslookup opm.gov
Server: serv3.opm.gov
Address: 198.78.46.3

Name: opm.gov
Address: 198.78.46.3

C:\>
```

Figure 5.14 Using nslookup in its non-interactive mode of operation

default name server and its address configured in the user's TCP/IP protocol stack, while the next two lines in the response repeat the domain name and the address of the name server for that domain.

nslookup supports a large number of options. Those options are considered to represent commands used with the nslookup command by some implementations of nslookup.

Figure 5.15 illustrates the commands and command options supported by the Microsoft Windows NT version of nslookup. Microsoft's version of nslookup

```
Commands: (identifiers are shown in uppercase, [  ] means optional)
NAME                        — print info about the host/domain NAME using
                              default server
NAME1 NAME2                 — as above, but use NAME2 as server
help or ?                   — print info on common commands
set OPTION                  — set an option
    all                     — print options, current server and host
    [no]debug               — print debugging information
    [no]d2                  — print exhaustive debugging information
    [no]defname             — append domain name to each query
    [no]recurse             — ask for recursive answer to query
    [no]search              — use domain search list
    [no]vc                  — always use a virtual circuit
    domain = NAME           — set default domain name to NAME
    srchlist = N1[/N2/.../N6] — set domain to N1 and search list to N1,N2, etc.
    root = NAME             — set root server to NAME
    retry = X               — set number of retries to X
    timeout = X             — set initial time-out interval to X seconds
    type = X                — set query type (ex. A,ANY,CNAME,MX,NS,PTR,SOA,SRV)
    querytype = X           — same as type
    class = X               — set query class (ex. IN (Internet), ANY)
    [no]msxfr               — use MS fast zone transfer
    ixfrver = X             — current version to use in IXFR transfer request
server NAME                 — set default server to NAME, using current default
                              server
lserver NAME                — set default server to NAME, using initial server
finger [USER]               — finger the optional NAME at the current default
                              host
root                        — set current default server to the root
ls [opt] DOMAIN [> FILE]    — list addresses in DOMAIN (optional: output to
                              FILE)
    -a                  — list canonical names and aliases
    -d                  — list all records
    -t TYPE             — list records of the given type (e.g. A,CNAME,MX,NS,PTR etc.)
view FILE                   — sort an 'ls' output file and view it with pg
exit                        — exit the program
```

Figure 5.15 nslookup commands and command options supported by Microsoft's Windows NT

Table 5.5 nslookup set querytype parameters values

Parameter value	Meaning
A	Computer ID address
ANY	All types of records
CNAME	Canonical name for an alias
GID	Group identifier of a group name
HINFO	Computer's CPU and operating system
MB	Mailbox domain name
MG	Mail group member
MINFO	Mailbox or mail list information
MR	Mail rename domain name
MX	Mail exchanger
NS	DNS name server for the named zone
PTR	Computer name if query is an IP address; otherwise the pointer to other information
SOA	DNS domain's start-of-authority record
TXT	Text information
UID	User ID
UINFO	User information
WKS	Well-known service description

runs in the DOS command prompt box and not as a graphic application. Note that you can obtain a screen display of information about each command by following the nslookup command with either the question mark (?) or the term help. Also note that you exit the interactive mode via the command exit.

One of the more important command options to note is its set querytype option, which is abbreviated set q=. You can use this option to specify a specific type of information you wish to query a DNS database for. Table 5.5 lists the values supported by the set querytype option. Note that if not specified A is the default.

A second important nslookup command option is ls which is used to list information for a DNS domain. The format of the ls option includes other options, as follows:

ls [option] dns domain [>filename] [>>filename]

The default for the ls option results in a list of computer names and their IP addresses being returned. When directed to a new file (>filename) or appended to an existing file (>>filename), hash marks will be printed for every 50 records received from the server. Table 5.6 lists the option parameters supported by the nslookup ls option.

Table 5.6 Option parameters supported by the nslookup ls command option

Option	Meaning
-t querytype	Lists all records of the specified querytype; see Table 5.5 for the listing of applicable querytypes
-a	Lists aliases of computers in the DNS domain
-d	Lists all records for the DNS domain
-h	Lists CPU operating system information for the DNS domain
-s	Lists well-known services of computers in the DNS domain

```
C:\>nslookup
Default Server:  serv3.opm.gov
Address:  198.78.46.3

> set  q = soa
> opm.gov
Server:  serv3.opm.gov
Address:  198.78.46.3

opm.gov
          primary name server = serv3.opm.gov
          responsible mail addr = clcheek.opm.gov
          serial = 98111303
          refresh = 10800 (3 hours)
          retry = 3600 (1 hour)
          expire = 10800 (3 hours)
          default TTL = 10800 (3 hours)
opm.gov nameserver = serv3.opm.gov
opm.gov nameserver = serv4.opm.gov
opm.gov nameserver = dcdhcp-dns1.opm.gov
serv3.opm.gov  internet address = 198.78.46.3
serv4.opm.gov  internet address = 198.78.46.18
dcdhcp-dns1.opm.gov  internet address = 172.16.8.3
> set q = ns
> opm.gov
Server:  serv3.opm.gov
Address:  198.78.46.3

opm.gov nameserver = serv4.opm.gov
opm.gov nameserver = dcdhcp-dns1.opm.gov
opm.gov nameserver = serv3.opm.gov
opm.gov nameserver = serv4.opm.gov
opm.gov nameserver = dcdhcp-dns1.opm.gov
opm.gov nameserver = serv3.opm.gov
serv4.opm.gov  internet address = 198.78.46.18
dcdhcp-dns1.opm.gov  internet address = 172.16.8.3
serv3.opm.gov  internet address = 198.78.46.3
>
```

Figure 5.16 Using the set querytype nslookup command option to display SOA and NS resource records for a domain

Because the best way to understand the capabilities of nslookup is by using the program, let us do this. In doing so, we will use the interactive mode of the program, as it provide us with a better capability for retrieving name server database records.

Figure 5.16 illustrates the use of the nslookup command to retrieve SOA and NS resource record listings for the domain opm.gov. Note that the set command was used to set the querytype to SOA (set q = soa), after which the domain opm.gov was entered. After the default name server and its address had been displayed, the nslookup program displayed the SOA resource record for the requested domain. This was followed by the entry of a second querytype, in this case for ns resource records (set q = ns) followed by the domain opm.gov, which resulted in the display of the name servers used by the denoted domain.

```
C:\>nslookup
Default Server:  serv3.opm.gov
Address:  198.78.46.3

> set q = ns
> opm.gov
Server:    serv3.opm.gov
Address:    198.78.46.3

opm.gov nameserver = dcdhcp-dns1.opm.gov
opm.gov nameserver = serv3.opm.gov
opm.gov nameserver = serv4.opm.gov
opm.gov nameserver = dcdhcp-dns1.opm.gov
opm.gov nameserver = serv3.opm.gov
opm.gov nameserver = serv4.opm.gov
dcdhcp-dns1.opm.gov       internet address = 172.16.8.3
serv3.opm.gov     internet address = 198.78.46.3
serv4.opm.gov     internet address = 198.78.46.18
> server    serv3.opm.gov
Default Server: serv3.opm.gov
Address:   198.78.46.3

> ls -t mx opm.gov
[serv3.opm.gov]
 bbs                        MX          10 mail.holonet.net
 bbs                        MX          20 altmail.holonet.net
> ls -t a opm.gov
[serv3.opm.gov]
 opm.gov.                   NS          server = dcdhcp-dns1.opm.gov
 opm.gov.                   NS          server = serv3.opm.gov
 opm.gov.                   NS          server = serv4.opm.gov
 opm.gov.                   A           198.78.46.33
 trusted.oer                A           172.16.200.11
 www.oer                    A           205.131.188.11
 jobentry                   A           205.131.175.16
 ipxgw                      A           198.78.46.27
 www2                       A           205.131.176.12
 www3                       A           205.131.176.13
 www5                       A           205.131.176.15
 search4.usajobs            A           205.131.175.34
 apps.usajobs               A           205.131.175.21
 search.usajobs             A           205.131.175.5
 webentry.usajobs           A           205.131.175.19
 www.usajobs                A           205.131.175.20
 resume.usajobs             A           205.131.175.24
 search2.usajobs            A           205.131.175.23
 search3.usajobs            A           205.131.175.10
 www.pmi                    A           205.131.175.33
 localhost                  A           127.0.0.1
 opm2                       A           205.131.175.102
 trusted.cplmr              A           172.16.200.12
 webtest                    A           198.78.46.121
 mcndns1                    A           198.78.46.249
 servfwDCint                A           198.78.46.251
 wfjic                      A           198.78.46.122
 mail                       A           198.78.46.20
 ccrmon.mail                A           198.78.46.41
```

```
fjob.mail             A    205.131.175.3
ftp.fjob.mail         A    205.131.175.4
www.usacareers        A    205.131.175.25
suzinttest            A    198.78.46.150
dcdhcp-dns1           A    172.16.8.3
serv2                 A    198.78.46.56
serv3                 A    198.78.46.3
serv4                 A    198.78.46.18
wdcoer-cs2            A    198.78.46.247
serv6                 A    198.78.46.72
cisco51001            A    198.78.46.147
eexp                  A    198.78.46.102
cisco51002            A    198.78.46.148
cisco51003            A    198.78.46.149
ftp.eex               A    205.131.175.53
dns.fipc              A    199.234.58.2
wdccplmr2-ws          A    205.131.188.12
maconims1             A    198.78.46.203
image_server          A    198.78.46.144
esalertpage           A    198.78.46.48
ei_tdw3               A    198.78.46.19
ntv5                  A    198.78.46.190
servfwDCext           A    205.131.175.100
www.nsep-net_bk       A    205.131.175.31
www.nsep-net          A    205.131.175.30
wdccplmr1             A    198.78.46.243
sbeachrmon            A    198.78.46.68
eex-customs_bk        A    198.78.46.206
ServicesOnline        NS   server = serv3.opm.gov
suzielab              A    198.78.46.139
www                   A    205.131.176.11
wdcoer-cs             A    198.78.46.117
mars                  A    198.78.46.125
fjob                  A    205.131.175.3
ftp.fjob              A    205.131.175.4
>
```

Figure 5.17 Using the nslookup ls command option

To continue our examination of the opm.gov domain, Figure 5.17 illustrates the use of the ls command option with the -t option to list all records of type MX for the domain opm.gov. Note that this request returned two MX records. Next, the ls command option was reentered with the -t option followed by the letter a to retrieve or list all A resource records for the domain. This action resulted in a listing of all A resource records maintained by the name server that supports the domain opm.gov. Note that an extensive list of A resource records as well as a few NS records were listed, which also indicates that the Windows NT nslookup program needs a bit of work.

Through the use of the nslookup program and its set querytype and ls options, you can list a variety of resource records. By doing so, you can easily verify if an inability to resolve a specific host address, a problem with a mail

```
UNIX(r) System V Release 4.0 (rrs5)

***************************************************************************
* – InterNIC Registration Services Center –
*
* For the *original* whois type:   WHOIS [search string]
<return>
* For referral whois type:         RWHOIS [search string]
<return>
*
* For user assistance call (703) 742-4777
# Questions/Updates on the whois database to
HOSTMASTER@internic.net
* Please report system problems to ACTION@internic.net
***************************************************************************
The InterNIC Registration Services database contains ONLY
non-military and non-US Government Domains and contacts.
Other associated whois servers:
     American Registry for Internet Numbers—whois.arin.net
     European IP Address Allocations       —whois.ripe.net
     Asia Pacific IP Address Allocations   —whois.apnic.net
     US Military                           —whois.nic.mil
     US Government                         —whois.nic.gov
Cmdinter Ver 1.3 Wed Nov 18 15:33:06 1998 EST
[ansi] InterNIC >Whois harvard.edu

Registrant:
Harvard University (HARVARD-DOM)
   Network Services Division
   Office for Information Technology
   10 Ware Street
   Cambridge, MA 02138

   Domain Name: HARVARD.EDU

   Administrative Contact:
      Tumas, Jay (JT929) jay_tumas@HARVARD.EDU
      617.495.8515 (FAX) 617.495.0914
   Technical Contact, Zone Contact:
      Donnelly, Leo (LD238) leo_donnelly@HARVARD.EDU
      617.496.0476
   Billing Contact:
      Tumas, Jay (JT929) jay_tumas@HARVARD.EDU
      617.495.8515 (FAX) 617.495.0914

   Record last updated on 31-Oct-97.
   Record created on 27-Jun-85.
   Database last updated on 18-Nov-98 05:04:20 EST.

   Domain servers in listed order:

NS1.HARVARD.EDU              128.103.200.101
NS2.HARVARD.EDU              128.103.1.1
CUNIXD.CC.COLUMBIA.EDU       128.59.35.142
```

Figure 5.18 Using Telnet to use the InterNIC Whois command to extract information about the domain *harvard.edu*

exchanger, or another resource record-related problem is related to a configuration problem in a name server.

The Whois command

In concluding our investigation of the DNS, we will turn our attention to the Whois command. This command is included in most versions of Unix. If you are using a version of Windows or a version of Unix that does not support Whois, you can obtain an equivalent capability by telneting to *rs.internic.net* and entering the command Whois.

Figure 5.18 illustrates the use of a Telnet session to the InterNIC and the entry of the Whois command to obtain information about the domain *harvard.edu*. Note that the result of the Whois command provides you with the ability to obtain a significant amount of information about a domain, including administrative contacts and domain servers for the domain. Thus, the use of Whois provides another valuable mechanism for obtaining information about a domain.

6

LAYER 2 MANAGEMENT

In the ISO Open System Interconnection (OSI) Reference Model the second layer of the model is the data link layer. At that layer frames are formed and passed to the lower physical layer to transport higher layer information, such as TCP/IP packets conveyed via an Ethernet or Token-Ring network. Although the TCP/IP protocol suite technically begins at the network layer, any book covering the management of the protocol would be remiss if it did not focus attention on layer 2 management issues. This is because the ability to transport TCP/IP on an end-to-end basis includes LAN delivery systems, and any problem or degradation in the capability or capacity of a LAN can adversely effect the flow of upper layer information.

In this chapter we will focus our attention upon Ethernet in its various 'flavors' as well as Token-Ring networks. In doing so we will examine the composition of Ethernet and Token-Ring frames, different types of error conditions applicable to each type of network, and parameters that affect the level of performance on each type of network. As we examine the operation of each distinct type of LAN, we will supplement text descriptions of events and conditions with visual images of screen displays generated by layer 2 monitoring programs. For details of demonstration software relevant to this chapter see Appendix B.

6.1 ETHERNET FRAME OPERATIONS

In this section we will focus attention upon the composition of different types of Ethernet frames. In reality, there is only one Ethernet frame, while the CSMA/CD frame format standardized by the IEEE is technically referred to as an 802.3 frame. As we will see later in this section, the physical 802.3 frame can have several logical formats. For consistency and ease of reference, we will refer to Carrier Sense Multiple Access/Collision Detection (CSMA/CD) operations collectively as Ethernet, and, when appropriate, indicate differences between Ethernet and the IEEE 802.3 Ethernet-based CSMA/CD standards. After describing the general composition of Ethernet and IEEE 802.3 frames, we will examine the function of the fields within each frame as well as the manner by which the placement of frames on the media is controlled — a process known as Media Access Control.

Preamble	Destination Address	Source Address	Type	Data	Frame Check Sequence
8 bytes	6 bytes	6 bytes	2 bytes	46—1500 bytes	4 bytes

IEEE 802.3

Preamble	Start-of-Frame Delimiter	Destination Address	Source Address	Length	Data	Frame Check Sequence
7 bytes	1 byte	2/6 bytes	2/6 bytes	2 bytes	46—1500 bytes	4 bytes

Figure 6.1 Ethernet and IEEE 802.3 frame formats

6.1.1 Ethernet frame composition

Figure 6.1 illustrates the general frame composition of Ethernet and IEEE 802.3 frames. You will note that they differ slightly. An Ethernet frame contains an 8-byte preamble, while the IEEE 802.3 frame contains a 7-byte preamble followed by a 1-byte start-of-frame delimiter field. A second difference between the composition of Ethernet and IEEE 802.3 frames concerns the 2-byte Ethernet type field. That field is used by Ethernet to specify the protocol carried in the frame, enabling several protocols to be carried independently of one another. Under the IEEE 802.3 frame format, the type field was replaced by a 2-byte Length field that specifies the number of bytes that follow that field as data. In addition, to enable different types of protocols to be carried in a frame and to be correctly identified, the 802.3 frame format subdivides the data field into subfields. Those subfields include a Destination Service Access Point (DSAP), a Source Service Access Point (SSAP), and a Control field that prefixes a reduced data field. The use of those fields defines a Logical Link Control (LLC) layer residing within an 802.3 frame, and will be discussed later in this chapter along with some common framing variations.

The differences between Ethernet and IEEE 802.3 frames, while minor, make the two incompatible with one another. This means that your network must contain all Ethernet-compatible Network Interface Cards (NICs), all IEEE 802.3-compatible NICs, or adapter cards that can examine the frame and automatically determine its type, a process described later in this chapter. Fortunately, the fact that the IEEE 802.3 frame format represents a standard means that most vendors market 802.3-compliant hardware and software. Although a few vendors continue to manufacture Ethernet or dual functioning Ethernet/IEEE 802.3 hardware, such products are primarily used to provide organizations with the ability to expand previously developed networks without requiring the wholesale replacement of NICs. Although the IEEE 802.3 frame does not directly support a type field within the frame, as we will note later in this chapter the IEEE defined a special type of frame to obtain compatibility with Ethernet LANs. That frame is referred to as an Ethernet

Subnetwork Access Protocol (Ethernet-SNAP) frame, which enables a type subfield to be included in the data field. While the IEEE 802.3 standard has essentially replaced Ethernet, because of their similarities and the fact that 802.3 was based upon Ethernet, we will consider both to be Ethernet.

Now that we have an overview of the structure of Ethernet and 802.3 frames, let us probe deeper and examine the composition of each frame field. We will take advantage of the similarity between Ethernet and IEEE 802.3 frames to examine the fields of each frame on a composite basis, noting the differences between the two when appropriate.

Preamble field

The Preamble field consists of 8 (Ethernet) or 7 (IEEE 802.3) bytes of alternating 1 and 0 bits. The Ethernet chip set contained on the network interface adapter places the Preamble and following start-of-frame delimiter on the front of each frame transmitted on the network. The purpose of the Preamble field is to announce the frame and to enable all receivers on the network to synchronize themselves to the incoming frame.

Start-of-Frame Delimiter field

This field is applicable only to the IEEE 802.3 standard, and can be viewed as a continuation of the preamble. In fact, the composition of this field continues in the same manner as the format of the preamble, with alternating 1 and 0 bits used for the first six bit positions of this 1-byte field. The last two bit positions of this field are 11: this breaks the synchronization pattern and alerts the receiver that frame data follows.

Both the Preamble field and the Start-of-Frame Delimiter field are removed by the Ethernet chip set or controller when it places a received frame in its buffer. Similarly, when a controller transmits a frame, it prefixes the frame with those two fields (if it is transmitting an IEEE 802.3 frame) or a Preamble field (if it is transmitting a true Ethernet frame).

Destination Address field

The Destination Address field identifies the recipient of the frame. Although this may appear to be a simple field, in reality its length can vary between IEEE 802.3 and Ethernet frames. In addition, each field can consist of two or more subfields, whose settings govern such network operations as the type of addressing used on the LAN, and whether the frame is addressed to a specific station or more than one station. To obtain an appreciation for the use of this field, let us examine how it is used under the IEEE 802.3 standard as one of the two field formats applicable to Ethernet.

Figure 6.2 illustrates the composition of the Source and Destination Address fields. As indicated, the 2-byte Source and Destination Address

A. 2-byte field (IEEE 802.3)

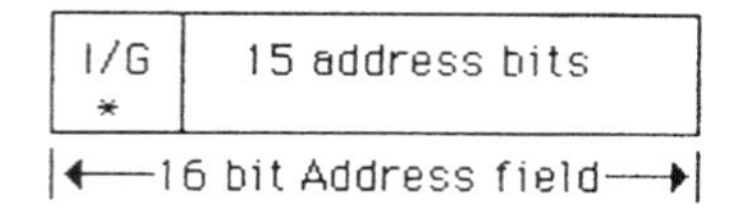

B. 6-byte field (Ethernet and IEEE 802.3)

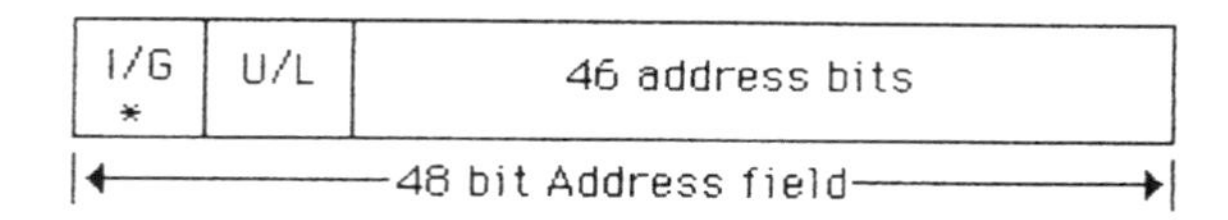

I/G bit subfield '0' = individual address '1' = group address
U/L bit subfield '0' = universally administrated addressing
'1' = locally administrated addressing

* Set to '0' in Source Address field

Figure 6.2 Source and Destination Address field formats

fields are applicable only to IEEE 802.3 networks, while the 6-byte Source and Destination Address fields are applicable to both Ethernet and IEEE 802.3 networks. A user can select either a 2- or 6-byte Destination Address field; however, with IEEE 802.3 equipment, all stations on the LAN must use the same addressing structure. Today, almost all 802.3 networks use 6-byte addressing, since the inclusion of a 2-byte field option was designed primarily to accommodate early LANs that use 16-bit Address fields. Both destination and source addresses are normally displayed by network monitors in hexadecimal, with the first three bytes separated from the last three by a colon (:) when 6-byte addressing is used. For example, the source address 02608C876543 would be displayed as 02608C:876543. As we will shortly note, the first three bytes identify the manufacturer of the adapter card, while the following three bytes identify a specific adapter manufactured by the vendor identified by the first three bytes or six hex digits.

I/G subfield The 1-bit I/G subfield is set to a 0 to indicate that the frame is destined to an individual station, or 1 to indicate that the frame is addressed to more than one station — a group address. One special example of a group address is the assignment of all 1s to the Address field. Hex FF-FF-FF-FF-FF-FF is recognized as a broadcast address, and each station on the network will receive and accept frames with that destination address.

When a destination address specifies a single station, the address is referred to as a unicast address. A group address that defines multiple stations is known as a multicast address, while a group address that specifies all stations on the network is, as previously mentioned, referred to as a broadcast address.

U/L subfield The U/L subfield is applicable only to the 6-byte Destination Address field. The setting of this field's bit position indicates whether the destination address is an address that was assigned by the IEEE (universally administered) or assigned by the organization via software (locally administered).

Universal versus locally administered addressing Each Ethernet Network Interface Card (NIC) contains a unique address burned into its read-only memory (ROM) at the time of manufacture. To ensure that this universally administered address is not duplicated, the IEEE assigns blocks of addresses to each manufacturer. These addresses normally include a 3-byte prefix, which identifies the manufacturer and is assigned by the IEEE, and a 3-byte suffix, which is assigned by the adapter manufacturer to its NIC. For example, the prefix hex 02-60-8C identifies an NIC manufactured by 3Com.

Table 6.1 lists the 3-byte identifiers associated with 15 manufacturers of Ethernet network interface cards. Note that many organizations beyond 3Com and Cisco listed in Table 6.1 were assigned two or more blocks of addresses by the IEEE. The entries in Table 6.1 represent only a small portion of 3-byte identifiers assigned by the IEEE to manufacturers of Ethernet adapter cards. For a comprehensive list of currently assigned 3-byte identifiers, readers should contact the IEEE at the following address:

IEEE Registration Authority
IEEE Standards Department
445 Hoes Lane, PO Box 1331
Piscataway, NJ 08844, USA

You can also obtain a full list of assigned vendor codes via FTP at *ftp.ieee.org* by retrieving the file *ieee/info/info.stds.ovi*. You should note that the list is limited to those companies that have agreed to make their vendor code assignment(s) public. Through the use of a table of 3-byte identifiers and associated manufacturer names, diagnostic hardware or software can be programmed to read the Source and Destination Address fields within frames and identify the manufacturer of the originating and destination adapter cards. Later in this section we will examine a few screen displays of one program whose use can focus attention upon different potential layer 2 problems.

Although the use of universally administered addressing eliminates the potential for duplicate network addresses, it does not provide the flexibility obtainable from locally administered addressing. For example, under locally administered addressing, you can configure mainframe software to work with a predefined group of addresses via a gateway PC. Then, as you add new stations to your LAN, you simply use your installation program to assign a locally administered address to the NIC instead of using its universally administered address. As long as your mainframe computer has a pool of locally administered addresses that includes your recent assignment, you do not have to modify your mainframe communications software configuration. Since the modification of mainframe communications software typically requires recompiling and reloading, the attached network must become

Table 6.1 Representative Ethernet Manufacturer IDs

Manufacturer	3-Byte Identifiers
3COM	02-60-8C
3COM	00-60-97
Cabletron	00-00-1D
Cisco Systems, Inc.	00-00-0C
Cisco Systems, Inc.	00-60-08
Cisco Systems, Inc.	00-60-09
Excelan	08-00-14
Hewlett-Packard	00-60-B0
NEC	00-00-4C
NeXT	00-00-0F
Novell	00-00-1B
Synoptics (Bay Networks)	00-0-81
Western Digital	00-00-C0
Xerox	00-00-AA
Xircom	00-80-C7

inoperative for a short period of time. Because a large mainframe may service hundreds to thousands of users, such changes are normally performed late in the evening or on a weekend. Thus, the changes required for locally administered addressing are more responsive to users than those required for universally administered addressing.

Source Address field

The Source Address field identifies the station that transmitted the frame. Like the Destination Address field, the source address can be either 2 or 6 bytes in length.

The 2-byte source address is supported only under the IEEE 802.3 standard, and requires the use of a 2-byte destination address; all stations on the network must use 2-byte addressing fields. The 6-byte source address field is supported by both Ethernet and the IEEE 802.3 standard. When a 6-byte address is used, the first three bytes represent the address assigned by the IEEE to the manufacturer for incorporation into each NIC's ROM. The vendor then normally assigns the last three bytes for each of its NICs.

Many software- and hardware-based network analyzers include the capability to identify each station on a LAN, and count the number of frames transmitted by the station and destined to the station, as well as to identify the manufacturer of the NIC used in the station. Concerning the latter capability, this is accomplished by the network analyzer containing a table of 3-byte identifiers assigned by the IEEE to each NIC manufacturer, along with the name of the manufacturer. Then the analyzer compares the 3-byte identifier read from frames flowing on the network and compares each identifier with the identifiers stored in its identifier table. By providing information concerning network statistics, network errors, and the vendor identifier for the NIC in each station, you may be able to isolate problems faster or better consider future decisions concerning the acquisition of additional NICs.

An example of the use of NIC manufacturer IDs can be obtained by examining two monitoring screen displays of the Triticom EtherVision network monitoring and analysis program. Figure 6.3 illustrates the monitoring screen during the program's autodiscovery process. During this process, the program reads the source address of each frame transmitted on the segment that the computer existing the program is connected to. Although obscured by the highlighted bar, the first three bytes of the adapter address covered by that bar is 00-60-8C, which represents a block of addresses assigned by the IEEE to 3Com Corporation. If you glance at the first column in Figure 6.3, you will note that the third through sixth and the next to last rows also have NIC addresses that commence with hex 00-60-8C. By pressing the F2 key, the program will display the manufacturer of each NIC encountered and for which statistics are being accumulated. This is indicated in Figure 6.4, which shows the first three bytes of each address replaced by the vendor assigned the appropriate manufacturer ID. Thus, rows 1, 3 through 6, and the next to last row commence with '3Com' to indicate the manufacturer of the NIC.

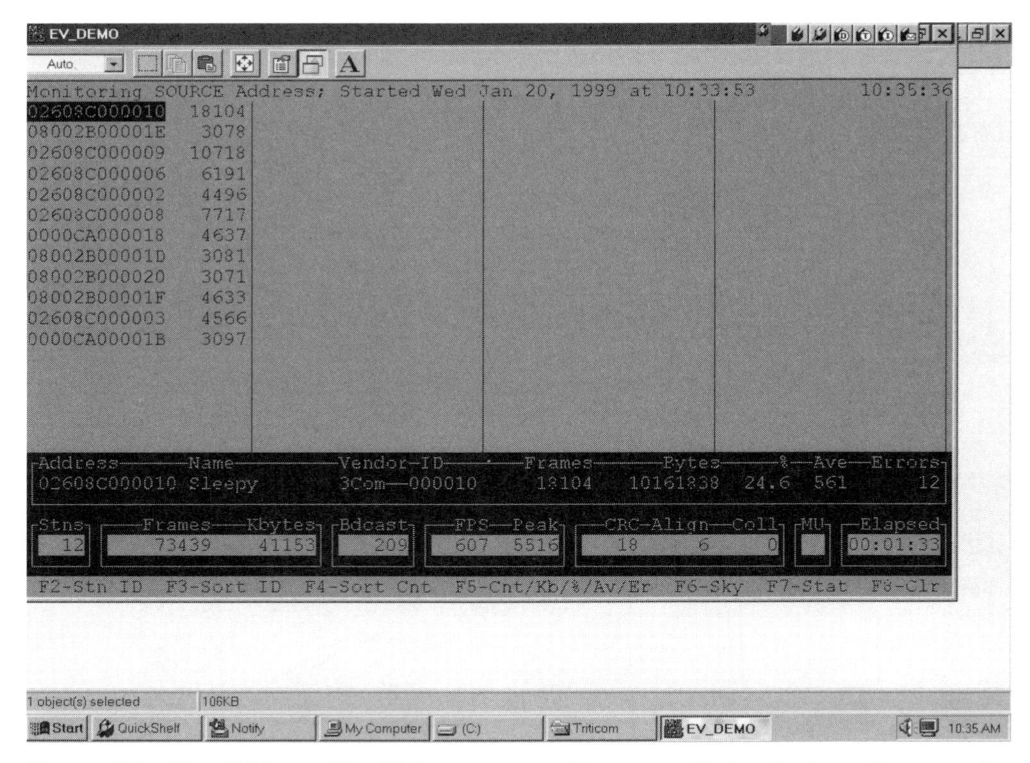

Figure 6.3 The Triticom EtherVision source address monitoring feature discovers the hardware address of each NIC. At the time this screen was captured, 12 stations were identified

In examining Figures 6.3 and 6.4, note that they show the number of frames transmitted by specific adapters at particular points in time. If your organization runs multiple protocols on a LAN, a layer 3 protocol analyzer restricted to monitoring TCP/IP would not indicate stations transmitting other protocols. Thus, it might be easy to overlook stations that heavily utilize the LAN transmitting IPX and other protocols, possibly causing network utilization problems. By using a layer 2 network monitoring program, you can note the total use of the network by each station on the network. This in turn provides you with the ability to note stations that might be adversely effecting the operation of the network. In addition, EtherVision and other layer 2 monitoring programs can be used to note certain types of network errors. You can use such information to correct certain types of problems, such as defective network adapter cards and faulty cabling.

Type field

The 2-byte Type field is applicable only to the Ethernet frame. This field identifies the higher-level protocol contained in the data field. Thus, this field tells the receiving device how to interpret the data field.

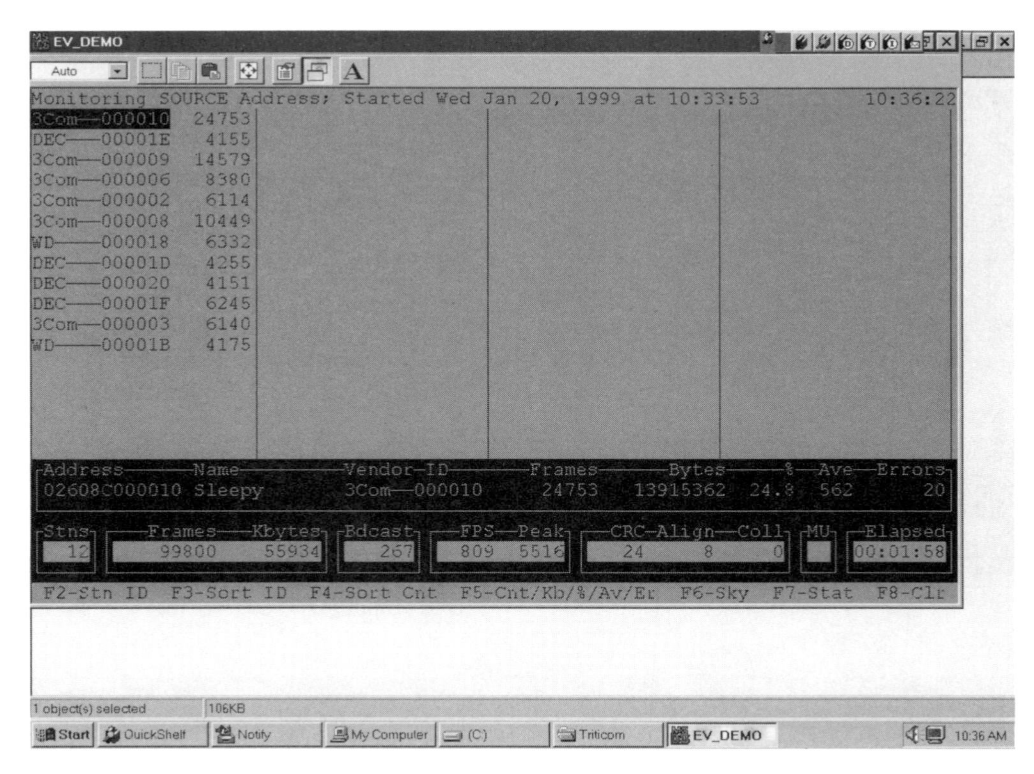

Figure 6.4 By pressing the F2 key, EtherVision will convert the 3-byte hex NIC manufacturer ID to the vendor name or an appropriate mnemonic

Under Ethernet, multiple protocols can exist on the LAN at the same time. Xerox served as the custodian of Ethernet address ranges licensed to NIC manufacturers and defined the protocols supported by the assignment of Type field values. Table 6.2 lists 31 of the more common Ethernet Type field identifiers to include their hex values. Note that the value of the Type field always exceeds decimal 1500 (hex 05-DC), and provides a mechanism for a receiving station to determine the type of frame on the network since a Length field (described next) cannot exceed decimal 1500. To illustrate the ability of Ethernet to transport multiple protocols, assume that a common LAN was used to connect stations to both UNIX and NetWare servers. Frames with hex value 0800 in their Type field would identify the transport of IP protocol data, while frames with the hex value 8137 in the Type field would identify the transport of IPX and SPX protocols. Thus, the placement of an appropriate hex value in the Ethernet Type field provides a mechanism to support the transport of multiple protocols on the LAN. Under the IEEE 802.3 standard, the Type field was replaced by a Length field, which precludes compatibility between pure Ethernet and 802.3 frames.

Length field

The 2-byte Length field, applicable to the IEEE 802.3 standard, defines the number of bytes contained in the Data field. Under both Ethernet and IEEE 802.3 standards, the minimum size frame must be 64 bytes in length from the Destination Address through FCS fields. The minimum length frame defined by the IEEE considers the frame prior to its placement onto the network. Thus, once placed onto the network, the preamble adds 8 bytes, resulting in a minimum length of the frame being 72 bytes. This minimum size frame ensures that there is sufficient transmission time to enable Ethernet NICs to detect collisions accurately, based on the maximum Ethernet cable length specified for a network and the time required for a frame to propagate the length of the cable. Based on the minimum frame length of 64 bytes and the possibility of using 2-byte addressing fields, this means that each Data field must be a minimum of 46 bytes in length.

The only exception to the preceding involves Gigabit Ethernet. At a 1000-Mbps operating rate, the original 802.3 standard would not provide a frame duration long enough to permit a 100-meter cable run. This is because at a 1000-Mbps data rate there is a high probability that a station could be in the middle of transmitting a frame before it becomes aware of any collision that might have occurred at the other end of the segment. Recognizing this problem resulted in the development of a carrier extension, which extends the minimum Ethernet frame to 512 bytes. The carrier extension is discussed in detail later in this chapter when we turn our attention to the Gigabit Ethernet carrier extension.

For all versions of Ethernet except Gigabit Ethernet, if data being transported is less than 46 bytes, the Data field is padded to obtain 46 bytes. However, the number of PAD characters is not included in the Length

Table 6.2 Ethernet Type field assignments

Protocol	Hex Value Assigned
Experimental	0101–DIFF
Xerox XNS	0600
IP	0800
X.75 Internet	0801
NBS Internet	0802
ECMA Internet	0803
CHAOSmet	0804
X.25 Level 3	0805
Address Resolution Protocol	0806
XNS Compatibility	0807
Banyan Systems	0BAD
BBN Simnet	5208
DEC MOP Dump/Load	6001
DEC MOP Remote Console	6002
DEC DECNET Phase IV Route	6003
DEC LAT	6004
DEC Diagnostic Protocol	6005
3Com Corporation	6010-6014
Proteon	7030
AT&T	8008
Excelan	8010
Tymshare	802E
DEC LANBridge	8038
DEC Ethernet Encryption	803D
AT&T	8046–8047
AppleTalk	809B
IBM SNA Service on Ethernet	80D5
AppleTalk ARP	80F3
IBM SNA Service on Ethernet	80D5
AppleTalk ARP	80F3
Wellfleet	80FF–8103
NetWare IPX/SPX	8137–8138
SNMP	814C

field value. NICs that support both Ethernet and IEEE 802.3 frame formats use the value in this field to distinguish between the two frames. That is, because the maximum length of the Data field is 1500 bytes, a value that exceeds hex 05DC indicates that instead of a Length field (IEEE 802.3), the field is a Type field (Ethernet).

Data field

As previously discussed, the Data field must be a minimum of 46 bytes in length to ensure that the frame is at least 64 bytes in length. This means that the transmission of 1 byte of information must be carried within a 46-byte Data field; if the information to be placed in the field is less than 46 bytes, the remainder of the field must be padded. Although some publications subdivide the Data field to include a PAD subfield, the latter actually represents optional fill characters that are added to the information in the Data field to ensure a length of 46 bytes. The maximum length of the Data field is 1500 bytes.

Frame Check Sequence field

The Frame Check Sequence field, applicable to both Ethernet and the IEEE 802.3 standard, provides a mechanism for error detection. Each chip set transmitter computes a Cyclic Redundancy Check (CRC) that covers both Address fields, the Type/Length field, and the Data field. The transmitter then places the computed CRC in the 4-byte FCS field.

The CRC treats the previously mentioned fields as one long binary number. The n bits to be covered by the CRC are considered to represent the coefficients of a polynomial $M(X)$ of degree $n-1$. Here, the first bit in the Destination Address field corresponds to the X^{n-1} term, while the last bit in the Data field corresponds to the X^0 term. Next, $M(X)$ is multiplied by X^{32}, and the result of that multiplication process is divided by the following polynomial:

$$G(X)=X^{32}+X^{26}+X^{23}+X^{22}+X^{16}+X^{12}+X^{11}+X^{10}+X^{8}+X^{7}+X^{5}+X^{4}+X^{2}+X+1$$

Note that the term X^n represents the setting of a bit to a 1 in position n. Thus, part of the generating polynomial $X^5+X^4+X^2+X^1$ represents the binary value 11011.

This division produces a quotient and remainder. The quotient is discarded, and the remainder becomes the CRC value placed in the four-byte FCS field. This 32-bit CRC reduces the probability of an undetected error to 1 bit in every 4.3 billion, or approximately 1 bit in $2^{32}-1$ bits.

Once a frame reaches its destination, the chip set's receiver uses the same polynomial to perform the same operation upon the received data. If the CRC computed by the receiver matches the CRC in the FCS field, the frame is accepted. Otherwise, the receiver discards the received frame, as it is considered to have one or more bits in error. The receiver will also consider a received frame to be invalid and discard it under two additional conditions. Those conditions occur when the frame does not contain an integral number of bytes, or when the length of the data field does not match the value contained in the Length field. The latter condition, obviously is only applicable to the 802.3 standard, since an Ethernet frame uses a Type field instead of a Length field.

Through the use of a layer 2 monitoring program that indicates the distribution of frames based upon frame length, you can obtain a general indication of the type of data being transported on a network. Figure 6.5 illustrates the Triticom EtherVision statistics display. Note that the upper right corner of the display illustrates the distribution of frames based on predefined frame lengths. At the point in time when the screen was captured there were very few frames under 128 bytes in length. This would indicate that a majority of traffic was for applications that tend to transmit more data per frame than short interactive queries such as file transfers or data base searches. At the bottom of Figure 6.5 note that EtherVision displays common error conditions as they occur as well as the station associated with the error. This information can be extremely valuable in isolating and fixing the cause of certain types of error conditions that can adversely effect the ability of all network users. For example, a very high number of CRC errors (covered next) results in an extensive amount of retransmissions, which adversely affects network performance. If a particular station is associated with a high level of CRC errors, it is possible that either the station's network adapter or cabling to the network is causing the problem and represents a good starting point to alleviate the problem.

6.2 ETHERNET MEDIA ACCESS CONTROL

Under the IEEE 802 series of standards, the data link layer of the OSI Reference Model was subdivided into two sublayers: Logical Link Control

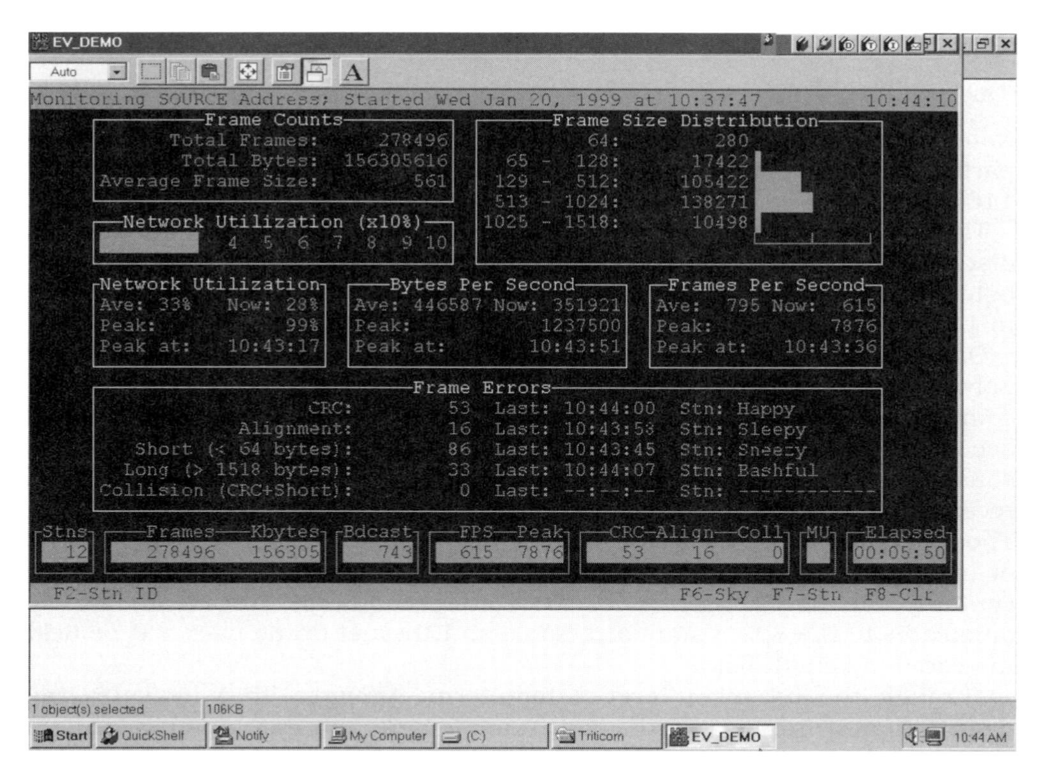

Figure 6.5 Through the EtherVision statistics display you can obtain information about how the network is being used as well as information about common types of frame errors

(LLC) and Medium Access Control (MAC). The frame formats previously examined represent the manner in which LLC information is transported. Directly under the LLC sublayer is the MAC sublayer.

6.2.1 Functions

The MAC sublayer, which is the focus of this section, is responsible for checking the channel and transmitting data if the channel is idle, checking for the occurrence of a collision, and taking a series of predefined steps if a collision is detected. Thus, this layer provides the required logic to control the network.

Figure 6.6 illustrates the relationship between the physical and LLC layers with respect to the MAC layer. The MAC layer is an interface between user data and the physical placement and retrieval of data on the network. To better understand the functions performed by the MAC layer, let us examine the four major functions performed by that layer: transmitting data operations, transmitting medium access management, receiving data operations, and receiving medium access management. Each of those four functions can be viewed as a functional area, since a group of activities is associated with each area.

Table 6.3 lists the four MAC functional areas and the activities associated with each area. Although the transmission and reception of data operations activities are self-explanatory, the transmission and reception of media access management require some elaboration. Therefore, let us focus our attention on the activities associated with each of these functional areas.

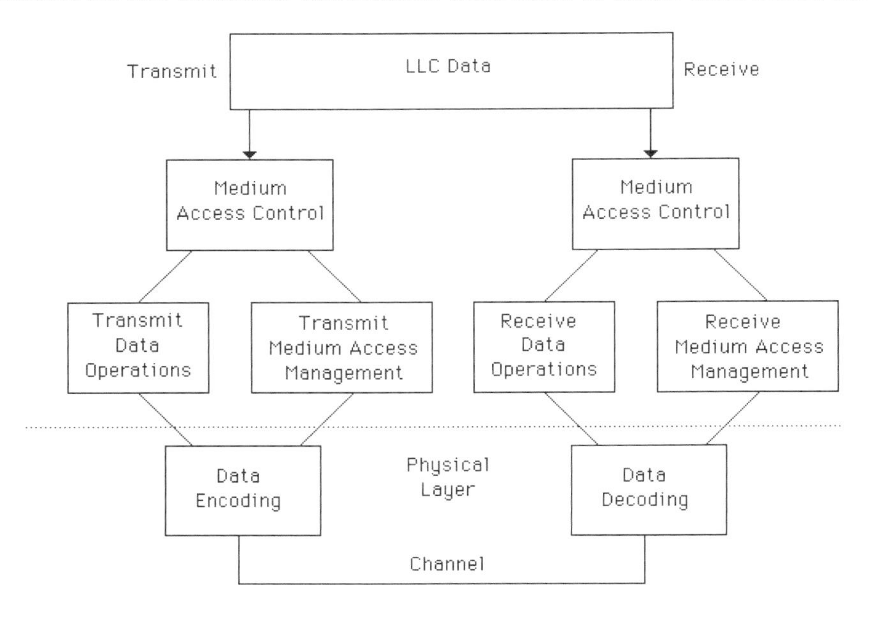

Figure 6.6 Medium access control. The Medium Access Control (MAC) layer can be considered an interface between user data and the physical placement and retrieval of data on the network

Table 6.3 MAC functional areas

Transmit data operations	• Accept data from the LC sublayer and construct a frame by appending preamble and start-of-frame by appending preamble and start-of-frame delimiter; insert destination and source address, length count; if frame is less than 64 bytes, insert sufficient PAD characters in the data field • Calculate the CRC and place in the FCS field
Transmit media access management	• Defer transmission if the medium is busy • Delay transmission for a specified interframe gap period • Present a serial bit stream to the physical layer for transmission • Halt transmission when a collision is detected • Transmit a jam signal to ensure that news of a collision propagates throughout the network • Reschedule retransmissions after a collision until successful, or specified retry limit is reached
Receive data operations	• Discard all frames not addressed to the receiving station • Recognize all broadcast frames and frames specifically addressed to station • Perform a CRC check • Remove preamble, start-of-frame delimiter, destination and source addresses, length count, FCS; if necessary, remove PAD fill characters • Pass data to LLC sublayer
Receive media access management	• Receive a serial bit stream from the physical layer • Verify byte boundary and length of frame • Discard frames not an even 8 bits in length or less than the minimum frame length

6.2.2 Transmit media access management

CSMA/CD can be described as a *listen-before-acting* access method. Thus, the first function associated with transmit media access management is to find out whether any data is already being transmitted on the network, and, if so, to defer transmission. During the listening process, each station attempts to sense the carrier signal of another station—hence the prefix *carrier sense* (*CS*) for this access method. Although broadband networks use RF modems that generate a carrier signal, a baseband network has no carrier signal in the conventional sense of a carrier as a periodic waveform altered to convey information. Thus, a logical question you may have is how the MAC sublayer on a baseband network can sense a carrier signal if there is no carrier. The answer to this question lies in the use of a digital signaling method known as *Manchester encoding* that a station can monitor to note whether another station is transmitting.

To understand the Manchester encoding signaling method used by baseband Ethernet LANs, let us first review the method of digital signaling used by computers and terminal devices. In this signaling method, a positive voltage is used to represent a binary 1, while the absence of voltage (0 volts) is used to represent a binary 0. If two successive 1 bits occur, two successive bit positions then have a similar positive voltage level or a similar zero voltage level. Since the signal goes from 0 to some positive voltage and does not return to 0 between successive binary 1s, it is referred to as a *unipolar non-return to zero signal* (*NRZ*). This signaling technique is illustrated at the top of Figure 6.7.

Although unipolar non-return to zero signaling is easy to implement, its use for transmission has several disadvantages. One of the major disadvantages associated with this signaling method involves determining where one bit ends and another begins. Overcoming this problem requires synchronization between a transmitter and receiver by the use of clocking circuitry, which can be relatively expensive.

To overcome the need for clocking, many baseband LANs use *Manchester* or *Differential Manchester* encoding. In Manchester encoding, a timing transition always occurs in the middle of each bit, while an equal amount of positive and negative voltage is used to represent each bit. This coding technique provides a good timing signal for clock recovery from received data, due to its timing transitions. In addition, since the Manchester code always maintains an equal amount of positive and negative voltage, it prevents direct current (DC) voltage buildup, enabling repeaters to be spaced farther apart from one another.

The lower portion of Figure 6.7 illustrates an example of Manchester coding. Note that a low-to-high voltage transition represents a binary 1, while a high-to-low voltage transition represents a binary 0. Although NRZI encoding is used on broadband networks, the actual data is modulated after it is encoded.

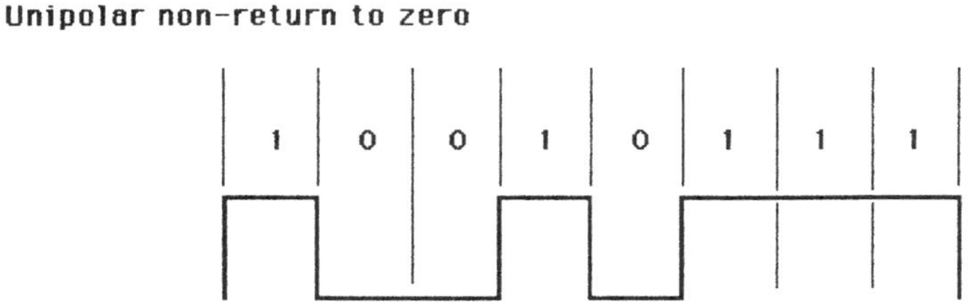

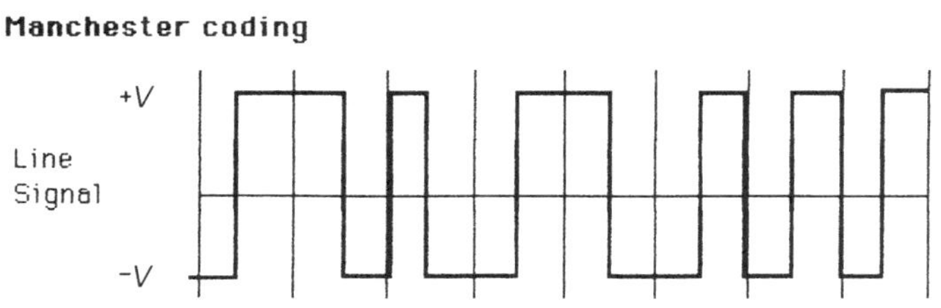

Figure 6.7 Unipolar non-return to zero signaling and Manchester coding. In Manchester coding, a timing transition occurs in the middle of each bit, and the line code maintains an equal amount of positive and negative voltage

Thus, the presence or absence of a carrier is directly indicated by the presence or absence of a carrier signal on a broadband network.

6.2.3 Collision detection

As previously discussed, under Manchester coding a binary 1 is represented by a high-to-low transition, while a binary 0 is represented by a low-to-high voltage transition. Thus, an examination of the voltage on the medium of a baseband network enables a station to determine whether a carrier signal is present.

If a carrier signal is found, the station with data to transmit will continue to monitor the channel. When the current transmission ends, the station will then transmit its data, while checking the channel for collisions. Since Ethernet and IEEE 802.3 Manchester encoded signals have a 1-volt average DC voltage level, a collision results in an average DC level of 2 volts. Thus, a transceiver or network interface card can detect collisions by monitoring the voltage level of the Manchester line signal.

Jam pattern

If a collision is detected during transmission, the transmitting station will cease transmission of data and initiate transmission of a jam pattern. The jam pattern consists of 32–48 bits. These bits can have any value other than the CRC value that corresponds to the partial frame transmitted before the jam. The transmission of the jam pattern ensures that the collision lasts long enough to be detected by all stations on the network.

When a repeater is used to connect multiple segments, it must recognize a collision occurring on one port and place a jam signal on all other ports. Doing so results in the occurrence of a collision with signals from stations that may have been in the process of beginning to transmit on one segment when the collision occurred on the other segment. In addition, the jam signal serves as a mechanism to cause non-transmitting stations to wait until the jam signal ends before attempting to transmit, alleviating additional potential collisions from occurring.

One of the most common causes of excessive collisions is improper cabling that violates Ethernet cabling distances. If you use EtherVision or a similar product and detect a high level of collisions, you should more than likely examine cabling distance between stations and hubs as well as the connectors on the cables. Upon occasion, a plug not fully mated into a pack can result in errors as persons walk through an area and cause minor vibrations that affect the connection.

Wait time

Once a collision has been detected, the transmitting station waits a random number of slot times before attempting to retransmit. Here the term *slot* represents 512 bits on a 10-Mbps network, or a minimum frame length of 64 bytes. The actual number of slot times the station waits is selected by a randomization process, formerly known as a *truncated binary exponential backoff*. Under this randomization process, a random integer r defines the number of slot times the station waits before listening to determine whether the channel is clear. If it is, the station begins to retransmit the frame, while listening for another collision.

If the station transmits the complete frame successfully and has additional data to transmit, it will again listen to the channel as it prepares another frame for transmission. If a collision occurs on a retransmission attempt, a slightly different procedure is followed. After a jam signal has been transmitted, the station simply doubles the previously generated random number and then waits the prescribed number of slot intervals prior to attempting a retransmission. Up to 16 retransmission attempts can occur before the station aborts the transmission and declares the occurrence of a multiple collision error condition.

Figure 6.8 illustrates the collision detection process by which a station can determine that a frame was not successfully transmitted. At time t_0 both stations A and B are listening and fail to detect the occurrence of a collision, and at time t_1 station A commences the transmission of a frame. As station A's frame begins to propagate down the bus in both directions, station B begins the transmission of a frame, since at time t_2 it appears to station B that there is no activity on the network.

Shortly after time t_2 the frames transmitted by stations A and B collide, resulting in a doubling of the Manchester-encoded signal level for a very short period of time. This doubling of the Manchester-encoded signal's voltage level is detected by station B at time t_3, since station B is closer to the collision than station A. Station B then generates a jam pattern that is detected by station A.

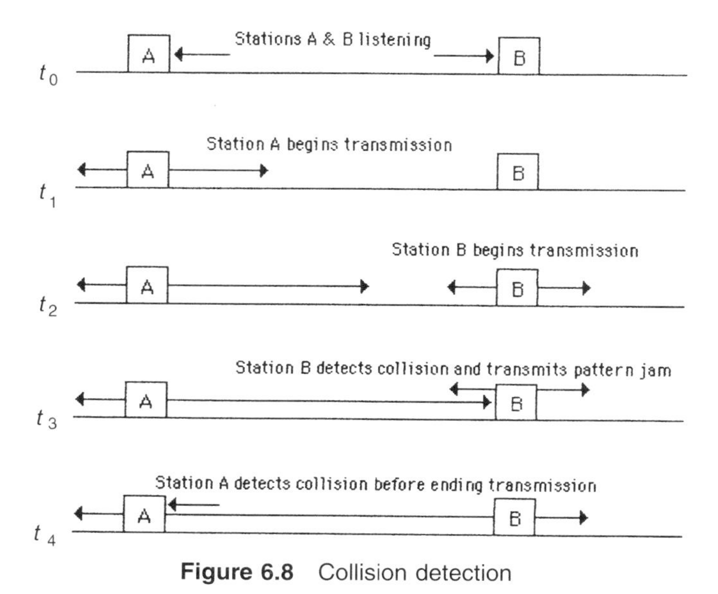

Figure 6.8 Collision detection

Late collisions

A late collision is a term used to reference the detection of a collision only after a station places a complete frame on the network. A late collision is normally caused by an excessive network segment cable length, resulting in the time for a signal to propagate from one end of a segment to another part of the segment being longer than the time required to place a full frame on the network. This results in two devices communicating at the same time never seeing the other's transmission until their signals collide.

A late collision is detected by a transmitter after the first slot time of 64 bytes and is applicable only for frames whose lengths exceed 65 bytes. The detection of a late collision occurs in exactly the same manner as a normal collision; however, it happens later than normal. Although the primary cause of late collisions is excessive segment cable lengths, an excessive number of repeaters, faulty connectors, and defective Ethernet transceivers or controllers can also result in late collisions. Many network analyzers provide information on late collisions, which can be used as a guide to check the previously mentioned items when late collisions occur.

6.3 ETHERNET LOGICAL LINK CONTROL

The Logical Link Control (LLC) sublayer was defined under the IEEE 802.2 standard to make the method of link control independent of a specific access method. Thus, the 802.2 method of link control spans Ethernet (IEEE 802.3), Token Bus (IEEE 802.4), and Token-Ring (IEEE 802.5) local area networks. Functions performed by the LLC include generating and interpreting commands to control the flow of data, including recovery operations for when a transmission error is detected.

Link control information is carried within the Data field of an IEEE 802.3 frame as an LLC Protocol Data Unit. Figure 6.9 illustrates the relationship between the IEEE 802.3 frame and the LLC Protocol Data Unit.

6.3.1 The LLC protocol data unit

Service Access Points (SAPs) function much like a mailbox. Since the LLC layer is bounded below the MAC sublayer and bounded above by the network layer, SAPs provide a mechanism for exchanging information between the LLC layer and the MAC and network layers. For example, from the network layer perspective, a SAP represents the place to leave messages about the services requested by an application.

There are two broad categories of SAPs: IEEE-administered and manufacturer-implemented. Table 6.4 provides six examples of each type of SAP. In examining the entries in Table 6.4, the hex value AA represents one of the more commonly used SAPs today. When that value is encoded in both DSAP and SSAP fields, it indicates a special type of Ethernet frame referred to as an Ethernet SNAP frame. The SNAP frame, as we will shortly note when we

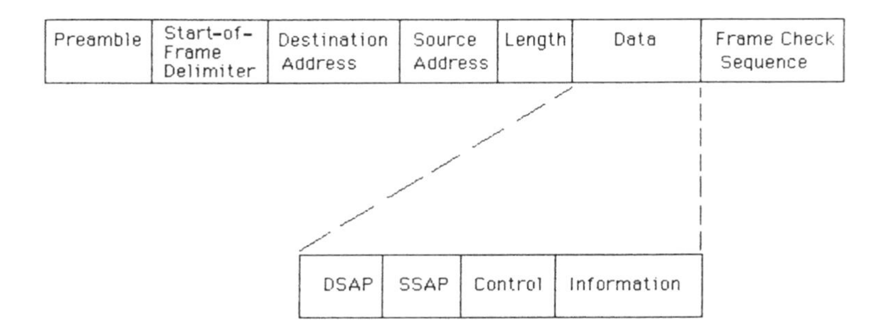

Figure 6.9 Formation of LLC protocol data unit. Control information is carried within a MAC frame

cover it in Section 6.4, unlike the Ethernet 802.3 frame enables several different protocols to be transported.

The Destination Services Access Point (DSAP) is one byte in length, and is used to specify the receiving network layer process, which is an IEEE term to denote the destination upper layer protocol. Because an IEEE 802.3 frame does not include a Type field, the DSAP field is used to denote the destination upper layer protocol carried in the frame. For example, the DSAP value hex 06 indicates the Data field is transporting IP. The Source Service Access Point (SSAP) is also one byte in length. The SSAP specifies the sending network layer process, which is in effect the source upper layer protocol. Both DSAP and SSAP addresses are either assigned by the IEEE or manufacture-implemented. They are always the same, since destination and source protocols must always be the same. For example, hex address 'FF' represents a DSAP broadcast address.

Table 6.4 Examples of SAP addresses

IEEE Administered Address (Hex)	Assignment
00	Null SAP
02	Individual LLC sublayer management functions
06	ARPANET Internet Protocol (IP)
42	IEEE 802.1 Bridge Spanning Tree Protocol
AA	Sub-Network Access Protocol (SNAP)
FE	ISO Network Layer Protocol
Manufacturer Implemented	
80	Xerox Network Systems
BC	Banyan Nines
EO	Novell NetWare
FO	IBM NetBIOS
F8	IBM Remote Program Load (RPL)
FA	Ungermann-Bass

The control field provides information that can indicate the type of service and protocol format. For example, if the frame is transporting NetWare data, the control field will contain the hex value 03, which indicates that the frame uses the unnumbered format for connectionless services. Prior to discussing the types and classes of service defined by the 802.2 standard, let us examine two additional IEEE 802.3 logical frame formats.

6.3.2 Types and classes of service

Under the 802.2 standard, there are three types of service available for sending and receiving LLC data. These types are discussed in the next three paragraphs. Figure 6.10 provides a visual summary of the operation of each LLC service type.

Type 1

Type 1 is an unacknowledged connectionless service. The term *connectionless* refers to the fact that transmission does not occur between two devices as if a logical connection were established. Instead, transmission flows on the channel to all stations; however, only the destination address acts upon the data. As the name of this service implies, there is no provision for the acknowledgment of frames. Neither are there provisions for flow control or for error recovery. Therefore, this is an unreliable service.

Despite those shortcomings, Type 1 is the most commonly used service, since most protocol suites use a reliable transport mechanism at the transport layer, thus eliminating the need for reliability at the link layer. In addition, by eliminating the time needed to establish a virtual link and the overhead of acknowledgments, a Type 1 service can provide a greater throughput than other LLC types of services.

Type 2 Connection-oriented service

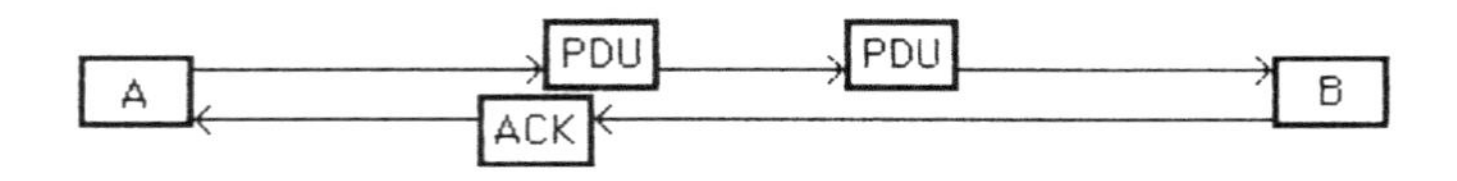

Type 3 Acknowledged connectionless source

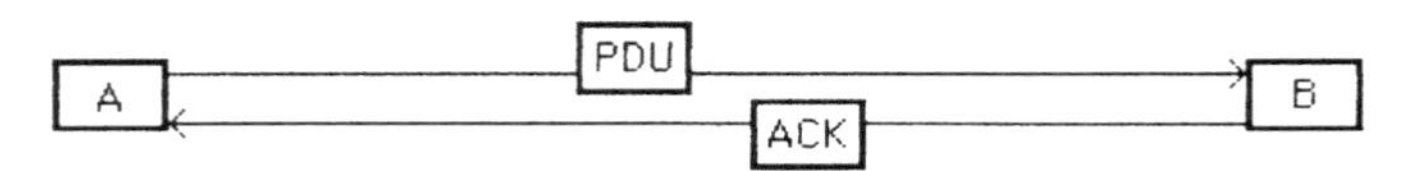

PDU: Protocol Data Unit
ACK: Acknowledgement
A,B : stations on the network

Figure 6.10 Logical Link Control service types

Type 2

The Type 2 connection-oriented service requires that a logical link be established between the sender and the receiver prior to information transfer. Once the logical connection has been established, data will flow between the sender and receiver until either party terminates the connection. During data transfer, a Type 2 LLC service provides all of the functions lacking in a Type 1 service, using a sliding window for flow control. When IBM's SNA data is transported on a LAN, it uses connection-oriented services. Type 2 LLC is also commonly referred to as LLC 2.

Type 3

The Type 3 acknowledged connectionless service contains provision for the setup and disconnection of transmission; it acknowledges individual frames using the stop-and-wait flow control method. Type 3 service is primarily used in an automated factory process-control environment, where one central computer communicates with many remote devices that typically have a limited storage capacity.

Classes of service

All LLC stations support Type 1 operations. This level of support is known as Class I service. The classes of service supported by LLC indicate the combinations of the three LLC service types supported by a station. Class I supports Type 1 service, Class II supports both Type 1 and Type 2, Class III supports Type 1 and Type 3 services, while Class IV supports all three service types. Since service Type 1 is supported by all classes, it can be considered a least common denominator, enabling all stations to communicate using a common form of service.

6.4 OTHER ETHERNET FRAME TYPES

Two additional frame types that warrant discussion are Ethernet-SNAP and Ethernet 802.3. In actuality, both types of frames represent a logical variation of the IEEE 802.3 frame in which the composition of the data field varies from the composition of the LLC protocol data unit previously illustrated in Figure 6.9.

6.4.1 Ethernet_SNAP frame

The Ethernet_SNAP (Subnetwork Access Protocol) frame provides a mechanism for obtaining a type field identifier associated with a pure Ethernet frame in an IEEE 802.3 frame, enabling it to be used to transport several protocols, such as AppleTalk, NetWare, and TCP/IP, at the same time. Thus, SNAP can be considered as an extension that permits vendors to create their own Ethernet protocol transports. To accomplish this, the data field is subdivided similarly to the previously illustrated LLC protocol data unit shown in Figure 6.9; however, two additional subfields are added after the Control field. Those fields are an organization code of 3 bytes and an Ethernet Type field of 2 bytes. Figure 6.11 illustrates the format of an Ethernet_SNAP frame. Although the format of this frame is based upon the IEEE 802.3 frame format, it does not use the DSAP and SSAP mailbox facilities. Instead, it places specific values in those fields to indicate that the frame is a SNAP frame. This results in a value of hex AA being placed in the DSAP and SSAP fields to indicate that the frame is an Ethernet_SNAP frame. The Control field functions similarly to the previously described LLC protocol data unit, indicating the type and class of service supported. This field is always fixed to a value of hex 03 to indicate a connectionless service unnumbered format, which is the only format supported by a SNAP frame.

The Organization Code field references the assigner of the value in the following Ethernet Type field. For most situations, a hex value of 00-00-00 is used to indicate that the Ethernet Type field value was assigned by Xerox. When the organization code is hex 00-00-00, the Ethernet Type field will contain one of the entries previously listed in Table 6.2.

6.4.2 NetWare Ethernet_802.3 frame

One additional logical variation of the IEEE 802.3 frame format that warrants an elaboration is known as the NetWare Ethernet_802.3 frame. Instead of using the IEEE 802.2 subfields to form a LLC protocol data unit, Novell places the IPX header immediately after the Length field, reducing the maximum Data field length by 30 bytes. The NetWare Ethernet_802.3 frame can only be used to transport NetWare IPX traffic, and represents a common level of frustration when an administrator attempts to use this frame format to transport a different protocol.

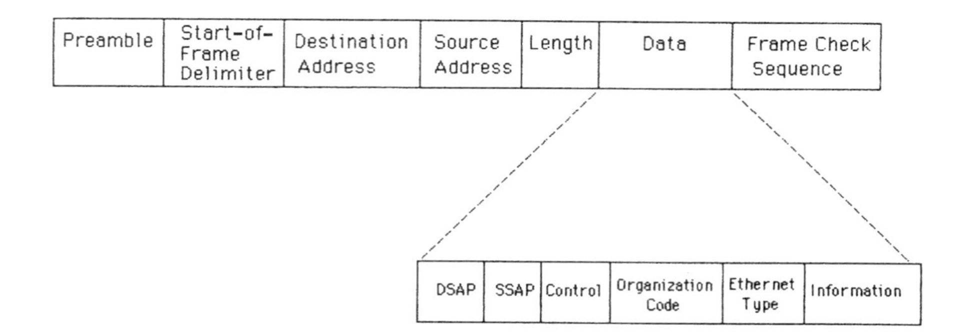

Figure 6.11 Ethernet_SNAP frame format

6.4.3 Receiver frame determination

A receiving station can distinguish between different types of Ethernet frames and correctly interpret data transported in those frames. To do so, it must examine the value of the field following the Source Address field, which is either a Type or Length field. If the field value exceeds 1500 decimals, the field must be a Type subfield. Thus, the frame is a 'raw' Ethernet frame. If the value is less than 1500, the field is a Length field and the two bytes following that field, which represent the first two bytes of an IEEE 802.3 frame's Data field, must be examined. If those two bytes have the value hex FF-FF, the frame is a NetWare Ethernet_802.3 frame used to transport IPX. If the value of the two bytes is hex AA-AA, the frame is an Ethernet_SNAP frame. Any other value in those bytes means that the frame is an IEEE_802.3 frame.

It is important during the LAN installation process to bind the appropriate protocol to the frame type capable of transporting the protocol. Table 6.5 lists several examples of protocols that can be bonded to different types of Ethernet frames.

6.5 FAST ETHERNET

The frame composition associated with Fast Ethernet is illustrated in Figure 6.12. In comparing the composition of the Fast Ethernet frame with Ethernet and IEEE 802.3 frame formats previously illustrated in Figure 6.1, you will note that other than the addition of starting and ending stream delimiters, the Fast Ethernet frame duplicates the older frames. Another difference between the two is not shown, as it is not actually observable from a comparison of frames, because this difference is associated with the time between frames. Ethernet and IEEE 802.3 frames are Manchester-encoded and have an interpacket gap of 9.6 μsec between frames. In comparison, the Fast Ethernet 100BASE-TX frame is transmitted using 4B5B encoding, and IDLE codes representing sequences of I (binary 11111) symbols are used to mark a 0.96-μs interpacket gap. Now that we have an overview of the differences between Ethernet/IEEE 802.3 and Fast Ethernet frames, let us focus upon the new fields associated with the Fast Ethernet frame format.

Table 6.5 Protocols vs. frame type

Frame Type	Protocols That Can be Bound
Ethernet	NetWare, AppleTalk, Phase I, TCP/IP
IEEE 802.3	NetWare, FTAM
Netware Ethernet_802.3	NetWare only
Ethernet_SNAP	NetWare, AppleTalk, Phase II, TCP/IP

SSD 1 byte	Preamble 7 bytes	SFD 1 byte	Destination Address 6 bytes	Source Address 6 bytes	L/T 2 bytes	Data 46–1500 bytes	FCS 1 byte	ESD 1 byte

SSD: Start-of-Stream Delimiter
SFD: Start-of-Frame Delimiter
L/T: Length (IEEE 802.3)/Type (Ethernet)
ESD: End-of-Stream Delimiter

Figure 6.12 Fast Ethernet frame. The 100BASE-TX frame differs from the IEEE 802.3 MAC frame through the addition of a byte at each end to mark the beginning and end of the stream delimiter

6.5.1 Start-of-Stream Delimiter

The Start-of-Stream Delimiter (SSD) is used to align a received frame for subsequent decoding. The SSD field consists of a sequence of J and K symbols, which define the unique code 11000 10001. This field replaces the first octet of the preamble in Ethernet and IEEE 802.3 frames, whose composition is 10101010.

6.5.2 End-of-Stream Delimiter

The End-of-Stream Delimiter (ESD) is used as an indicator that data transmission terminated normally, and a properly formed stream was transmitted. This 1-byte field is created by the use of T and R codes whose bit composition is 01101 00111. The ESD field lies outside of the Ethernet/IEEE 802.3 frame and for comparison purposes can be considered to fall within the interframe gap of those frames.

6.6 GIGABIT ETHERNET

Earlier in this chapter it was briefly mentioned that the Ethernet frame was extended for operations at 1 Gbps. In actuality the Gigabit Ethernet standard resulted in two modifications to conventional CSMA/CD operations. The first modification, which is referred to as carrier extension, is only applicable for half-duplex links and was required to maintain an approximate 200 meter topology at gigabit speeds. Instead of actually extending the frame, as we will shortly note, the time the frame is on the wire is extended. A second modification, referred to as packet burst, enables gigabit compatible network devices to transmit bursts of relatively short packets without having to relinquish control of the network. Both carrier extension and packet bursting represent modifications to the CSMA/CD protocol to extend the collision domain and enhance the efficiency of Gigabit Ethernet, respectively. Both topics are covered in detail in this section.

6.6.1 Carrier extension

In an Ethernet network the attachment of workstations to a hub creates a segment. That segment or multiple segments interconnected via the use of one or more repeaters forms a collision domain. The latter term is formally defined as a single CSMA/CD network in which a collision will occur if two devices attached to the network transmit at or approximately the same time. The reason we can say approximately the same time is that there is a propagation delay time associated with the transmission of signals on a conductor. Thus, if one station is relatively close to another, the propagation delay time is relatively short, requiring both stations to transmit data at nearly the same time for a collision to occur. If two stations are at opposite ends of the network, the propagation delay for a signal placed on the network by one station to reach the other station is much greater. This means that one station could initiate transmission and actually transmit a portion of a frame while the second station might 'listen' to the network, hear no activity, and begin to transmit, resulting in a collision.

Figure 6.13 illustrates the relationship between a single collision domain and two collision windows. Note that as stations are closer to one another the collision window that represents the propagation delay time during which one station could transmit and another would assume there is no network activity decreases.

Ethernet requires that a station should be able to hear any resulting collision for the frame it is transmitting before it completes the transmission of the entire frame. This means that the transmission of the next to last bit of a frame that results in a collision should allow the transmitting station to 'hear' the collision voltage increase before it transmits the last bit. Thus, the maximum allowable cabling distance is limited by the bit duration associated with the network operating rate and the speed of electrons on the wire.

When Ethernet operates at 1 Gbps, the allowable cabling distance would be reduced to approximately 10 meters or 33 feet. Clearly this would be a major restriction on the ability of Gigabit Ethernet to be effectively used in a shared media half-duplex environment. To overcome this transmission distance limitation, Sun Microsystems, Inc. suggested the carrier extension scheme, which became part of the Gigabit Ethernet standard for half-duplex operations.

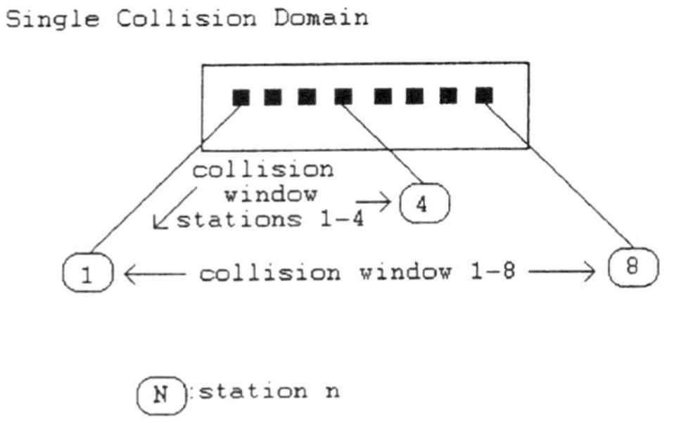

Figure 6.13 Relationship between a collision domain and collision windows

Under the carrier extension scheme, the original Ethernet frame is extended by increasing the time the frame is on the 'wire.' The timing extension occurs after the end of the standard CSMA/CD frame as illustrated in Figure 6.14. The carrier extension extends the frame timing to guarantee at least a 512-byte slot time for half-duplex Ethernet. Note that Ethernet's slot time is considered as the time from the first bit of the Destination Address field reaching the wire through the last bit of the Frame Check Sequence field. The increase in the minimum length frame does not change the frame size and only alters the time the frame is on the wire. Due to this, compatibility is maintained between the original Ethernet frame and the Gigabit Ethernet frame.

Although the carrier extension scheme enables the cable length of a half-duplex Gigabit network to be extended to a 200-meter diameter, that extension is not without a price. This price is one of overhead, since extension symbols attached to a short frame wastes bandwidth. For example, a frame with a 64-byte Data field would have 448 bytes of wasted carrier extension symbols attached to it. To further complicate bandwidth utilization, when the Data field is less than 46 bytes in length, nulls are added to produce a 64-byte minimum length Data field. Thus, a simple query to be transported by Ethernet, such as 'Enter your age' consisting of 44 data characters, would be padded with 32 null characters when transported by Ethernet to ensure a minimum 72-byte length frame. Under Gigabit Ethernet the minimum 512-byte time slot would require the use of 448 carrier extension symbols to ensure that the time slot from destination address through any required extension is at least 512 bytes in length.

In examining Figure 6.14, it is important to note that the carrier extension scheme does not extend the Ethernet frame beyond a 512-byte time slot. Thus, Ethernet frames with a time slot equal to or exceeding 512 bytes have no carrier extension. Another important item to note concerning the carrier extension scheme is that it has no relationship to a 'Jumbo Frames' feature that is proprietary to a specific vendor. That feature is supported by a switch manufactured by Alteon Networks and is used to enhance data transfers between servers, permitting a maximum frame size of up to 9 Kbytes to be supported. Since 'Jumbo Frames' are not part of the

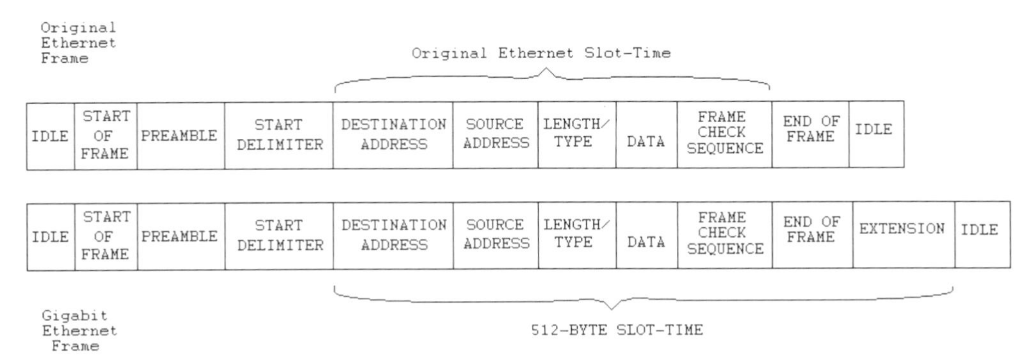

Figure 6.14 Half-duplex Gigabit Ethernet uses a carrier extension scheme to extend timing so that the slot time consists of at least 512 bytes

Gigabit Ethernet standard, you must disable that feature to obtain interoperability between that vendor's 1 Gbps switch and other vendors' Gigabit Ethernet products.

6.6.2 Packet bursting

Packet bursting represents a scheme added to Gigabit Ethernet to counteract the overhead associated with transmitting relatively short frames. This scheme was proposed by NBase Communications and is included in the Gigabit Ethernet standard as an addition to carrier extension.

Under packet bursting, each time the first frame in a sequence of short frames successfully passes the 512-byte collision window using the carrier extension scheme previously described, subsequent frames are transmitted without including the carrier extension. The effect of packet bursting is to average the wasted time represented by the use of carrier extension symbols over a series of short frames. The limit on the number of frames that can be burst is a total of 1500 bytes for the series of frames, which represents the longest Data field supported by Ethernet. To inhibit other stations from initiating transmission during a burst carrier extension, signals are inserted between frames in the burst.

In addition to enhancing network utilization and minimizing bandwidth overhead, packet bursting also reduces the probability of collisions occurring. This is because the burst of frames are only susceptible to a collision during the first frame in the sequence. Thereafter, carrier extension symbols between frames followed by additional short frames are recognized by all other stations on the segment, and inhibit those stations from initiating a transmission that would result in the occurrence of a collision.

6.7 TOKEN-RING FRAME OPERATIONS

In this section we will examine Token-Ring frame operations, enabling us to understand the manner in which different frame fields are used for such functions as access control, error checking, routing of data between interconnected networks, and other Token-Ring network functions.

A Token-Ring network consists of ring stations representing devices that attach to a ring and an attaching medium. Concerning the latter, the attaching medium can be shielded, twisted-pair, or fiber optic cable, each having constraints concerning transmission distance and number of stations allowed on the network.

Frames are transmitted sequentially from one station to another physically active station in a clockwise direction. The next active station is referred to as downstream neighbor, which regenerates the frame, as well as performing Media Access Control (MAC) address checking and other functions. In performing a MAC address check, the station compares its address with the destination address contained in the frame. If the two match or if the station has a functional address that matches the frame destination's address, the station copies the data contained in the frame. While performing the previously described operations, the station performs a number of error checks based upon the composition of data in the frame and reports errors via the generation of different types of error reporting frames. Thus, it is important to understand the composition of the fields within the Token-Ring frames as they govern the operation of a Token-Ring network.

6.7.1 Transmission formats

Three types of transmission formats are supported on a Token-Ring network: token, abort, and frame. The token format as illustrated in the top of Figure 6.15 is the mechanism by which access to the ring is passed from one computer attached to the network to another device connected to the network. Here the token format consists of three bytes, of which the starting and ending delimiters are used to indicate the beginning and end of a token frame. The middle byte of a token frame is an access control byte. Three bits

a. **Token format**

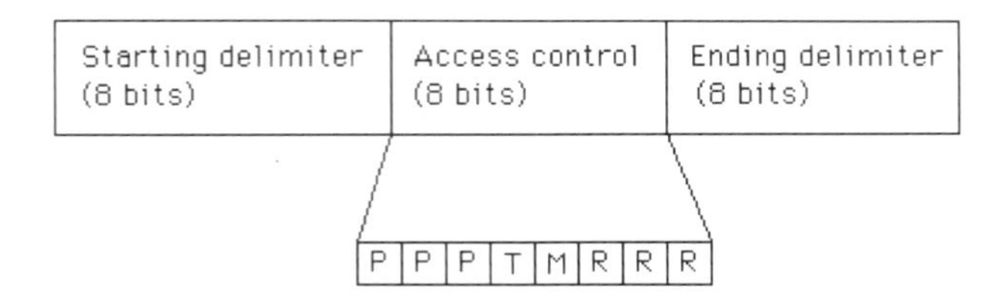

b. **Abort token format**

Starting delimiter	Ending Delimiter

c. **Frame format**

Starting delimiter (8 bits)	Access control (8 bits)	Frame control (8 bits)	Destination address (48 bits)	Source address (48 bits)	Routing information (optional)

Information variable	Frame check sequence (32 bits)	Ending delimiter (8 bits)	Frame status (8 bits)

Figure 6.15 Token, abort, and frame formats (P: priority bits; T: token bit; M: monitor bit; R: reservation bits)

are used as a priority indicator, three bits are used as a reservation indicator, while one bit is used for the token bit, and another bit position functions as the monitor bit.

When the token bit is set to a binary 0, it indicates that the transmission is a token. When it is set to a binary 1, it indicates that data in the form of a frame is being transmitted.

The second Token-Ring frame format signifies an abort token. In actuality there is no token, since this format is indicated by a starting delimiter followed by an ending delimiter. The transmission of an abort token is used to abort a previous transmission. The format of an abort token is illustrated in Figure 6.15b.

The third type of Token-Ring frame format occurs when a station seizes a free token. At that time the token format is converted into a frame that includes the addition of frame control, addressing data, an error detection field, and a frame status field. The format of a Token-Ring frame is illustrated in Figure 6.15c. At any given point in time, only one token can reside on a ring, represented either as a token format, abort token format, or frame. By examining each of the fields in the frame, we will also examine the token and token abort frames due to the commonality of fields between each frame.

Starting/ending delimiters

The starting and ending delimiters mark the beginning and ending of a token or frame. Each delimiter consists of a unique code pattern that identifies it to the network. In order to understand the composition of the starting and ending delimiter fields, we must review the method by which data is represented on a Token-Ring network using Differential Manchester encoding.

Differential Manchester encoding Figure 6.16 illustrates the use of Differential Manchester encoding, comparing its operation with non-return to zero (NRZ) and conventional Manchester encoding.

At the top of Figure 6.16, NRZ coding illustrates the representation of data by holding a voltage low ($-V$) to represent a binary 0 and high ($+V$) to represent a binary 1. This method of signaling is called non-return to zero since there is no return to a 0-V position after each data bit is coded.

To avoid the necessity of building clocking circuitry into devices, a mechanism is required for encoded data to carry clocking information. One method by which encoded data carries clocking information is obtained from the use of Manchester encoding, which is illustrated in Figure 6.16 and which represents the signalling method used by Ethernet. In Manchester encoding, each data bit consists of a half-bit time signal at a low voltage ($-V$) and another half-bit time signal at the opposite positive voltage ($+V$). Every binary 0 is represented by a half-bit time at a low voltage and the remaining bit time at a high voltage. Every binary 1 is represented by a half-bit time at a high voltage followed by a half-bit time at a low voltage. By changing the voltage for every binary digit, Manchester encoding ensures that the signal carries self-clocking information.

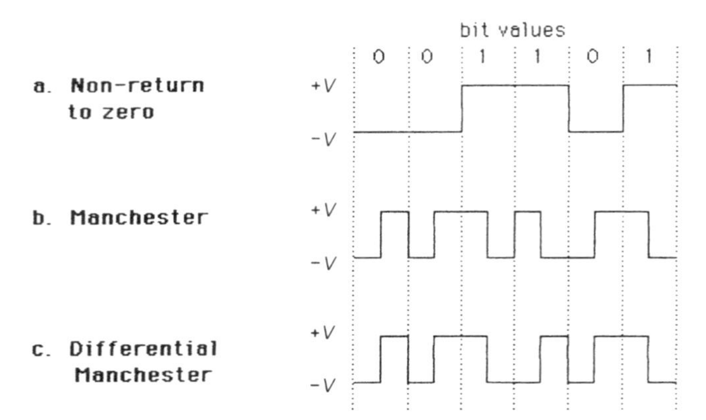

Figure 6.16 Differential Manchester encoding: in this, the direction of the signal's voltage transition changes whenever a binary 1 is transmitted, but remains the same for a binary 0

In Figure 6.16c, Differential Manchester encoding is illustrated. The difference between Manchester encoding and Differential Manchester encoding occurs in the method by which binary 1s are encoded. In Differential Manchester encoding, the direction of the signal's voltage transition changes whenever a binary 1 is transmitted, but remains the same for a binary 0. The IEEE 802.5 standard specifies the use of Differential Manchester encoding, and this encoding technique is used on Token-Ring networks at the physical layer to transmit and detect four distinct symbols: a binary 0, a binary 1, and two non-data symbols.

Non-data symbols Under Manchester and Differential Manchester encoding there are two possible code violations that can occur. Each code violation produces what is known as a non-data symbol, and is used in the Token-Ring frame to denote starting and ending delimiters similar to the use of the flag in an HDLC frame. However, unlike the flag, whose bit composition 01111110 is uniquely maintained by inserting a 0 bit after every sequence of five set bits and removing a 0 following every sequence of five set bits, Differential Manchester encoding maintains the uniqueness of frames by the use of non-data J and non-data K symbols. This eliminates the bit stuffing operations required by HDLC.

The two non-data symbols each consist of two half-bit times without a voltage change. The J symbol occurs when the voltage is the same as that of the last signal, while the K symbol occurs when the voltage becomes opposite of that of the last signal. Figure 6.17 illustrates the occurrence of the J and K non-data symbols based upon different last bit voltages. Readers will note in comparing Figure 6.17 with Figure 6.17c that the J and K non-data symbols are distinct code violations that cannot be mistaken for either a binary 0 or a binary 1.

Now that we have an understanding of the operation of Differential Manchester encoding and the composition of the J and K non-data symbols, we can focus our attention upon the actual format of each frame delimiter.

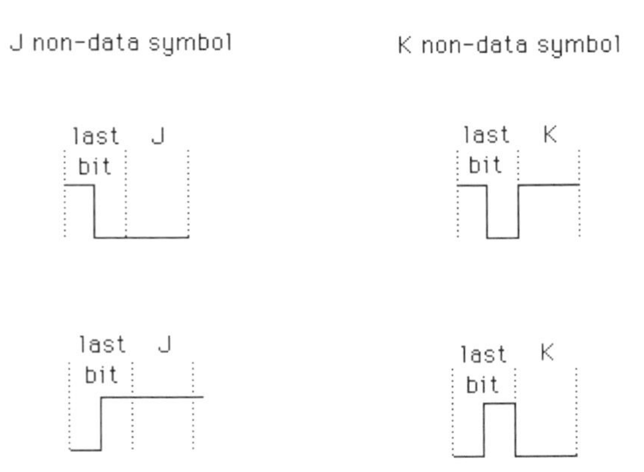

Figure 6.17 J and K non-data symbol composition. J and K non-data symbols are distinct code violations that cannot be mistaken for data

The start delimiter field marks the beginning of a frame. The composition of this field is the bits and non-data symbols JK0JK000. The end delimiter field marks the end of a frame as well as denoting whether or not the frame is the last frame of a multiple frame sequence using a single token or if there are additional frames following this frame.

The format of the end delimiter field is JK1JK1IE, where I is the intermediate frame bit. If I is set to 0, this indicates that it is the last frame transmitted by a station. If I is set to 1, this indicates that additional frames follow this frame.

E is an Error-Detected bit. The E bit is initially set to 0 by the station transmitting a frame, token, or abort sequence. As the frame circulates the ring, each station checks the transmission for errors. Upon detection of a Frame Check Sequence (FCS) error, inappropriate non-data symbol, illegal framing, or another type of error, the first station detecting the error will set the E bit to a value of 1. Since stations keep track of the number of times they set the E bit to a value of 1, it becomes possible to use this information as a guide to locating possible cable errors. For example, if one workstation accounted for a very large percentage of E bit settings in a network, there is a high degree of probability that there is a problem with the lobe cable to that workstation. The problem could be a crimped cable or a loose connector and represents a logical place to commence an investigation in an attempt to reduce E bit errors.

Access control field

The second field in both token and frame formats is the access control byte. As illustrated at the top of Figure 6.15, this byte consists of four subfields and serves as the controlling mechanism for gaining access to the network. When a free token circulates the network, the access control field represents one-third of the length of the frame since it is prefixed by the start delimiter and suffixed by the end delimiter.

The lowest priority that can be specified by the priority bits in the access control byte is 0 (000), while the highest is seven (111), providing eight levels of priority. Table 6.6 lists the normal use of the priority bits in the access control field. Workstations have a default priority of three, while bridges have a default priority of four.

To reserve a token, a workstation will attempt to insert its priority level in the priority reservation subfield. Unless another station with a higher priority bumps the requesting station, the reservation will be honored and the requesting station will obtain the token. If the token bit is set to 1, this serves as an indication that a frame follows instead of the ending delimiter.

A station that needs to transmit a frame at a given priority can use any available token that has a priority level equal to or less than the priority level of the frame to be transmitted. When a token of equal or lower priority is not available, the ring station can reserve a token of the required priority through the use of the reservation bits. In doing so, the station must follow two rules. First, if a passing token has a higher priority reservation than the reservation level desired by the workstation, the station will not alter the reservation field contents. Secondly, if the reservation bits have not been set or indicate a lower priority than that desired by the station, the station can now set the reservation bits to the required priority level.

Once a frame is removed by its originating station, the reservation bits in the header will be checked. If those bits have a non-zero value, the station must release a non-zero priority token, with the actual priority assigned based upon the priority used by the station for the recently transmitted frame, the reservation bit settings received upon the return of the frame, and any stored priority.

On occasion, the Token-Ring protocol will result in the transmission of a new token by a station prior to that station having the ability to verify the settings of the Access Control field in a returned frame. When this situation arises, the token will be issued according to the priority and reservation bit settings in the Access Control field of the transmitted frame.

Figure 6.18 illustrates the operation of the Priority (P) and Reservation (R) Bit fields in the Access Control field. In this example, the prevention of a high-priority station from monopolizing the network is illustrated by station A entering a Priority-Hold state. This occurs when a station originates a token

Table 6.6 Priority bit settings

Priority Bits	Priority
000	Normal user priority, MAC frames that do not require a token and response type MAC frames
001	Normal user priority
010	Normal user priority
011	Normal user priority and MAC frames that require tokens
100	Bridge
101	Reserved
110	Reserved
111	Specialized Station Management

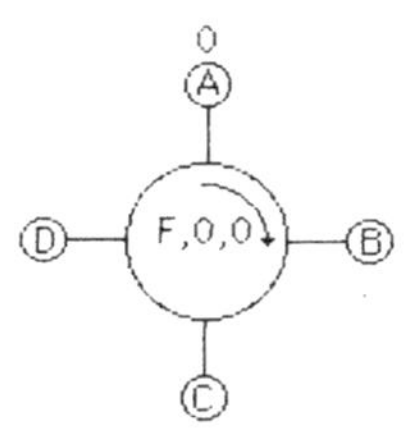

a. Station A generates a frame using a non-priority token P,R=0,0

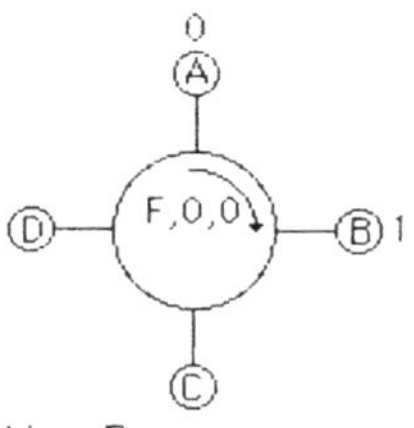

b. Station B reserves a priority 1 in the reservation bits in the frame P,R=0,1; Station A enters a Priority-Hold state

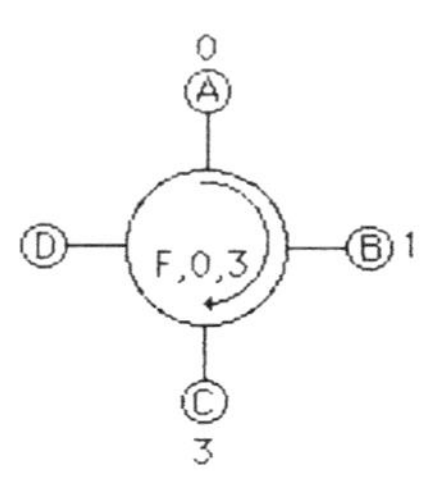

c. Station C reserves a priority of 3 overriding B's reservation of 1; P,R=0,3

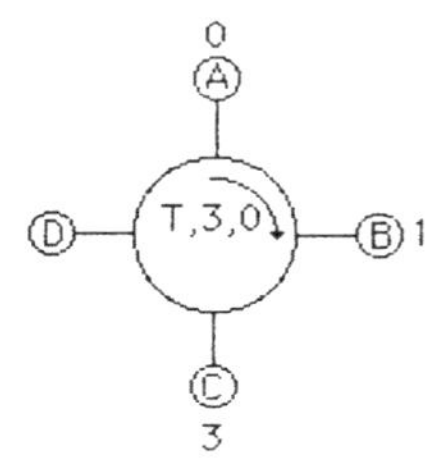

d. Station A removes its frame and generates a token at reserved priority level 3; P,R=3,0

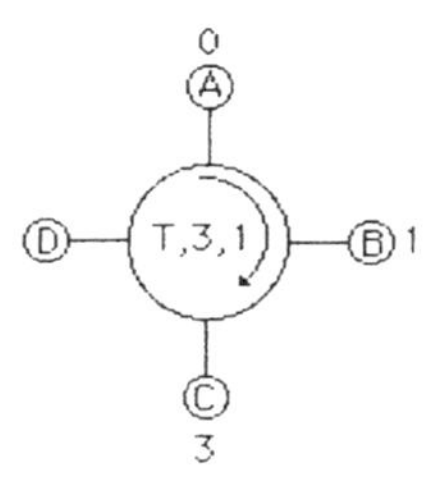

e. Station B repeats priority token and makes a new reservation of priority level 1; P,R=3,1

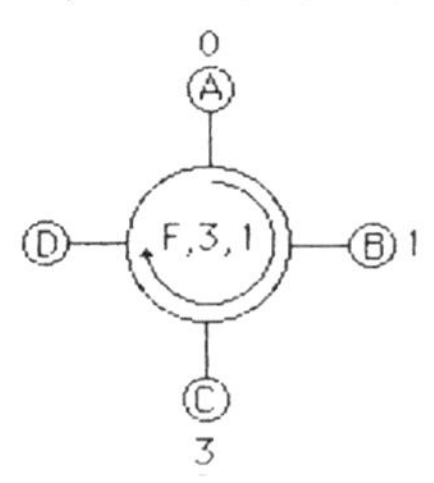

f. Station C grabs token and transmits a frame with a priority of 3; P,R=3,1

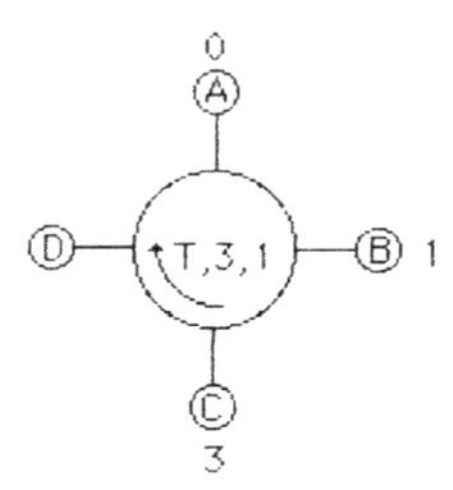

g. Upon return of frame to Station C it is removed. Station C generates a token at the priority just used; P,R=3,1

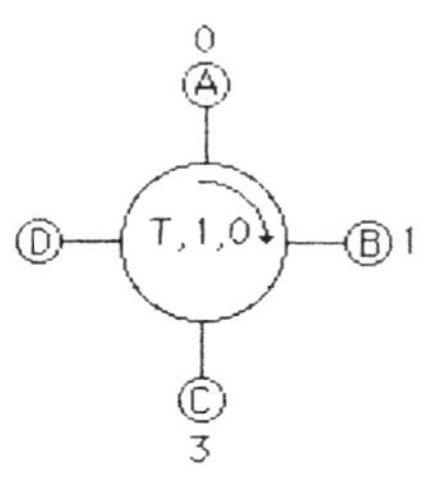

h. Station A in a Priority-Hold state grabs token and changes its priority to 1; P,R=1,0. Station A stays in Priority-Hold state until priority is reduced to 0

Ⓐ, Ⓑ, Ⓒ, Ⓓ = stations

Numeric outside station identifier indicates priority level

Figure 6.18 Priority and Reservation field utilization

at a higher priority than the last token it generated. Once in a Priority-Hold state, the station will issue tokens that will bring the priority level eventually down to zero as a mechanism to prevent a high-priority station from monopolizing the network.

The monitor bit The monitor bit is used to prevent a token with a priority exceeding zero or a frame from continuously circulating on the Token-Ring. This bit is transmitted as a 0 in all tokens and frames, except for a device on the network that functions as an active monitor and thus obtains the capability to inspect and modify that bit.

When a token or frame is examined by the active monitor, it will set the monitor bit to a 1 if it was previously found to be set to 0. If a token or frame is found to have the monitor bit already set to 1, this indicates that the token or frame has already made at least one revolution around the ring and an error condition has occurred, usually caused by the failure of a station to remove its transmission from the ring or the failure of a high-priority station to seize a token. When the active monitor finds a monitor bit set to 1, it assumes that an error condition has occurred. The active monitor then purges the token or frame and releases a new token onto the ring. Now that we have an understanding of the role of the monitor bit in the Access Control field and the operation of the active monitor on that bit, let us focus our attention upon the active monitor.

The active monitor The active monitor is the device that has the highest address on the network. All other stations on the network are considered as standby monitors and watch the active monitor.

As previously explained, the function of the active monitor is to determine if a token or frame is continuously circulating the ring in error. To accomplish this, the active monitor sets the monitor count bit as a token or frame goes by. If a destination workstation fails or has its power turned off, the frame will circulate back to the active monitor, where it is then removed from the network. In the event that the active monitor should fail or be turned off, the standby monitors watch the active monitor by looking for an active monitor frame. If one does not appear within seven seconds, the standby monitor that has the highest network address then takes over as the active monitor.

In addition to detecting and removing frames that might otherwise continue to circulate the ring, the active monitor performs several other ring management functions. These functions include the detection and recovery of multiple tokens and the loss of a token or frame on the ring, as well as initiation of a token when a ring is started. The loss of a token or frame is detected by the expiration of a timer whose time-out value exceeds the time required for the longest possible frame to circulate the ring. The active monitor restarts this time, and each time it transmits a starting delimiter that precedes every frame and token. Thus, if the timer expires without the appearance of a frame or token, the active monitor will assume the frame or token was lost and initiate a purge operation, which is described later in this section.

Frame Control field

The Frame Control field informs a receiving device on the network of the type of frame that was transmitted and how it should be interpreted. Frames can be either Logical Link Control (LLC) or reference physical link functions according to the IEEE 802.5 Media Access Control (MAC) standard. A MAC frame carries network control information and responses, while an LLC frame carries data.

The eight-bit frame control field has the format FFZZZZZZ, where FF are frame definition bits. The top of Table 6.7 indicates the possible settings of the frame bits and the assignment of those settings. The ZZZZZZ bits convey MAC buffering information when the FF bits are set to 00. When the FF bits are set to 01 to indicate an LLC frame, the ZZZZZZ bits are split into two fields, designated rrrYYY. Currently, the rrr bits are reserved for future use and are set to 000. The YYY bits indicate the priority of the LLC data. The lower portion of Table 6.7 indicates the value of the Z bits when used in MAC frames to notify a Token-Ring adapter that the frame is to be expressed buffered.

Destination Address field

Although the IEEE 802.5 standard supports both 16-bit and 48-bit Address fields, IBM's implementation requires the use of 48-bit Address fields. IBM's Destination Address field is made up of five subfields as illustrated in Figure 6.19. The first bit in the destination address identifies the destination as an individual station (bit set to 0) or as a group (bit set to 1) of one or more stations. The latter provides the capability for a message to be broadcast to a group of stations.

Table 6.7 Frame Control Field subfields

Frame Type Field	
F bit settings	Assignment
00	MAC frame
01	LLC frame
10	Undefined (reserved for future use)
11	Undefined (reserved for future use)
Z bit settings	Assignment*
000	Normal buffering
001	Remove ring station
010	Beacon
011	Claim token
100	Ring purge
101	Active monitor present
110	Standby monitor present

*When F bits are set to 00, Z bits are used to notify an adapter that the frame is to be expressed buffered.

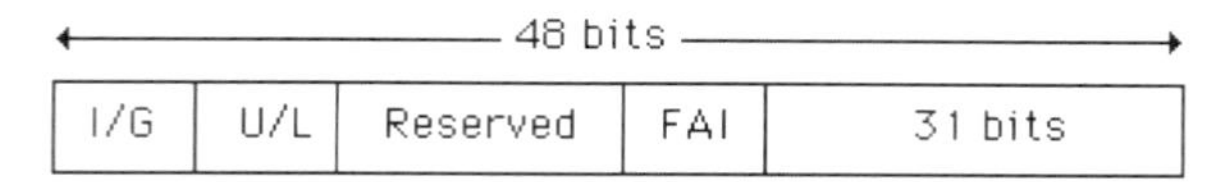

Figure 6.19 Destination Address subfields (I/G: individual or group bit address identifier; U/L: universally or locally administered bit identifier; FAI: functional address indicator). The Reserved field contains the manufacturer's identification in 22 bits represented by 6 hex digits

Universally administered address Similar to an Ethernet universally administered address, a Token-Ring universally administered address is a unique address permanently encoded into an adapter's ROM. Because it is placed into ROM, it is also known as a burned-in address. The IEEE assigns blocks of addresses to each vendor manufacturing Token-Ring equipment, which ensures that Token-Ring adapter cards manufactured by different vendors are uniquely defined. Token-Ring adapter manufacturers are assigned universal addresses that contain an organizationally unique identifier. This identifier consists of the first six hex digits of the adapter card address, and is also referred to as the manufacturer identification. For example, cards manufactured by IBM will begin with the hex address 08-00-5A or 10-00-5A, whereas adapter cards manufactured by Texas Instruments will begin with the address 40-00-14.

Locally administered address A key problem with the use of universally administered addresses is the requirement to change software coding in a mainframe computer whenever a workstation connected to the mainframe via a gateway is added or removed from the network. To avoid constant software changes, locally administered addressing can be used. This type of addressing functions similarly to its operation on an Ethernet LAN, temporarily overriding universally administered addressing; however, the user is now responsible for ensuring the uniqueness of each address. To accomplish locally administered addressing, a statement is inserted into a configuration file that sets the adapter's address at adapter-open time, normally when a station is powered on or a system reset operation is performed.

Functional address indicator The functional address indicator subfield in the destination address identifies the function associated with the destination address, such as a bridge, active monitor, or configuration report server.

The functional address indicator indicates a functional address when set to 0 and the I/G bit position is set to a 1 — the latter indicating a group address. This condition can only occur when the U/L bit position is also set to a 1, and results in the ability to generate locally administered group addresses that are called functional addresses. Table 6.8 lists the functional addresses defined by the IEEE. Currently, 21 functional addresses have been defined out of a total of 31 that are available for use, with the remaining addresses available for user definitions or reserved for future use.

Address values The range of addresses that can be used on a Token-Ring primarily depends upon the settings of the I/G, U/L, and FAI bit positions. When the I/G and U/L bit positions are set to 00, the manufacturer's universal

Table 6.8 IEEE functional addresses

Active Monitor	C0-00-00-00-00-01
Ring Parameter Server	C0-00-00-00-00-02
Network Server Heartbeat	C0-00-00-00-00-04
Ring Error Monitor	C0-00-00-00-00-08
Configuration Report Server	C0-00-00-00-00-10
Synchronous Bandwidth Manager	C0-00-00-00-00-20
Locate-Directory Server	C0-00-00-00-00-40
NETBIOS	C0-00-00-00-00-80
Bridge	C0-00-00-00-01-00
IMPL Server	C0-00-00-00-02-00
Ring Authorization Server	C0-00-00-00-04-00
LAN Gateway	C0-00-00-00-08-00
Ring Wiring Concentrator	C0-00-00-00-10-00
LAN Manager	C0-00-00-00-20-00
User-defined	C0-00-00-00-80-00 through C0-00-40-00-00-00
ISO OSI ALL ES	C0-00-00-00-40-00
ISO OSI ALL IS	C0-00-00-00-80-00
IBM discovery non-server	C0-00-00-01-00-00
IBM resource manager	C0-00-00-02-00-00
TCP/IP	C0-00-00-04-00-00
6611-DECnet	C0-00-20-00-00-00
LAN Network Manager	C0-00-40-00-00-00

address is used. When the I/G and U/L bits are set to 01, individual locally administered addresses are used in the defined range listed in Table 6.8. When all three bit positions are set, this situation indicates a group address within the range contained in Table 6.9. If the I/G and U/L bits are set to 11 but the FAI bit is set to 0, this indicates that the address is a functional address. In this situation the range of addresses is bit-sensitive, permitting only those functional addresses previously listed in Table 6.8.

A number of destination ring stations can be identified through the use of a group address. Table 6.10 lists a few of the standard group addresses that have been defined when the I/G, U/L and FAI bits are set to one.

In addition to the previously mentioned addresses, there are two special destination address values that are defined. An address of all 1s (FF-FF-FF-FF-FF-FF) identifies all stations as destination stations. If a null address is used in which all bits are set to 0 (00-00-00-00-00), the frame is not addressed to any workstation. In this situation it can only be transmitted but not received, enabling you to test the ability of the active monitor to purge this type of frame from the network.

Source Address field

The Source Address field always represents an individual address that specifies the adapter card responsible for the transmission. The Source

Table 6.9 Token-Ring addresses

	Bit Settings			
	I/G	U/L	FAI	Address/Address Range
Individual, universal administered	0	0	0/1	Manufacturer's serial no.
Individual, locally administrated	0	1	0	40-00-00-00-00-00 - 40-00-7F-FF-FF-FF
Group address	1	1	1	40-00-80-00-00-00 - 49-00-FF-FF-FF-FF
Functional address	1	1	0	C0-00-00-00-00-01 - C0-00-FF-FF-FF-FF (bit-sensitive)
All stations broadcast	1	1	1	FF-FF-FF-FF-FF-FF
Null address	0	0	0	00-00-00-00-00-00

Table 6.10 Representative standardized group addresses

Bridge	80-02-43-00-00-00
Bridge Management	80-01-43-00-00-08
Novell IPX	90-00-72-00-00-40
Hewlett-Packard probe	90-00-90-00-00-80
Vitalink gateway	90-00-3C-A0-00-80
Customer use	D5-00-20-00-XX-XX
DECnet phase IV station addresses	55-00-20-00-XX-XX

Address field consists of three major subfields as illustrated in Figure 6.20. When locally administrated addressing occurs, only 24 bits in the Address field are used since the 22 manufacturer identification bit positions are not used.

The routing information bit identifier identifies the fact that routing information is contained in an optional Routing Information Field. This bit is set when a frame is routed across a bridge using IBM's source routing technique.

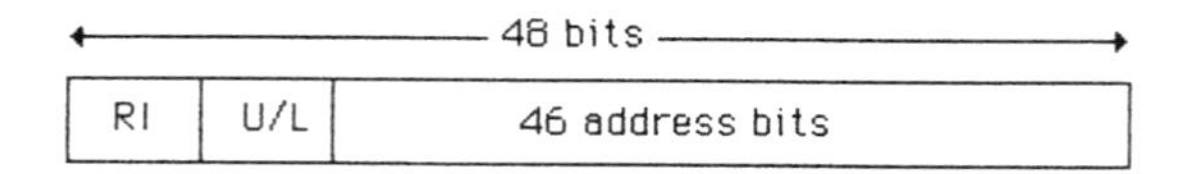

The 46 address bits consist of 22 manufacturer identification bits and 24 universally administered bits when the U/L bit is set to 0. If set to 1, a 31-bit locally administered address is used, with the manufacturer's identification bit set to 0.

Figure 6.20 Source Address field (RI: routing information bit identifier; U/L: universally or locally administered bit identifier)

Routing Information field

The Routing Information Field (RIF) is optional and is included in a frame when the RI bit of the source address field is set. Figure 6.21 illustrates the format of the optional RIF. If this field is omitted, the frame cannot leave the ring it was originated on under IBM's source routing bridging method. Under transparent bridging, the frame can be transmitted onto another ring. The RIF is of variable length and contains a Control subfield and one or more two-byte Route Designator fields when included in a frame, as the latter are required to control the flow of frames across one or more bridges.

The maximum length of the RIF is 18 bytes. Since each RIF must contain a 2-byte Routing Control field, this leaves a maximum of 16 bytes available for use by up to eight route designators. As illustrated in Figure 6.21, each two-byte route designator consists of a 12-bit ring number and a 4-bit bridge

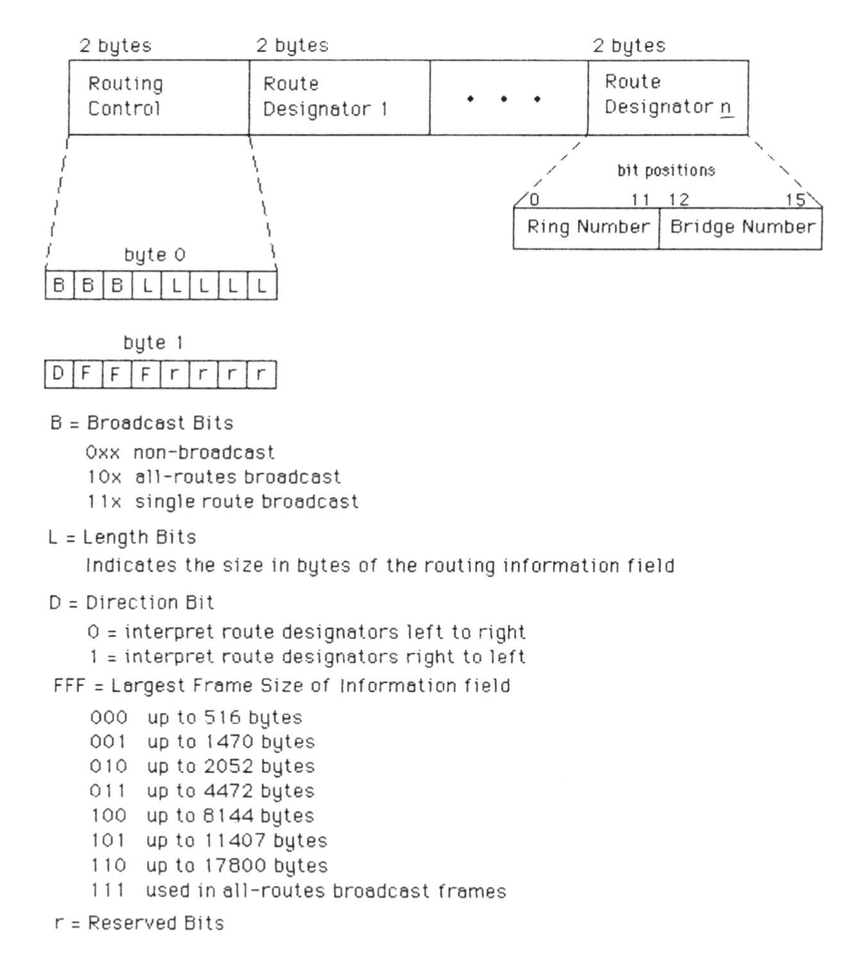

Figure 6.21 Routing Information field

number. Thus, a maximum total of 16 bridges can be used to join any two rings in an Enterprise Token-Ring network.

Information field

The Information field is used to contain Token-Ring commands and responses as well as to carry user data. The type of data carried by the Information field depends upon the F bit settings in the Frame Type field. If the F bits are set to 00, the information field carries Media Access Control (MAC) commands and responses that are used for network management operations. If the F bits are set to 01, the Information field carries Logical Link Control (LLC) or user data. Such data can be in the form of portions of a file being transferred on the network or an electronic mail message being routed to another workstation on the network. The Information field is of variable length and can be considered to represent the higher level protocol enveloped in a Token-Ring frame.

In the IBM implementation of the IEEE 802.5 Token-Ring standard the maximum length of the Information field depends upon the Token-Ring adapter used and the operating rate of the network. Token-Ring adapters with 64 Kbytes of memory can handle up to 4.5 Kbytes on a 4-Mbps network and up to 18 Kbytes on a 16-Mbps network.

Frame Check Sequence field

The Frame Check Sequence field contains four bytes that provide the mechanism for checking the accuracy of frames flowing on the network. The cyclic redundancy check data included in the Frame Check Sequence field covers the Frame Control, Destination Address, Source Address, Routing Information, and Information fields. If an adapter computes a cyclic redundancy check that does not match the data contained in the Frame Check Sequence field of a frame, the destination adapter discards the frame information and sets an error bit (E bit) indicator. This error bit indicator, as previously discussed, actually represents a ninth bit position of the ending delimiter and serves to inform the transmitting station that the data was received in error.

Frame Status field

The Frame Status field serves as a mechanism to indicate the results of a frame's circulation around a ring to the station that initiated the frame. Figure 6.22 indicates the format of the Frame Status field. The Frame Status field contains three subfields that are duplicated for accuracy purposes since they reside outside of CRC checking. One field (A) is used to denote whether an address was recognized, while a second field (C) indicates whether the frame was copied at its destination. Each of these fields is 1 bit in length. The

A	C	r	r	A	C	r	r

A: Address-Recognized Bits
B: Frame-Copied Bits
r: Reserved Bits

Figure 6.22 Frame Status field. This field denotes whether the destination address was recognized and whether the frame was copied. Since this field is outside of CRC checking, its subfields are duplicated for accuracy

third field, which is 2 bit positions in length (rr), is currently reserved for future use.

Similar to Ethernet layer 2 monitoring products, you can also select from a range of Token-Ring monitoring products to obtain an appreciation of the activity occurring on a Token-Ring network. Figure 6.23 illustrates a screen display resulting from the use of the Triticom TokenVision monitoring program. In the example shown in this figure the program is monitoring source addresses and counting the errors associated with each station on the network.

The manner in which TokenVision operates is similar to the manner in which EtherVision operates, looking into the Source or Destination Address field of each frame to discover active stations on the network. If the F2 key

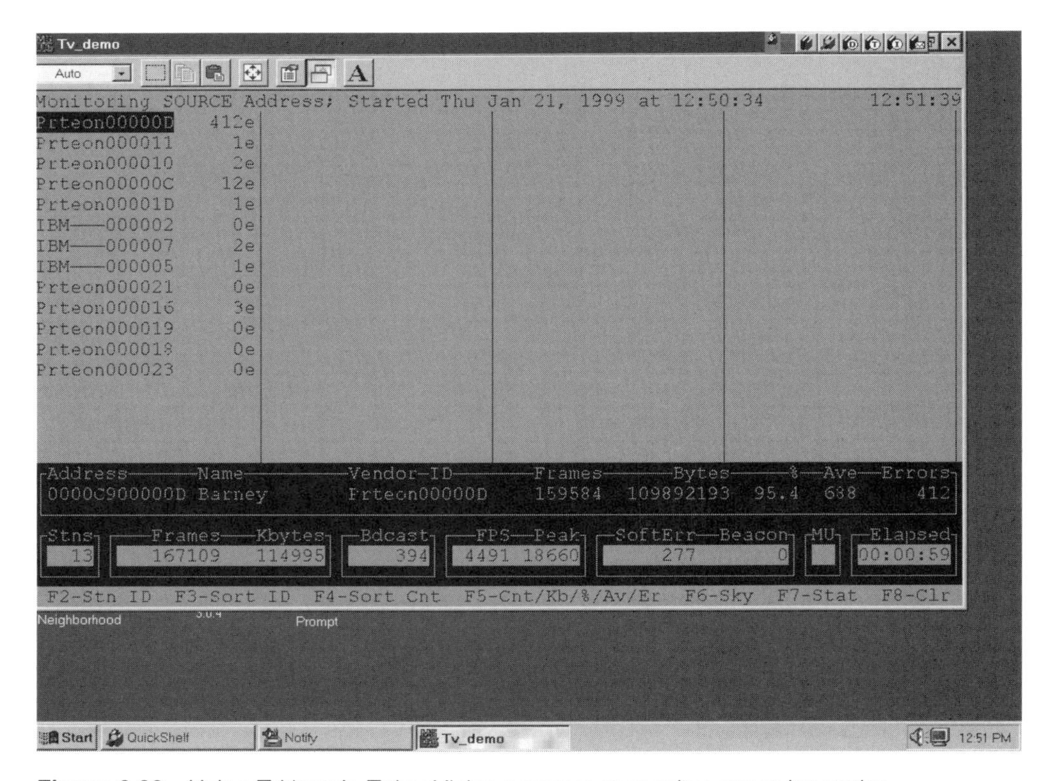

Figure 6.23 Using Triticom's TokenVision program to monitor errors by station

had been pressed instead of the F5 key, the display would be similar to the one shown in Figure 6.4, where instead of an error count the number of frames would be displayed.

By first using the TokenVision statistics display and then selecting the program's MAC option, you can obtain detailed information about the state of MAC activity on the network. This is illustrated in Figure 6.24, which shows the display of the program's MAC screen. In examining Figure 6.24, note that the top portion of the display presents information about the active monitor, including the station that is currently the active monitor and the number of active monitor present frames recorded. The middle portion of the screen display shows ring recovery information, including claimed tokens, ring purges, and beacons, topics discussed later in this section. The lower third of Figure 6.24 displays soft errors and frames associated with different error conditions. Thus, you can use this layer 2 monitoring program to obtain an insight into the functioning of the Token-Ring data link layer used to transport various protocols to include TCP/IP on a Token-Ring network.

6.8 TOKEN-RING MEDIUM ACCESS CONTROL

As previously discussed, a MAC frame is used to transport network commands and responses. As such, the MAC layer controls the routing of

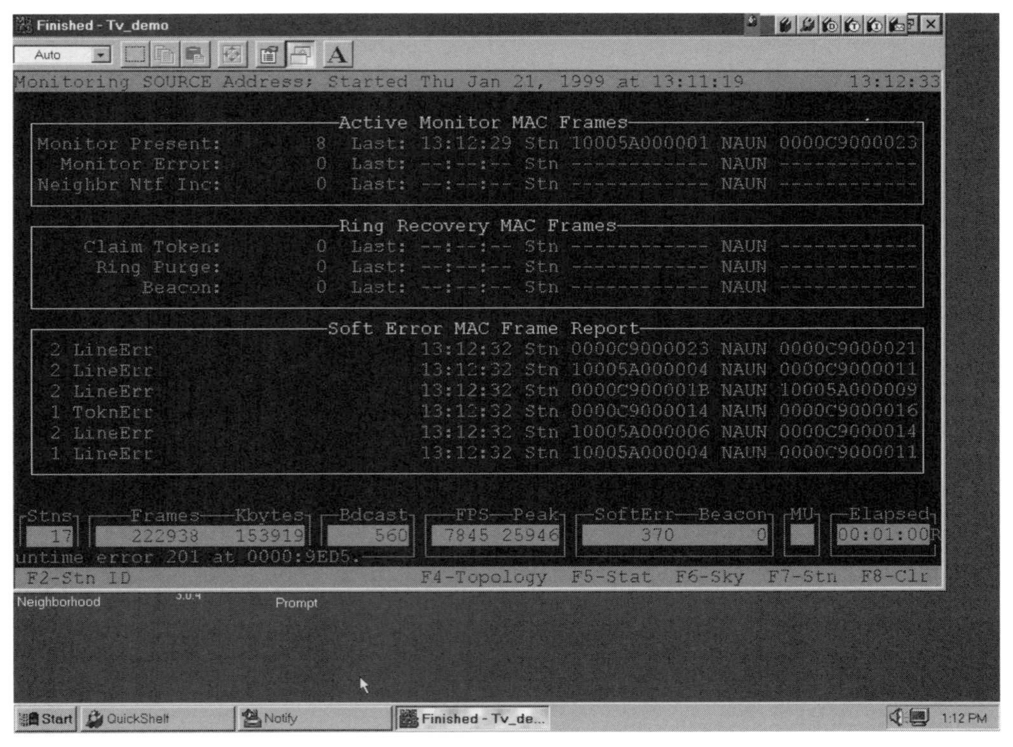

Figure 6.24 The TokenVision MAC display provides MAC layer statistic information, including active monitor, ring recovery, and soft error frame data

information between the LLC and the physical network. Examples of MAC protocol functions include the recognition of adapter addresses, physical medium access management, and message verification and status generation. A MAC frame is indicated by the setting of the first two bits in the frame control field to 00. When this situation occurs, the content of the Information field that carries MAC data is known as a vector.

6.8.1 Vectors and subvectors

Only one vector is permitted per MAC frame. That vector consists of a major Vector Length (VL), a major Vector Identifier (VI), and zero or more subvectors.

As indicated in Figure 6.25, there can be multiple subvectors within a vector. VL is a 16-bit number that gives the length of the vector, including the VL subfield in bytes. VL can vary between decimal 4 and 65 535 in value. The minimum value that can be assigned to VL results from the fact that the smallest information field must contain both VL and VI subfields. Since each subfield is 2 bytes in length, the minimum value of VL is 4.

When one or more subvectors are contained in a MAC information field, each subvector contains three fields. The Subvector Length (SVL) is an 8-bit number that indicates the length of the subvector. Since an 8-bit number has a maximum value of 255 and cannot indicate a length exceeding 256 bytes (0–255), a method was required to accommodate Subvector Values (SVV) longer than 254 bytes. The method used is the placement of hex FF in the

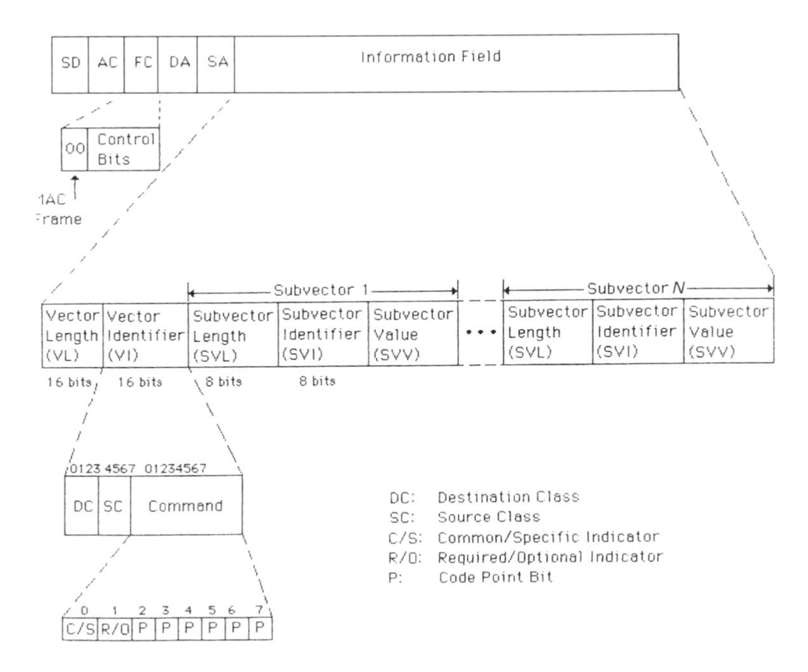

Figure 6.25 MAC Frame Information field format

SVL field to indicate that SVV exceeds 254 bytes. Then, the actual length is placed in the first two bytes following SVL. Finally, each SVV contains the data to be transmitted. The Command field within the major vector identifier contains bit values referred to as code points that uniquely identify the type of MAC frame. Figure 6.25 illustrates the format of the MAC Frame Information field, while Table 6.11 lists currently defined vector identifier codes for six MAC control frames defined under the IEEE 802.5 standard.

6.8.2 MAC control

As previously discussed, each ring has a station known as the active monitor, which is responsible for monitoring tokens and taking action to prevent the endless circulation of a token on a ring. Other stations function as standby monitors, and one such station will assume the functions of the active monitor if that device should fail or be removed from the ring. For the standby monitor with the highest network address to take over the functions of the active monitor, the standby monitor needs to know there is a problem with the active monitor. If no frames are circulating on the ring but the active monitor is operating, the standby monitor might falsely presume the active monitor has failed. Thus, the active monitor will periodically issue an Active Monitor Present (AMP) MAC frame. This frame must be issued every 7 seconds to inform the standby monitors that the active monitor is operational. Similarly, standby monitors periodically issue a Standby Monitor Present (SMP) MAC frame to denote they are operational.

If an active monitor fails to send an AMP frame within the required time interval, the standby monitor with the highest network address will continuously transmit Claim Token (CL_TK) MAC frames in an attempt to become the active monitor. The standby monitor will continue to transmit CL_TK MAC frames until one of three conditions occurs:

- a MAC CL_TK frame is received and the sender's address exceeds the standby monitor's station address;
- a MAC Beacon (BCN) frame is received;
- a MAC Purge (PRG) frame is received.

If one of the preceding conditions occurs, the standby monitor will cease its transmission of CL_TK frames and resume its standby function.

Table 6.11 Vector identifier codes

Code Value	MAC Frame Meaning
010	Beacon (BCN)
011	Claim Token (CL_TK)
100	Purge MAC frame (PRG)
101	Active Monitor Present (AMP)
110	Standby Monitor Present (SMP)
111	Duplicate Address Test (DAT)

Purge frame

If a CL_TK frame issued by a standby monitor is received back without modification and neither a BCN nor a PRG frame is received in response to the CL_TK frame, the standby monitor becomes the active monitor and transmits a MAC PRG frame. The PRG frame is also transmitted by the active monitor each time a ring is initialized or if a token is lost. Once a PRG frame has been transmitted, the transmitting device will place a token back on the ring.

Beacon frame

In the event of a major ring failure, such as a cable break or the continuous transmission by one station (known as jabbering), a BCN frame will be transmitted. The transmission of BCN frames can be used to isolate ring faults. For an example of the use of a BCN frame, consider Figure 6.26, in which a cable fault results in a ring break. When a station detects a serious problem with the ring, such as the failure to receive a frame or token, it transmits a BCN frame. That frame defines a failure domain, which consists of the station reporting the failure via the transmission of a beacon and its Nearest Active Upstream Neighbor (NAUN), as well as everything between the two.

If a BCN frame makes its way back to the issuing station, that station will remove itself from the ring and perform a series of diagnostic tests to determine if it should attempt to reinsert itself into the ring. This procedure ensures that a ring error caused by a beaconing station can be compensated for by having that station remove itself from the ring. Since BCN frames indicate a general area where a failure occurred, they also initiate a process known as auto-reconfiguration. The first step in the auto-reconfiguration process is the diagnostic testing of the beaconing station's adapter. Other steps in the auto-reconfiguration process include diagnostic tests performed by other nodes located in the failure domain in an attempt to reconfigure a ring around a failed area.

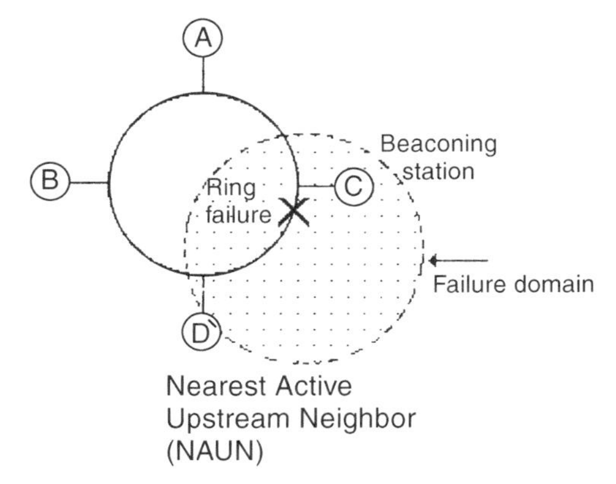

Figure 6.26 Beaconing. A beaconing frame indicates a failure occurring between the beaconing station and its nearest active upstream neighbor — an area referred to as a failure domain

Duplicate Address Test frame

The last type of MAC command frame is the Duplicate Address Test (DAT) frame. This frame is transmitted during a station initialization process when a station joins a ring. The station joining the ring transmits a MAC DAT frame with its own address in the frame's destination address field. If the frame returns to the originating station with its address-recognized (A) bit in the Frame Control field set to 1, this means that another station on the ring is assigned that address. The station attempting to join the ring will send a message to the ring network manager concerning this situation and will not join the network.

6.8.3 Station insertion

Depending upon the type of LAN adapter installed in your workstation, you may observe a series of messages at the format 'Phase X' followed by the message 'Completed' or 'Passed' when you power-on your computer. Those messages reference a five-phase ring insertion process during which your workstation's Token-Ring adapter attempts to become a participant on the ring. Table 6.12 lists the steps in the ring insertion process.

During the lobe testing phase, the adapter transmits a series of Lobe Media Test MAC frames to the Multistation Access Unit (MAU). Those frames should be wrapped at the MAU, resulting in their return to the adapter. Assuming that the returned frames are received correctly, the adapter sends a 5-volt DC current, which opens a relay at the MAU port and results in an attachment to the ring.

After the station has attached to the ring, it sets a value in a timer known as the Insert timer and watches for an AMP, SMP or Purge MAC frame prior to the timer expiring. If the timer expires, a token claiming process is initiated. If the station is the first station on the ring, it then becomes the active monitor.

Once the Monitor Check Phase has been completed, the station transmits a Duplicate Address Test frame during which the destination and source address fields are set to the station's universal address. If a duplicate address is found when the A bit is set to 1, the station cannot become a participant on the ring and detaches itself from the ring.

Assuming the station has a unique address, it next begins the neighbor notification process. During this ring insertion phase, the station learns the address of its Nearest Active Upstream Neighbor (NAUN) and reports its address to its Nearest Active Downstream Neighbor.

The address learning process begins when the active monitor transmits an

Table 6.12 Ring station insertion process

Phase 0: Lobe testing
Phase 1: Monitor check
Phase 2: Duplicate address check
Phase 3: Participation in neighbor notification
Phase 4: Request initialization

AMP frame. The first station that receives the frame and is able to copy it sets the address-recognized (A) and frame-copied (C) bits to '1'. The station then saves the source address from the copied frame as the NAUN address and initiates a Notification-Response timer. As the frame circulates the ring, other active stations only repeat it as its A and C bits were set.

When the Notification-Response timer of the first station downstream from the active monitor expires, it broadcasts an SMP frame. The next station downstream copies its NAUN address from the Source Address field of the SMP frame and sets the A and C bits in the frame to '1'. Then, it starts its own Notification-Response time, which, upon expiration, results in that station transmitting its SMP frame. As the SMP frames originate from different stations, the notification process proceeds around the ring until the active monitor copies its NAUN address from an SMP frame. At this point, the active monitor sets its Neighbor-Notification Complete flag to 1, which indicates that the neighbor notification process was successfully completed.

The final phase in the ring insertion process occurs after the neighbor notification process has been completed. During this phase, the station's adapter transmits a Request Initialization frame to the ring parameter server. The server responds with an Initialize-Ring-Station frame, which contains values that enable all stations on the ring to use the same ring number and soft error report time value, thereby completing the insertion process.

6.9 TOKEN-RING LOGICAL LINK CONTROL

In concluding this chapter, we will examine the flow of information within a Token-Ring network at the Logical Link Control (LLC) sublayer. Similar to Ethernet, the Token-Ring LLC sublayer is responsible for performing routing, error control, and flow control. In addition, this sublayer is responsible for providing a consistent view of a LAN to upper OSI layers, regardless of the type of media and protocols used on the network.

Figure 6.27 illustrates the format of an LLC frame that is carried within the Information field of the Token-Ring frame. As discussed earlier in this chapter, the setting of the first two bits in the Frame Control field of a Token-Ring frame to 01 indicates that the Information field should be interpreted as an LLC frame. The portion of the Token-Ring frame that carries LLC information is known as a protocol data unit, and consists of either three or four fields, depending upon the inclusion or omission of an optional information field. The Control field is similar to the Control field used in the HDLC protocol and defines three types of frames: Information (I-frames) are used for sequenced messages, Supervisory (S-frames) are used for status and flow control, while Unnumbered (U-frames) are used for unsequenced, unacknowledged messages.

6.9.1 Service Access Points

Service Access Points (SAPs) can be considered as interfaces to the upper layers of the OSI Reference Model, such as the network layer protocols. A

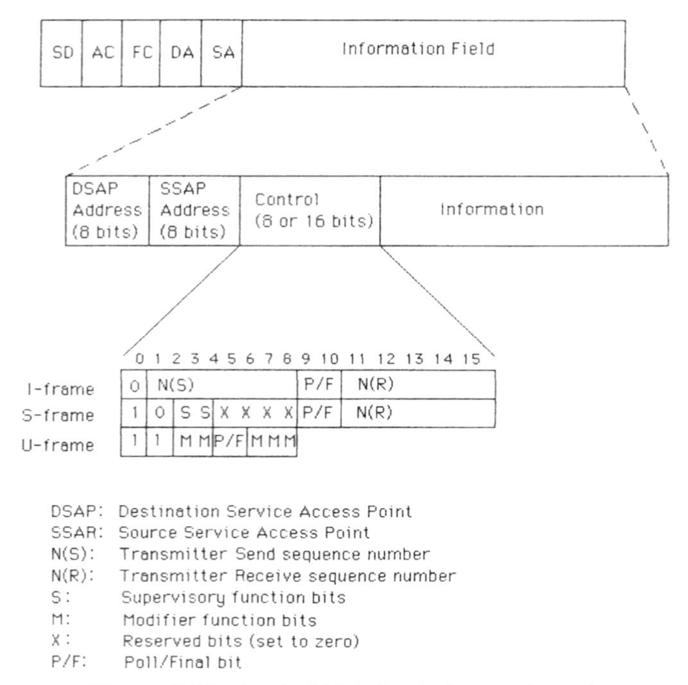

Figure 6.27 Logical Link Control frame format

station can have one or more SAPs associated with it for a specific layer and can have one or more active sessions initiated through a single SAP. Thus, we can consider a SAP to function similarly in scope to a mailbox, containing an address that enables many types of mailings to reach the box. However, instead of mail, SAP addresses identify different network layer processes or protocols, and function as locations where messages can be left concerning desired network services.

DSAP

The first field in the LLC protocol data unit is the Destination Services Access Point (DSAP). The DSAP address field identifies one or more service access points for which information is to be delivered.

SSAP

The second field in the LLC protocol data unit is the Source Services Access Point (SSAP). The SSAP address field identifies the service access point that transmitted the frame. Both DSAP and SSAP addresses are assigned to vendors by the IEEE to ensure that each is unique.

Both DSAPs and SSAPs are 8-bit fields; however, only 7 bits are used for addressing, which results in a maximum of 128 distinct addresses available for each service access point. The eighth DSAP bit indicates whether the destination is an individual or a group address, while the eighth SSAP bit indicates whether the PDU contains a request or a response.

The Control field contains information that defines how the LLC frame will be handled. U-frames are used for what is known as connectionless service in which frames are not acknowledged, while I-frames are used for connection-oriented services in which frames are acknowledged.

6.9.2 Types and classes of service

The types and classes of service supported by Token-Ring are the same as those supported by Ethernet, which were described in Section 6.3. Thus, readers are referred to Section 6.3 for information concerning the types and classes of service supported by a Token-Ring LAN.

6.10 SUMMARY

Regardless of whether your organization operates an Ethernet or Token-Ring network infrastructure, it is important to recognize the manner in which the layer 2 network operates. As previously mentioned, most layer 2 networks transport a variety of protocols, including TCP/IP. For example, consider Figure 6.28, which illustrates the TokenVision protocol display screen. To

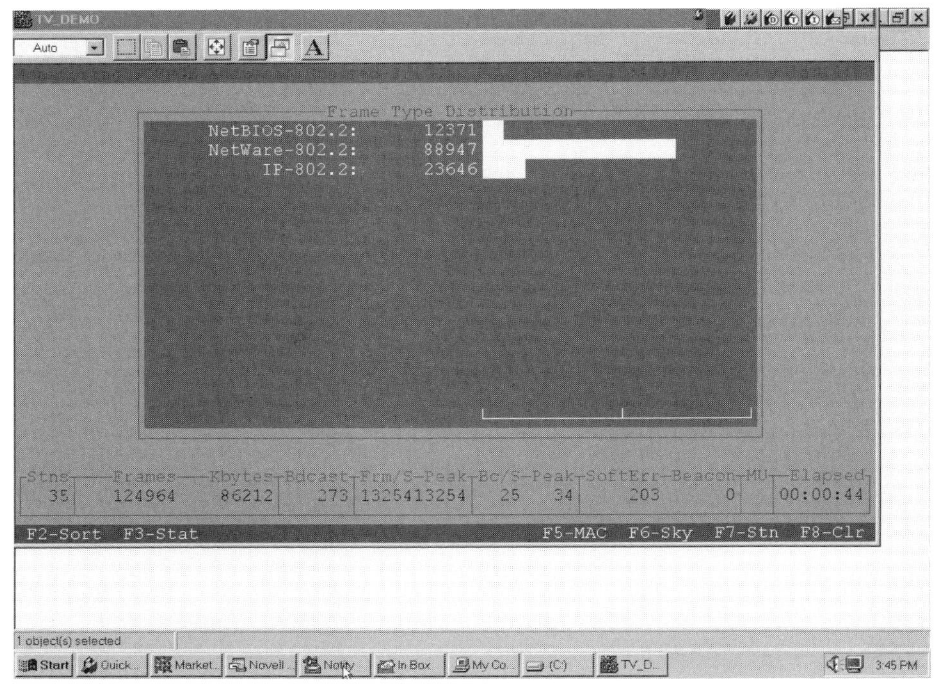

Figure 6.28 The TokenVision protocol display screen indicates the distribution of frames transporting different network layer protocols

generate this display, TokenVision literally looks further into each frame to determine the protocol being transported in each frame. Note that at the time the screen shown in Figure 6.28 was captured, the number of frames transporting IP was significantly less than the number of frames transporting NetWare. Thus, this illustrates why other protocols being transported can adversely affect the performance of TCP/IP and the reason why you should examine the flow of frames on a LAN to determine the possible or actual cause of bottlenecks or error conditions that may or may not be associated with TCP/IP traffic.

7

LAYER 3 AND LAYER 4 MANAGEMENT

In the first five chapters in this book we focused our attention upon obtaining a detailed understanding of the TCP/IP protocol suite. In Chapter 6 we moved our management analysis up the protocol stack, turning our attention to tools and techniques we can use to effect layer 3 and layer 4 management.

Because it is very difficult to divorce the transport layer from the network layer, we will focus on both in this chapter. In doing so, we will examine the use of several network analysis programs that can be used to examine layer 3 and layer 4 operations as well as to obtain information that will be used as a mechanism for discussing techniques that can be used to facilitate TCP/IP network management.

Although many SNMP and RMON-compatible network management programs generate layer 3 and layer 4 statistics, we will defer a discussion of standardized network management tools and techniques until Chapter 8. While standardized network management products considerably facilitate interoperability, many times the ability to use a stand-alone program can be sufficient to provide information required to successfully manage a TCP/IP network.

The two programs used in this chapter were selected as they include features and capabilities representative of a number of stand-alone layer 3 and layer 4 network management products. While I do not recommend any particular vendor product, it should be noted that the commercial programs used to illustrate layer 3 and layer 4 management methods were acquired as a result of their capability. While this does not mean—nor should it imply—that there are no other similar capable programs, my experience with these programs resulted in their use as a mechanism to illustrate concepts and management techniques.

7.1 USING WebXRay

The first program that will be used to illustrate layer 3 and layer 4 management is WebXRay. This program was originally developed by Cinco

Systems. Cinco Systems was acquired by Network General, which in turn was acquired by Network Associates.

7.1.1 Overview

WebXRay represents a network monitoring and analysis program that includes the ability to capture and decode data flowing on a network. Although you can use WebXRay to discover the addresses and services supported by computers on a distant network, the primary use of this program is to monitor and analyze the flow of data on a local network. To do so, you would install WebXRay on a PC connected to the network whose traffic you wish to monitor and analyze.

7.1.2 Operation

By default, when WebXRay is initiated, the program displays a three-gauge dashboard that provides a visual summary of packet flow, network utilization, and ICMP activity on the networks being monitored. Figure 7.1 illustrates the WebXRay dashboard, which can be toggled between its gauge and detail display modes by clicking on one of the two tabs located at the bottom of the display.

In examining the three gauges in the dashboard, note that the first two, Packets/s and Utilization %, have dual numbers. Because more than one

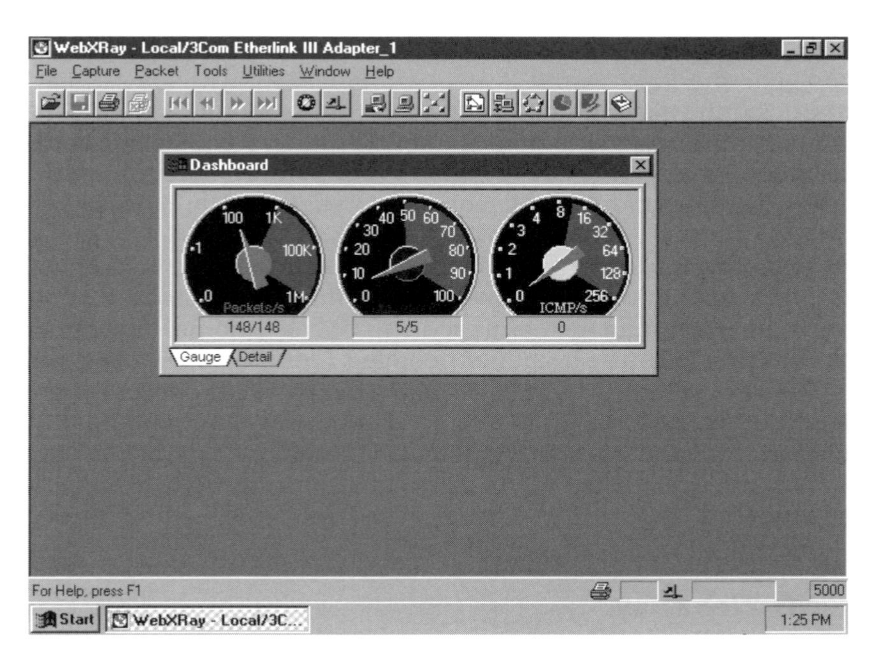

Figure 7.1 The WebXRay dashboard contains three gauges that provide a visual indication of total network and TCP/IP-related activity

protocol can be flowing on a network, WebXRay counts both total packets and TCP/IP packets flowing on a network. The first number displayed represents the total packet flow, while the second number represents the flow of TCP/IP packets on the network. If the monitored network only uses the TCP/IP protocol, then the two numbers will be the same; however, they can also be the same when TCP/IP represents the protocol used by the vast majority of network traffic.

The middle gauge is similar to the first in that the first number represents the percent utilization of the network to include TCP/IP, while the second metric represents the utilization of the network attributable to TCP/IP. Because ICMP is only relevant to the TCP/IP protocol, the third gauge only displays one value.

By focusing your attention upon the WebXRay Dashboard display, it becomes possible to note certain conditions that could adversely affect network operations. For example, if the network is experiencing a high level of utilization, this could explain the reason why network users are beginning to complain that interactive sessions and file transfers are requiring more time than normal. Even if the network is being used for a limited amount of TCP/IP transmission, a high level of network activity will adversely affect TCP/IP transmission. Thus, it is important to examine both the level of TCP/IP traffic and the overall network activity.

Another gauge that warrants attention is the Packets/s gauge. This gauge indicates the flow of data on the network and is commonly indirectly used as a measurement of network activity. However, by clicking on the Detail tab of the Dashboard display, you obtain the ability to directly view a breakdown of the data flow on the network, which can facilitate determining several possible reasons for an inordinate amount of network activity.

Figure 7.2 illustrates the Dashboard's Detail display. Note that this display can be used to identify the amount of IP, TCP, UDP, ICMP, broadcast, and multicast packets. Thus, if a high level of network utilization is caused by the transmission of a specific type of packet, the Detail Dashboard can be used to identify this. However, some readings may require an additional level of analysis. For example, because both voice over IP and SNMP are transported by UDP, a high level of UDP traffic might not directly point to a specific application. Similarly, a high level of broadcast or multicast traffic could result from a number of causes, and would require further investigation. In spite of the previously mentioned problems, WebXRay's Dashboard and similar dashboards used by other programs represent a good starting point for examining activity on a network. If network traffic is light and users are complaining of delays, then the use of a dashboard would indicate that the delay might be attributable to a server. If the server operates Windows NT, you might then consider using the Windows NT (or Windows 2000) Performance Monitor to examine processor and disk utilization levels.

Autodiscovery

One of the more interesting features of WebXRay is its Topology Discovery capability, which is actually not quite correctly named. This is because this

Dashboard

Packets:		Bytes:		Utilization:	
Network	80135757	Network	332403628	Network	7
IP	7949107[illegible]	IP	266789497	IP	7
TCP	7919354[illegible]	TCP	224223955		
UDP	221992	UDP	36310159		
ICMP	75314	ICMP	6235153		
Broadcast	196595				
Multicast	9				

Gauge Detail

Figure 7.2 Viewing the WebXRay Dashboard's Detail display

feature discovers devices and the services they provide instead of a network topology.

Figure 7.3 illustrates the WebXRay Topology Discovery dialog box with its default Address tab shown displayed. To automatically discover the active stations on a network and the services they support, you would first enter the network portion of a Class C address, such as 205.131.176, in the previously referenced illustration. Then you would enter a range of host addresses, such as 1 and 254, which represents all possible addresses on a Class C network. The last metric you would enter is a Ping timeout value to change the default setting of 300 milliseconds.

The WebXRay Topology Discovery program operates by cycling through the IP network plus host address range, pinging each address. The number of pings performed depends upon the setting of items in the Service tab, so let us turn our attention to that tab.

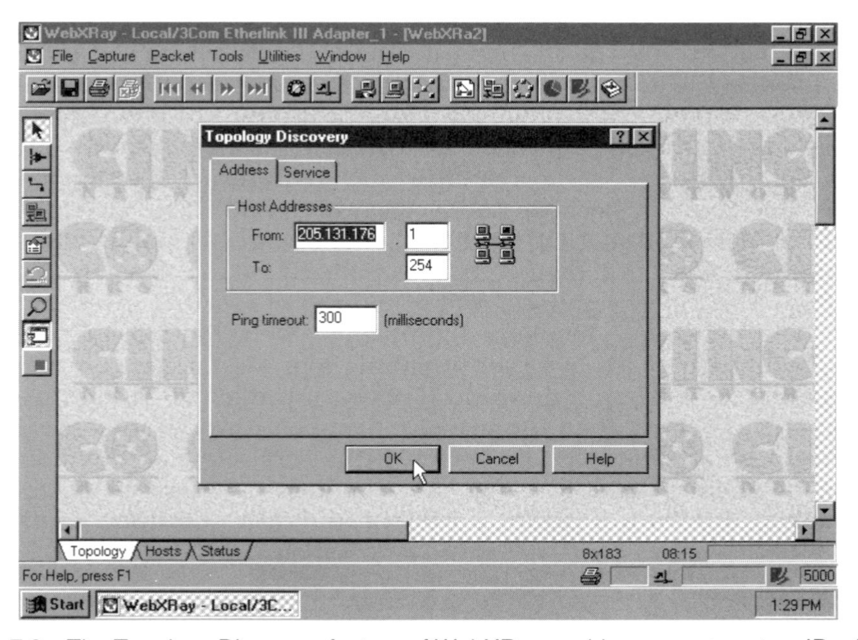

Figure 7.3 The Topology Discovery feature of WebXRay enables a user to set an IP address range for the program to use to determine active stations and the services they support

Service selection Figure 7.4 illustrates the Topology Discovery dialog box with its Service tab displayed in the foreground. In this example the 'Select All' button was clicked on to select all services. This means that each ping operation will cycle through TCP and UDP ports associated with each service as a mechanism to determine if the service is available.

Topology discovery The actual operation of the Topology Discovery process is illustrated in Figure 7.5. Note that since an IP network could be operating on an

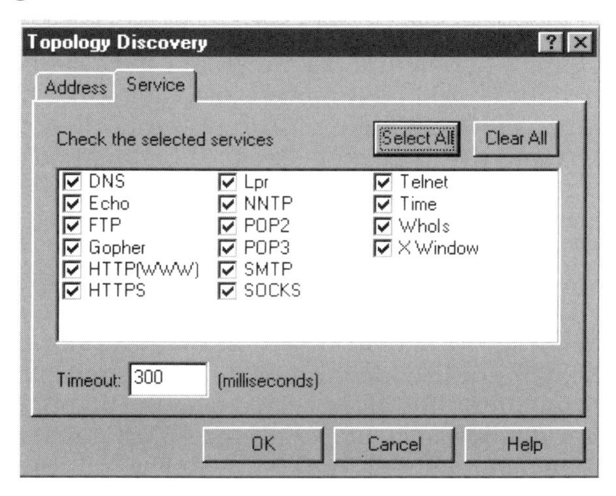

Figure 7.4 Through the Service tab in the Topology Discovery dialog box, you can define the services on the range of IP addresses the program should check for

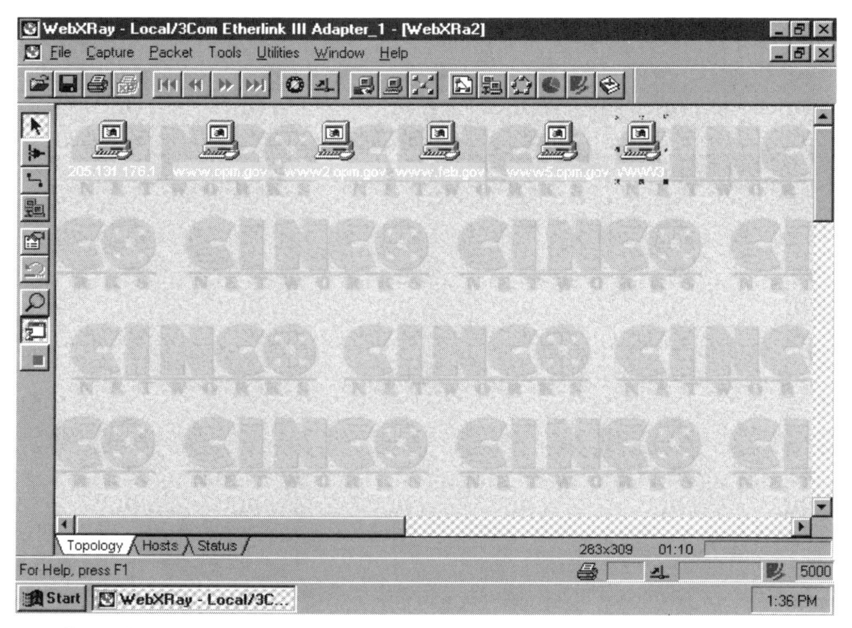

Figure 7.5 The Topology Discovery process results in the display of an icon for each station discovered on a network

Ethernet or Token-Ring LAN infrastructure, in my opinion the program is performing an autodiscovery process and not a topology discovery process.

The previously described autodiscovery process can be performed by a number of programs in addition to the use of a menu item in WebXRay. Regardless of which program performs the autodiscovery process, it is important to note that the computer running the program does not have to be connected to the network it is attempting to perform the autodiscovery operation on. However, if a firewall or router access list is configured to block pings, then you will not be able to conduct a remote autodiscovery process.

In examining Figure 7.5, note that as a result of the autodiscovery process WebXRay denotes both IP and host names, displaying the host name when possible. The display shown in Figure 7.5 represents the program's default Topology tab foreground display and is one of three methods you can use to view the results of the autodiscovery process. The other two viewing methods result from the selection of the Hosts and Status tabs shown in the lower left portion of the screen display.

Hosts information Figure 7.6 illustrates the display of the Hosts tab from the WebXRay Topology Discovery screen. Note that the resulting split window display consists of the name or IP address of hosts discovered (which is displayed in the left window) and the status of the services for the highlighted host (which is displayed in the right window). In examining Figure 7.6, note that the host *www.feb.gov* is shown highlighted. The right window indicates that that host presently supports FTP, HTTP, and HTTPS services. Thus, the Topology Discovery feature included in WebXRay provides the ability to note what hosts are

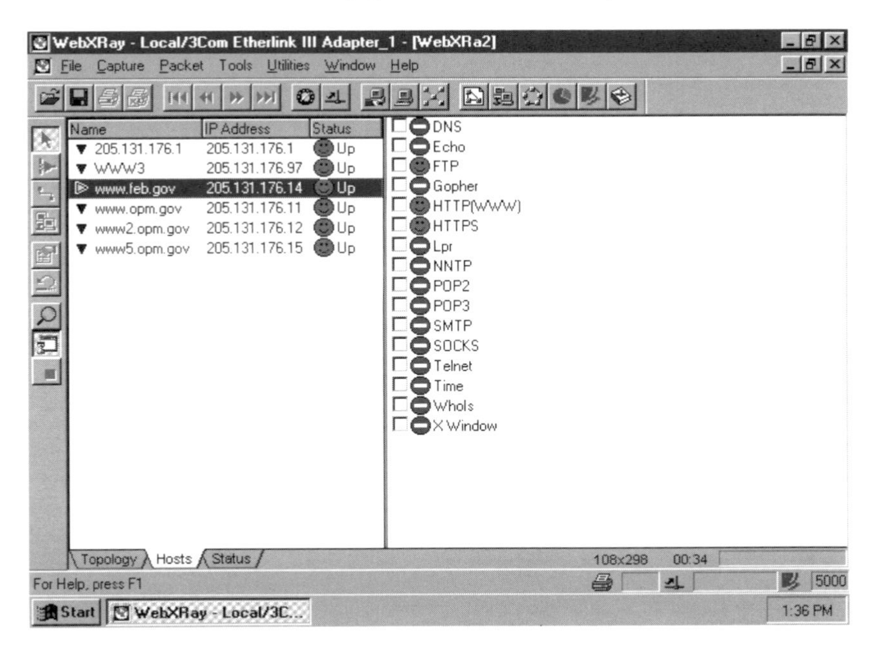

Figure 7.6 The Topology Discovery Hosts tab provides a mechanism to visually note services supported by each active host

currently operating the TCP/IP protocol stack as well as the services operating on the stack. This information can be extremely valuable when network users or customers call complaining about the inability to access a particular device or service provided by a host.

Services information The third option in the WebXRay Topology Discovery display occurs when the Status tab is selected. This option results in the display of available services for each active host discovered on the network. Figure 7.7 illustrates an example of the use of the Status tab. Note that because the program displays the status of a large number of services for each host within a common window, you must scroll to the right to observe the status of many services for each active host located on the network.

Now that we have an appreciation of the use of the discovery feature of WebXRay, let us turn our attention to the program's traffic measuring features. Those features are only applicable to the network to which a computer operating the program is connected, as the monitoring computer must read each frame and process various field values in some manner.

Traffic measuring

WebXRay includes several built-in statistical-generating displays that count various metrics and display the counts in a table. Examples of such tables include a Server Host Table, Server and Client Matrix Table, IP Host Table, and IP Matrix Table. Data obtained from the previously mentioned tables

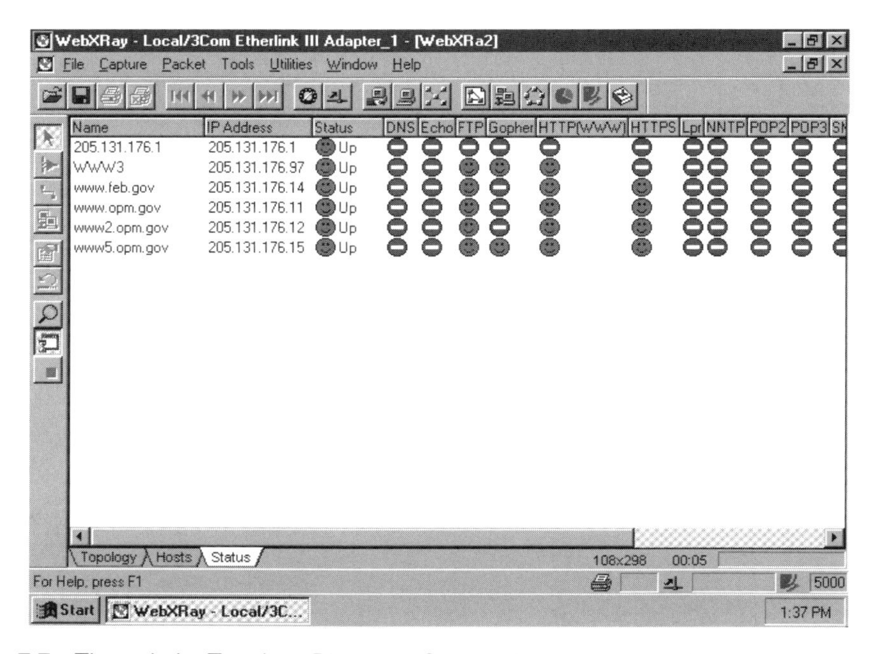

Figure 7.7 Through the Topology Discovery Status tab, you can view the services associated with each host through a common scrollable window

generally corresponds to tabular information generated through the use of an SNMP management station accessing a RMON; however, unlike a RMON probe, which can be located on any network and provide statistical data retrieved by the SNMP management station, the computer operating WebXRay must be physically connected to the network to be monitored.

Server Host Table The Server Host Table results in a display of inbound and outbound traffic whose packets are flowing on the monitored network. Figure 7.8 illustrates an example of the WebXRay Server Host Table at a particular point in time.

In addition to counting inbound and outbound packets and bytes, the Server Host Table accumulates the number of connections attempted, accepted, and rejected. Such information is displayed when a user scrolls to the right in Figure 7.8, and can be valuable for ascertaining actual or potential problems. For example, if a server's acceptance count is well below the connection attempts or if the reject count is relatively high, this would indicate a potential server overload condition and not a network problem. In addition to the general monitoring performed by the Server Host Table, a WebXRay user can set the display to break down traffic by several categories, such as HTTP, FTP, POP3, SMTP, Telnet, DNS, and other applications. Thus, the Server Host Table can be used to identify not only data flow but, in addition, data flow by application as well as information concerning connection attempts.

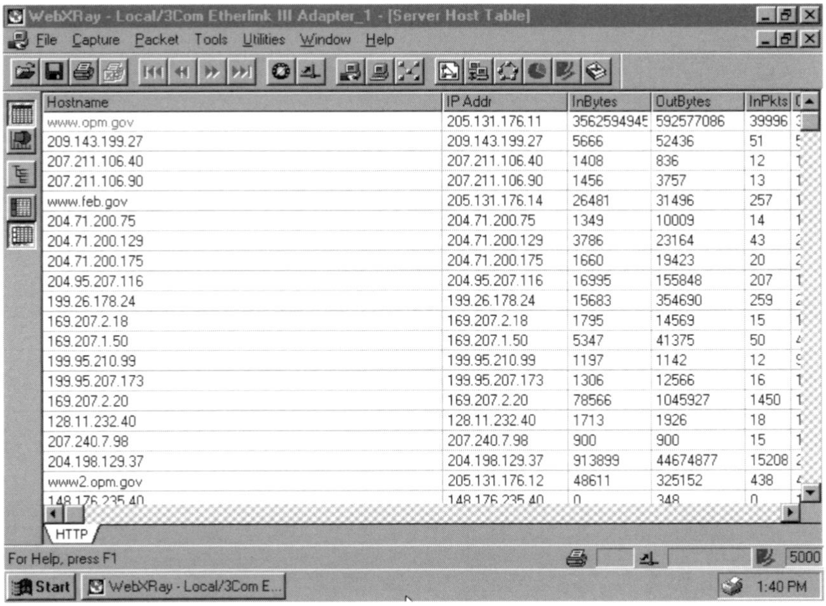

WebXRay - Local/3Com Etherlink III Adapter_1 - [Server Host Table]

File Capture Packet Tools Utilities Window Help

Hostname	IP Addr	InBytes	OutBytes	InPkts
www.opm.gov	205.131.176.11	3562594945	592577086	39996
209.143.199.27	209.143.199.27	5666	52436	51
207.211.106.40	207.211.106.40	1408	836	12
207.211.106.90	207.211.106.90	1456	3757	13
www.feb.gov	205.131.176.14	26481	31496	257
204.71.200.75	204.71.200.75	1349	10009	14
204.71.200.129	204.71.200.129	3786	23164	43
204.71.200.175	204.71.200.175	1660	19423	20
204.95.207.116	204.95.207.116	16995	155848	207
199.26.178.24	199.26.178.24	15683	354690	259
169.207.2.18	169.207.2.18	1795	14569	15
169.207.1.50	169.207.1.50	5347	41375	50
199.95.210.99	199.95.210.99	1197	1142	12
199.95.207.173	199.95.207.173	1306	12566	16
169.207.2.20	169.207.2.20	78566	1045927	1450
128.11.232.40	128.11.232.40	1713	1926	18
207.240.7.98	207.240.7.98	900	900	15
204.198.129.37	204.198.129.37	913899	44674877	15208
www2.opm.gov	205.131.176.12	48611	325152	438
148.176.235.40	148.176.235.40	0	348	0

HTTP

For Help, press F1 5000

Start WebXRay - Local/3Com E... 1:40 PM

Figure 7.8 The Server Host Table provides information concerning traffic flow by host as well as connection and traffic distribution by application

Server–Client Matrix Table The Server–Client Matrix Table provides a distribution of traffic between hosts based upon application. Figure 7.9 illustrates an example of the display of the WebXRay Server–Client Matrix Table.

In examining Figure 7.9, it should be noted that during the monitoring period the only traffic observed on the network was HTTP. Thus, Figure 7.9 only has a tab labeled HTTP in the bottom left of the display. Users can sort this table to determine which client–server pair consumes large portions of network bandwidth. Thus, the Server–Client Matrix Table provides another traffic monitoring tool that can be extremely useful for locating unusual network activity that may require remedial action.

IP Host Table The IP Host Table represents a collection of statistics for each IP address noted on the monitored network. Those statistics include traffic expressed in terms of packets and bytes for inbound and outbound packets as well as for specific protocols and packet types, such as TCP, UDP, ICMP, and broadcast and multicast packets. By sorting the entries in the IP Host Table, you can determine which stations are the most or least active based upon different categories of IP statistics. The IP Host Table is illustrated in Figure 7.10.

IP Matrix Table The fourth WebXRay table that we will examine is the IP Matrix Table. As its name implies, this table provides statistics between pairs of network nodes, indicating the data flow between those pairs. Figure 7.11 illustrates an example of the display of the WebXRay IP Matrix Table.

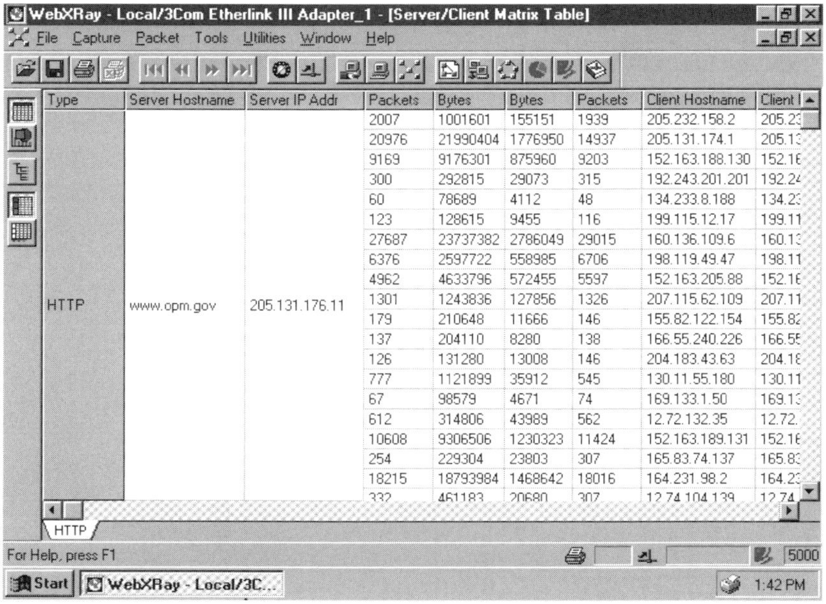

WebXRay - Local/3Com Etherlink III Adapter_1 - [Server/Client Matrix Table]

File Capture Packet Tools Utilities Window Help

Type	Server Hostname	Server IP Addr	Packets	Bytes	Bytes	Packets	Client Hostname	Client I
HTTP	www.opm.gov	205.131.176.11	2007	1001601	155151	1939	205.232.158.2	205.23
			20976	21990404	1776950	14937	205.131.174.1	205.13
			9169	9176301	875960	9203	152.163.188.130	152.16
			300	292815	29073	315	192.243.201.201	192.24
			60	78689	4112	48	134.233.8.188	134.23
			123	128615	9455	116	199.115.12.17	199.11
			27687	23737382	2786049	29015	160.136.109.6	160.13
			6376	2597722	558985	6706	198.119.49.47	198.11
			4962	4633796	572455	5597	152.163.205.88	152.16
			1301	1243836	127856	1326	207.115.62.109	207.11
			179	210648	11666	146	155.82.122.154	155.82
			137	204110	8280	138	166.55.240.226	166.55
			126	131280	13008	146	204.183.43.63	204.18
			777	1121899	35912	545	130.11.55.180	130.11
			67	98579	4671	74	169.133.1.50	169.13
			612	314806	43989	562	12.72.132.35	12.72.
			10608	9306506	1230323	11424	152.163.189.131	152.16
			254	229304	23803	307	165.83.74.137	165.83
			18215	18793984	1468642	18016	164.231.98.2	164.23
			332	461183	20680	307	12.74.104.139	12.74

HTTP

For Help, press F1 5000

Start WebXRay - Local/3C... 1:42 PM

Figure 7.9 The WebXRay Server–Client Matrix Table denotes the flow of data between pairs of computers by application

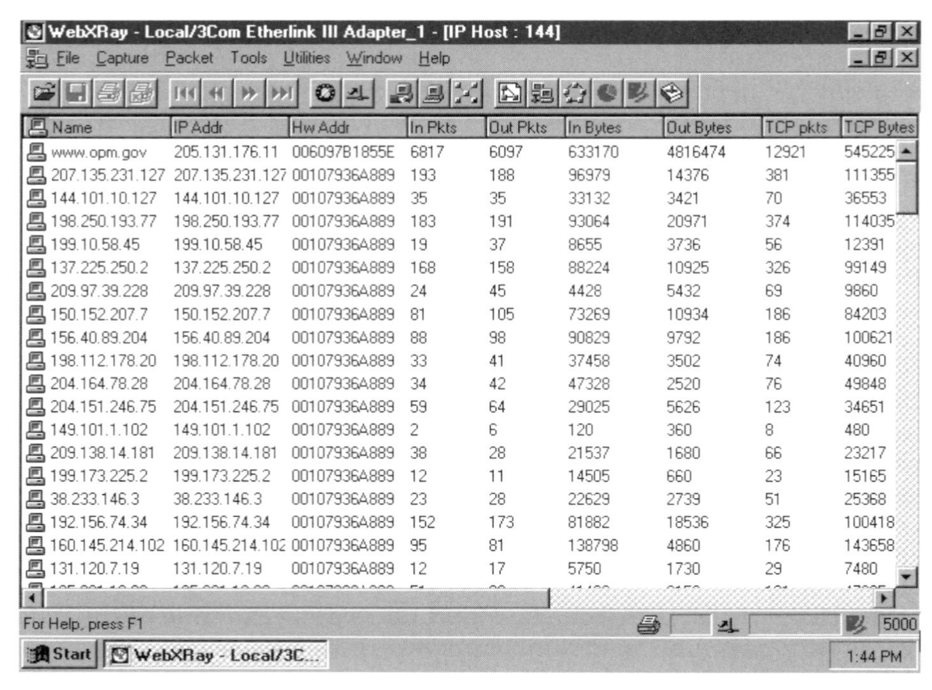

WebXRay - Local/3Com Etherlink III Adapter_1 - [IP Host : 144]

Name	IP Addr	Hw Addr	In Pkts	Out Pkts	In Bytes	Out Bytes	TCP pkts	TCP Bytes
www.opm.gov	205.131.176.11	006097B1855E	6817	6097	633170	4816474	12921	545225
207.135.231.127	207.135.231.127	00107936A889	193	188	96979	14376	381	111355
144.101.10.127	144.101.10.127	00107936A889	35	35	33132	3421	70	36553
198.250.193.77	198.250.193.77	00107936A889	183	191	93064	20971	374	114035
199.10.58.45	199.10.58.45	00107936A889	19	37	8655	3736	56	12391
137.225.250.2	137.225.250.2	00107936A889	168	158	88224	10925	326	99149
209.97.39.228	209.97.39.228	00107936A889	24	45	4428	5432	69	9860
150.152.207.7	150.152.207.7	00107936A889	81	105	73269	10934	186	84203
156.40.89.204	156.40.89.204	00107936A889	88	98	90829	9792	186	100621
198.112.178.20	198.112.178.20	00107936A889	33	41	37458	3502	74	40960
204.164.78.28	204.164.78.28	00107936A889	34	42	47328	2520	76	49848
204.151.246.75	204.151.246.75	00107936A889	59	64	29025	5626	123	34651
149.101.1.102	149.101.1.102	00107936A889	2	6	120	360	8	480
209.138.14.181	209.138.14.181	00107936A889	38	28	21537	1680	66	23217
199.173.225.2	199.173.225.2	00107936A889	12	11	14505	660	23	15165
38.233.146.3	38.233.146.3	00107936A889	23	28	22629	2739	51	25368
192.156.74.34	192.156.74.34	00107936A889	152	173	81882	18536	325	100418
160.145.214.102	160.145.214.102	00107936A889	95	81	138798	4860	176	143658
131.120.7.19	131.120.7.19	00107936A889	12	17	5750	1730	29	7480

For Help, press F1

Figure 7.10 The IP Host Table provides a summary of network statistics for all IP addresses that have traffic on the monitored network

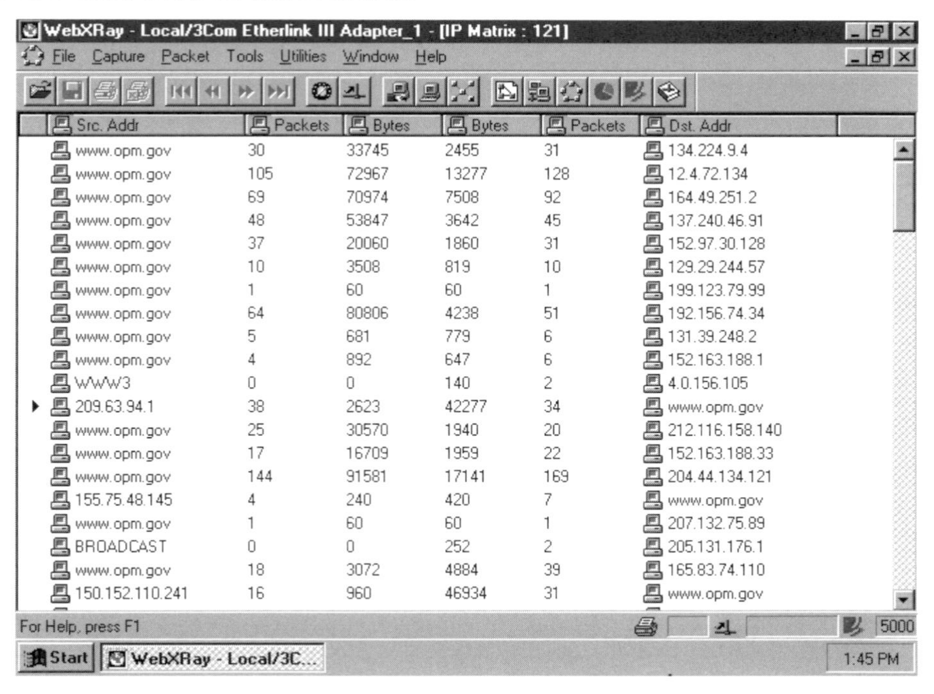

WebXRay - Local/3Com Etherlink III Adapter_1 - [IP Matrix : 121]

Src. Addr	Packets	Bytes	Bytes	Packets	Dst. Addr
www.opm.gov	30	33745	2455	31	134.224.9.4
www.opm.gov	105	72967	13277	128	12.4.72.134
www.opm.gov	69	70974	7508	92	164.49.251.2
www.opm.gov	48	53847	3642	45	137.240.46.91
www.opm.gov	37	20060	1860	31	152.97.30.128
www.opm.gov	10	3508	819	10	129.29.244.57
www.opm.gov	1	60	60	1	199.123.79.99
www.opm.gov	64	80806	4238	51	192.156.74.34
www.opm.gov	5	681	779	6	131.39.248.2
www.opm.gov	4	892	647	6	152.163.188.1
WWW3	0	0	140	2	4.0.156.105
209.63.94.1	38	2623	42277	34	www.opm.gov
www.opm.gov	25	30570	1940	20	212.116.158.140
www.opm.gov	17	16709	1959	22	152.163.188.33
www.opm.gov	144	91581	17141	169	204.44.134.121
155.75.48.145	4	240	420	7	www.opm.gov
www.opm.gov	1	60	60	1	207.132.75.89
BROADCAST	0	0	252	2	205.131.176.1
www.opm.gov	18	3072	4884	39	165.83.74.110
150.152.110.241	16	960	46934	31	www.opm.gov

For Help, press F1

Figure 7.11 The IP Matrix table can be used to track statistics concerning network activity between pairs of network nodes

From examining Figure 7.11, you will note that it appears that source–destination address pairs are displayed as encountered, which is correct. However, you can click on the tables 'Packets' column to obtain a sorted display, which can be useful to identify top talkers. In addition, note that the 'Packets' and 'Bytes' columns on the left of the table denote traffic from source to destination, while traffic from destination to source is identified by the second column labeled 'Bytes' and the second column labeled 'Packets.'

Now that we have an appreciation of the general use of different tables, let us turn our attention to other features of WebXRay. These features include a protocol distribution capability as well as the ability to filter and decode the contents of packets flowing on a monitored network.

Protocol distribution

Like other network monitoring programs, WebXRay includes the ability to display in real time the percentage of network usage based upon the network layer protocol. Figure 7.12 illustrates an example of the IP protocol distribution for a network primarily consisting of Web servers accessed via the Internet. As might be expected, a vast majority of traffic involves HTTP.

While a preponderance of HTTP traffic is not too meaningful, if you found a significant amount of FTP traffic and your network was on the verge of being

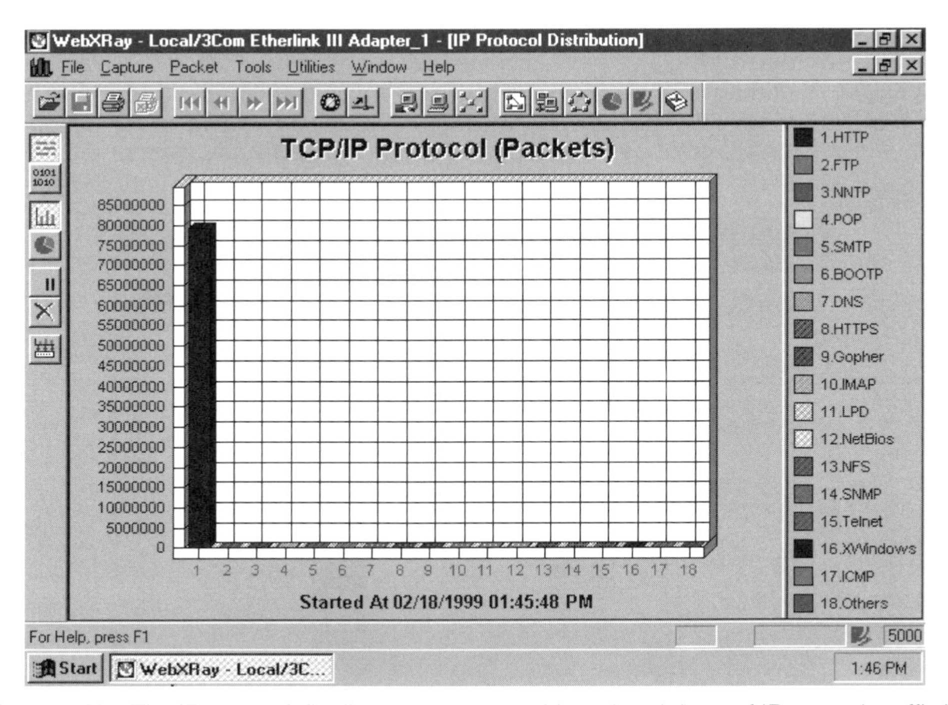

Figure 7.12 The IP protocol distribution screen provides a breakdown of IP network traffic by applications

saturated, you could consider the possibility of invoking timed FTP transfers in the evening. If possible, this could alleviate a costly network upgrade, and illustrates one possible use of the protocol distribution display.

Filtering and packet decoding

In concluding our examination of WebXRay, we will turn our attention to the program's filtering and packet decoding capability. Figure 7.13 illustrates the program's Filter Settings dialog box with its Address tab selected for foreground display. Note you can select filtering by MAC or IP address type as well as by an include or exclude mode. You can also create rather fancy filters through the use of the Data Pattern and Advanced Filter tabs. In the example shown in Figure 7.13 we will create a rather simple filter, telling the program to capture packets addressed to the IP address 205.131.176.11 from any source address.

Figure 7.14 illustrates the first of two packet decoding screens. In examining the top portion of Figure 7.14, you will note five packets, with packet one having a check mark and highlighted. A portion of the packet decoding is shown in the middle portion of Figure 7.14, while the bottom portion indicates the contents of the packet in hex. From the middle portion of Figure 7.14, you will note that the packet was transported in an Ethernet Version II frame and the MAC addresses of the source and destination. The program then looks further into the frame and decodes the contents of the Ethernet Data field. Thus, you will note that the Ver field in the IP header has the value 4, identifying the packet as an IPv4 packet.

Because decoding of a packet provides a significant number of field values, you can either scroll through the middle screen or expand it for scrolling. Figure 7.15 illustrates an expanded packet decoding window that was

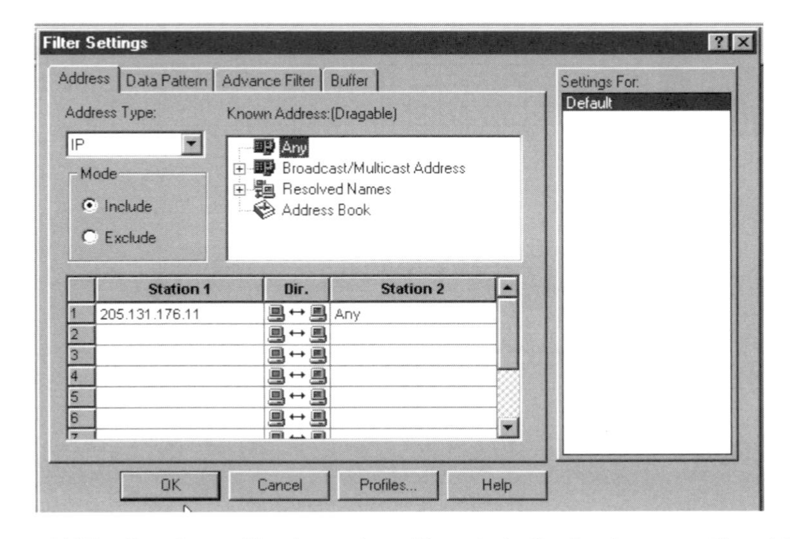

Figure 7.13 Creating a filter to capture IP packets flowing to a specific address

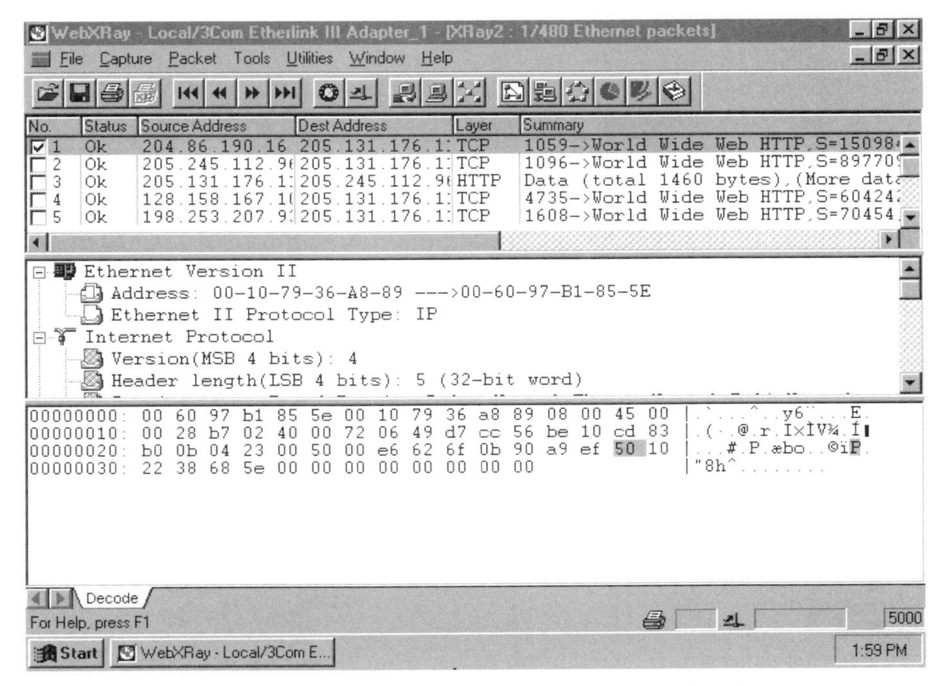

Figure 7.14 Viewing the decoding of a selected packet

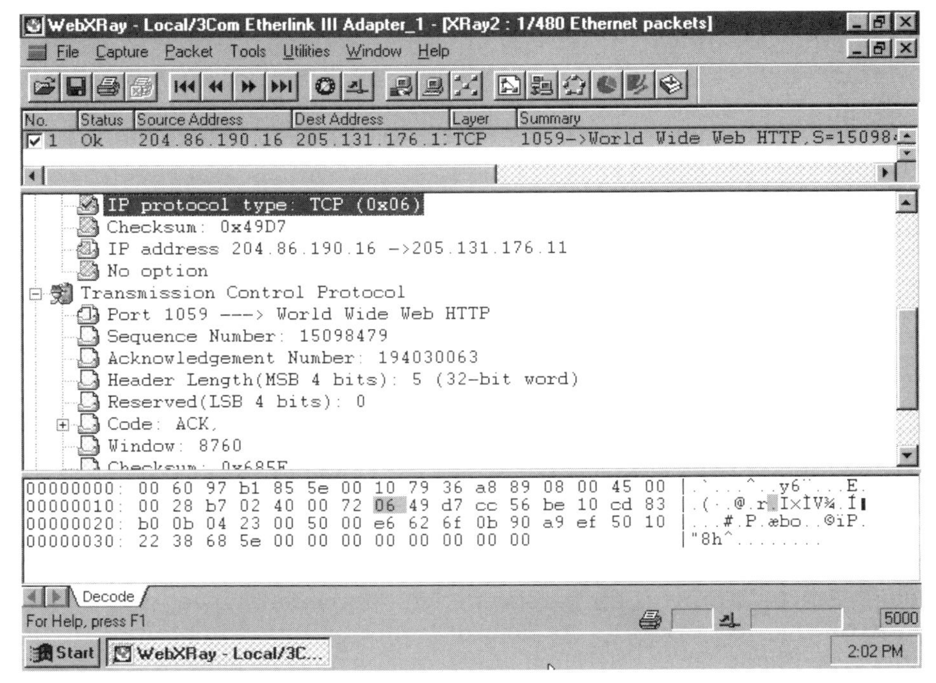

Figure 7.15 You can make the WebXRay protocol decode window larger by dragging upper and lower bars to expand the middle window

partially scrolled to the IP Protocol Type field. Note that the value of 6 identifies a TCP header following the IP header. Also note that as you move the highlighted bar, the bar in the hex decode portion of the lower window changes. Thus, the program automatically indicates where you are in the hex composition of the packet as you scroll through the decode middle window. By focusing attention upon the values of certain fields, you may be able to isolate a variety of potential problems. Thus, the filtering and packet decoding capability of WebXRay can be an important tool for troubleshooting IP-related problems.

7.2 USING EtherPeek

The second program that will be used in this chapter to illustrate layer 3 and layer 4 management is EtherPeek, a product of the AG Group. As its name implies, EtherPeek is a monitoring and analysis program that is restricted to operating on Ethernet LANs. However, since approximately 80–90% of all TCP/IP networks operate on an Ethernet infrastructure, this restriction will only affect persons that operate TCP/IP on another physical network infrastructure.

7.2.1 Operation

EtherPeek represents a very sophisticated packet capture and analysis program that includes the ability to generate and display a number of network-related statistics. Similar to WebXRay, EtherPeek operates by using a network adapter card in its promiscuous mode of operation, examining the contents of each frame that flows on the network to which the adapter is connected.

Although there are a variety of ways to employ EtherPeek, one of the more common methods using this program is to examine the composition of packets when normal activities appear abnormal, such as a situation where traffic may not be recognized by its destination or traffic does not reach its intended destination, and similar network problems. In such situations you would probably want to use EtherPeek's packet capture capability. Therefore, let us first turn our attention to that feature.

Packet capture

Figure 7.16 illustrates the EtherPeek main window, showing its packet Capture window located within the main window. By clicking on the button labeled 'Start Capture,' the program will immediately copy every frame flowing on the network until its default buffer of approximately 2 Mbytes of memory is filled. On a 10-Mbps LAN that averages only a 25% level of utilization, the default buffer could become filled in approximately 3 seconds. On a 100-Mbps LAN, the default buffer size would become filled within a

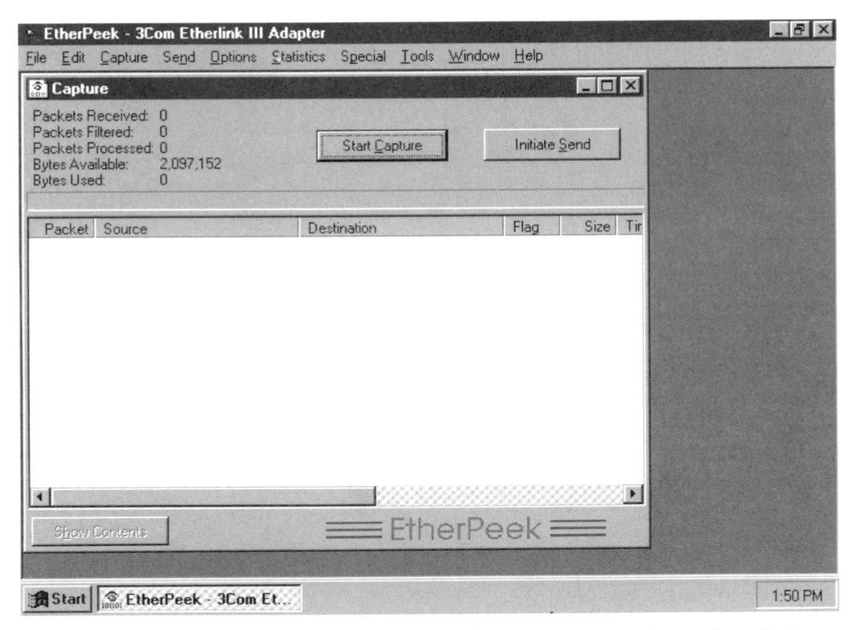

Figure 7.16 Through the EtherPeek Capture window, you can capture all packets or packets that satisfy a predefined filtering criteria

fraction of a second. Even if the computer operating EtherPeek has 64 Mbytes of memory the majority of which you could assign to the capture buffer, you would only be able to capture less than a few minutes of network activity at best. Due to this, you would more than likely want to capture network traffic based upon the use of one or more filters. In this area EtherPeek excels, providing a comprehensive filtering capability that can be used to explicitly control the capture of predefined network activity based upon protocols, addresses, and even specific bytes within a frame. Therefore, let us turn our attention to the filtering capability of EtherPeek to obtain an appreciation for the manner by which you can use this feature to selectively capture and decode the contents of specific packets in an attempt to better manage a TCP/IP network at layers 3 and 4 of the protocol suite.

Filtering Figure 7.17 illustrates the EtherPeek Filters window. Note that the upper portion of the window contains three buttons whose selection defines the manner by which filters will be used. In this example the middle button is shown selected, which will result in network activity that matches the filter we will create being captured and copied off the network and into the computer's capture buffer. The other two filter modes will result in either all filters being ignored or the capture of packets not matching the filter criteria you establish. As an example of the latter, assume you wanted to concentrate your efforts upon TCP/IP and the network transported that protocol as well as AppleTalk and NetWare. By selecting the button labeled 'Capture packets not matching checked filters' and checking those two protocol filter boxes, you would ensure anything other than AppleTalk and NetWare packets were captured.

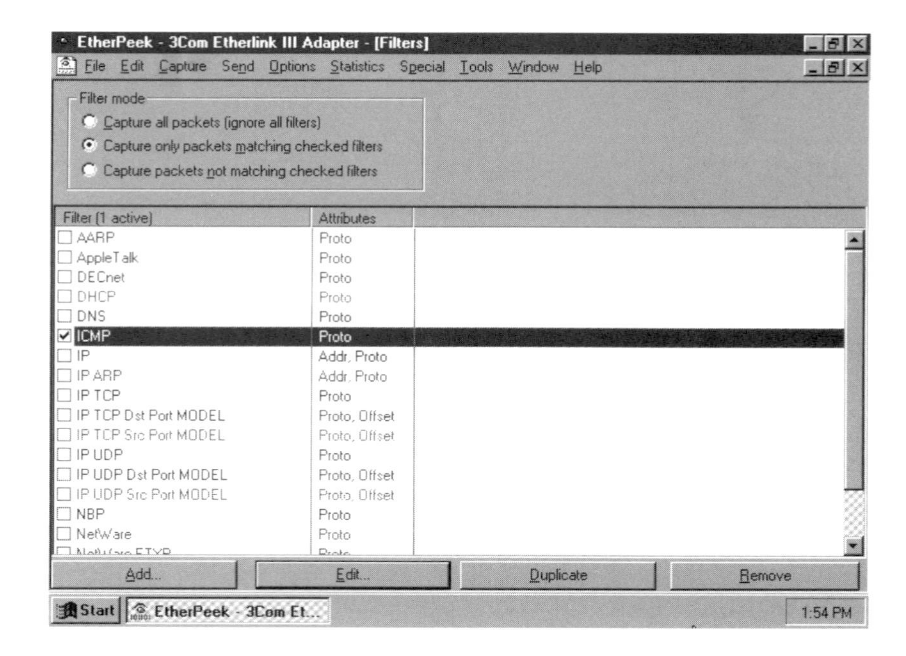

Figure 7.17 Using the EtherPeek Filters screen to initiate filtering based upon ICMP packets

In the example illustrated in Figure 7.17 we will assume that data being transported to a destination off the network is not reaching its intended destination and we want to decode ICMP packets in an attempt to gain a possible insight into the cause of the problem. Thus, we would check the box next to the ICMP entry to filter based upon ICMP packets.

If we left the entry in Figure 7.17 as it is, this would result in the capture of all ICMP packets. If we are encountering a problem with packets flowing off the network, we might wish to examine packets addressed to the router that connects the local network to a distant network or to the Internet. Thus, we would probably want to consider more narrowly defining the ICMP filter. Using EtherPeek, you would double-click on the ICMP filter entry shown in Figure 7.17. This action would result in the display of the program's Filter Settings dialog box, which is illustrated in Figure 7.18.

The Filter Settings dialog box illustrated in Figure 7.18 provides EtherPeek users with the ability to quantify filtering based upon source and destination address, packet flow direction, protocol, and an offset. In addition, because EtherPeek operates beginning at layer 2 of the protocol stack, it provides an error filter capability that allows capturing to occur based upon CRC and frame alignment errors as well as on runt and oversized packets. In our use of the Filter Settings dialog box, we will maintain the use of ICMP as the protocol filter and add an address filter such that all ICMP frames generated to or from IP address 205.131.176.1 will be captured. Thus, the address entry in Figure 7.18 shows the 'Any address' button selected for address 1 and that button not selected for address 2, with the latter assigned a specific address.

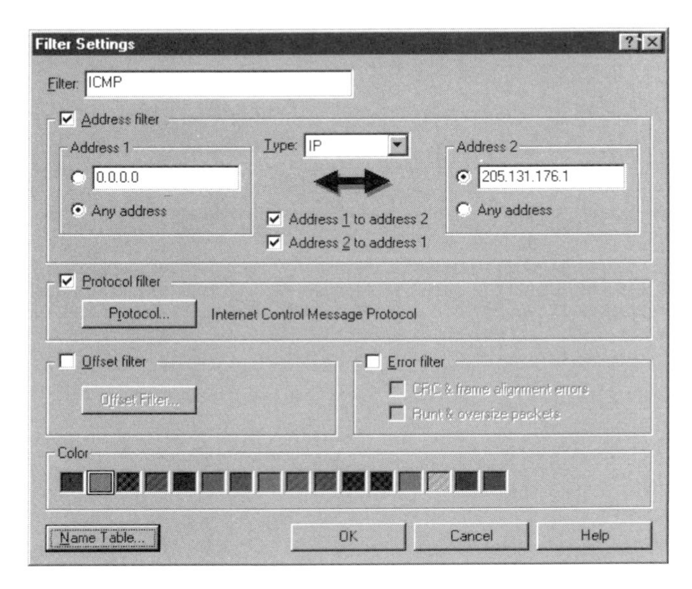

Figure 7.18 Through the Filter Settings dialog box, you can filter based upon address, protocol, an offset value, and one of two specific layer 2 error conditions

Selective packet capture Assume that we have just completed our filter and clicked on the Start Capture button previously shown in Figure 7.16. Figure 7.19 illustrates the display of the EtherPeek Capture window at a period of time after 30 730 packets have been examined and six have been found to match our previously created filter criteria. In examining the upper portion of the Capture window shown in Figure 7.19, note that only 540 bytes in the capture buffer have been used to capture the six packets that meet our filter criteria. If we did not use a filter, each packet flowing on the network would automatically be placed into the capture buffer and it would have been completely filled, probably after only one or a few packets of interest were captured. Then we would be looking for a 'needle in a haystack' in attempting to locate the few packets among many that were of specific interest. Thus, the ability to filter packets is a most important feature for any monitoring and analysis program you may wish to consider using.

Returning to Figure 7.19, the white background area in the display lists summary information about each packet captured by the program. The program numbers each packet for reference purposes, as well as denoting the source and destination address contained in the packet, any packet flag settings, and other information. By clicking on a captured packet, you can invoke EtherPeek's powerful packet decoder facility. Thus, let us click on the first packet and attempt to determine what the ICMP packet might tell us about a frame that could not leave the local network.

Packet decoding

Figure 7.20 illustrates the first of two packet decode windows generated by EtherPeek that we will examine. In actuality, both Figures 7.20 and 7.21

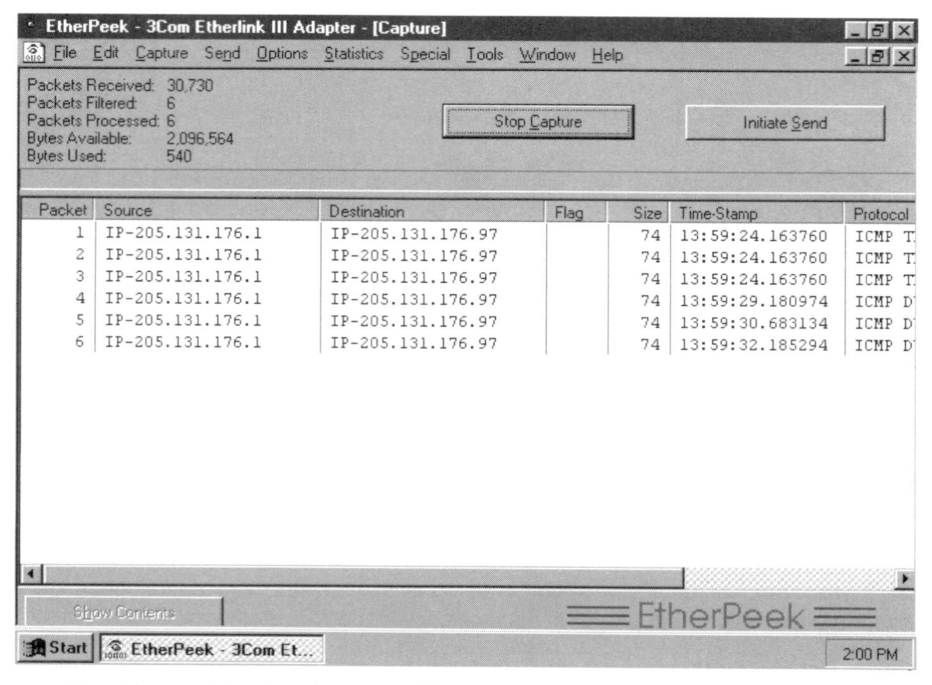

Figure 7.19 Once a filter has been established, the Capture window will display summary information about each captured packet that satisfies the filter criteria

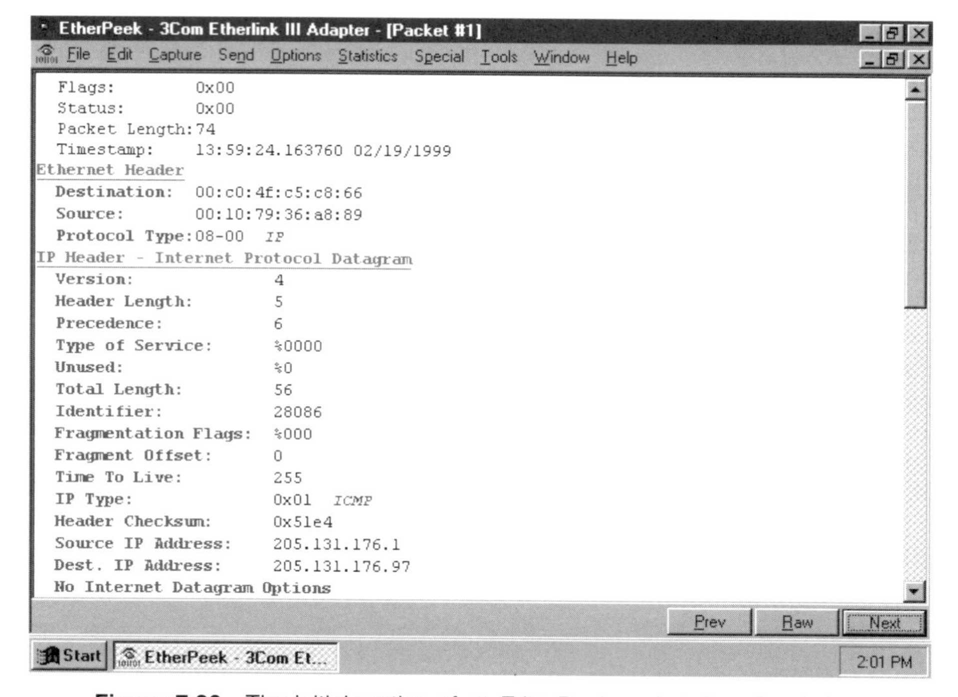

Figure 7.20 The initial portion of an EtherPeek packet decode window

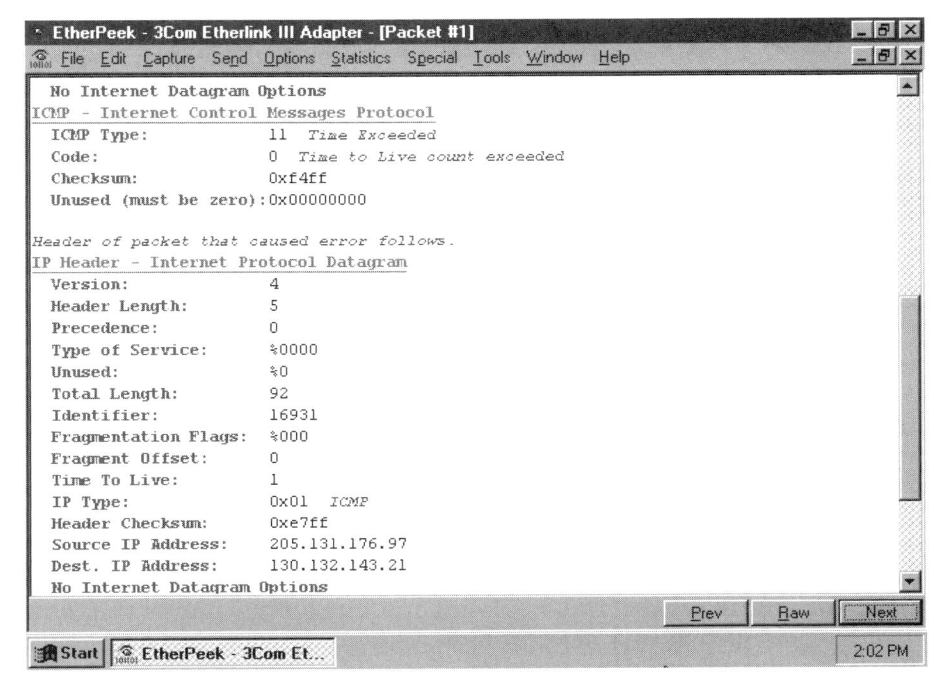

Figure 7.21 Viewing additional packet decoding by scrolling through the decode window

represent the same packet decode window, with the second screen illustrating additional details after we scroll through an initial portion of the packet decode window.

Figure 7.20 illustrates the initial portion of the packet decoding for the first packet captured in Figure 7.19. In examining Figure 7.20, note that because EtherPeek begins operations at layer 2 in the protocol stack it provides us with information about the Ethernet frame header transporting the ICMP packet. Such information includes the destination and source MAC addresses as well as the value of the Protocol Type field, which is set to hex 80 for IP. Because ICMP messages are transported by IP, this is normal.

Ethernet information is followed by a decoding of the IP header. On examining the IP header decoding shown in Figure 7.20, it appears that everything so far is normal, with the router at IP address 205.131.176.1 transmitting an ICMP message to a network station at IP address 205.131.176.97. Thus, let us scroll further down the packet decode window to view additional decoding details.

If we continue our scrolling through the packet decoding window the screen display would appear similar to that shown in Figure 7.21. Note that the decoding of the ICMP Type field indicates 'Time Exceeded,' while the ICMP code value is zero, indicating that the time-to-live count was exceeded. If we focus our attention upon the IP header that caused the error, we note that the Time-to-Live (TTL) field value is set to 1 for the datagram sent from source address 205.131.176.97 to the router for forwarding off the network to IP address 130.132.143.21. Because the router decrements the TTL value and discards packets whose TTL value reaches zero, it is doing what it is programmed to do. Thus, the problem is in the application or protocol stack at IP address 205.131.176.97, which is placing a value of 1 in the TTL field of the IP datagram! Now that we have an appreciation for the use of EtherPeek's packet capture and decoding capability, let us complete our examination of the use of this program by turning our attention to a few of its statistical screens.

7.2.2 Network statistics

By now you have probably noticed that EtherPeek has a menu labeled Statistics. Through the use of that menu, you can display a range of network-related statistical information. In concluding our examination of EtherPeek, we will turn our attention to two of several statistical displays the program is capable of generating.

Figure 7.22 illustrates the EtherPeek Network Statistics display. Note that this screen provides both a bar chart and numeric indication of network utilization, and allows you to adjust both the sample period whose default is 60 seconds and the scale of the display. Thus, the top portion of the Network Statistics screen provides a general indication of network activity similar to the WebXRay dashboard. The lower portion of the Network Statistics screen

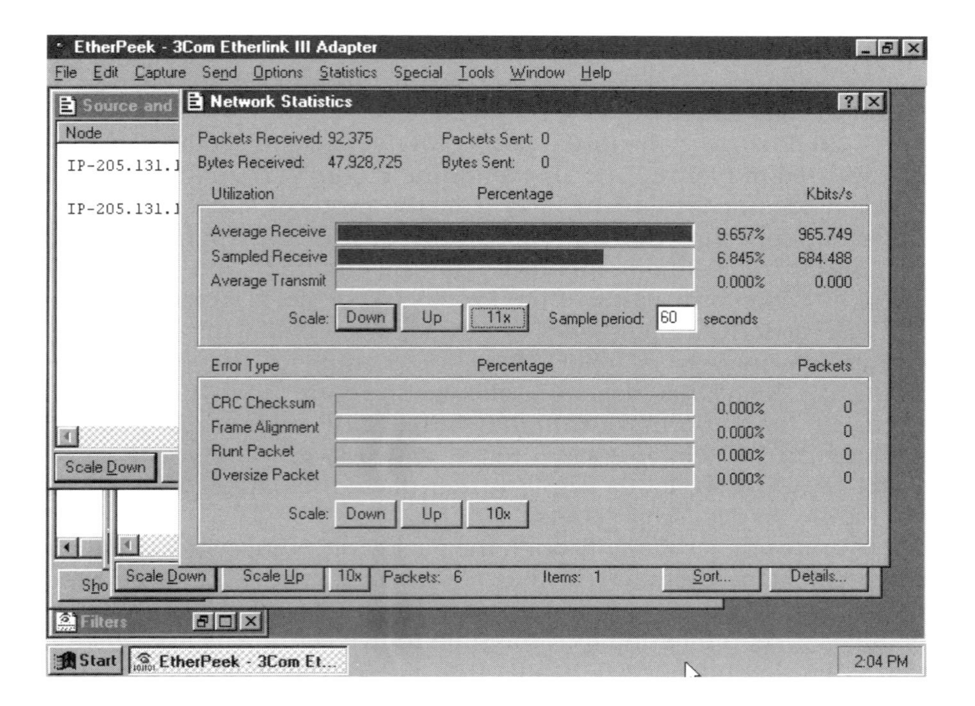

Figure 7.22 The EtherPeek Network Statistics display provides information on network utilization and layer 2 errors

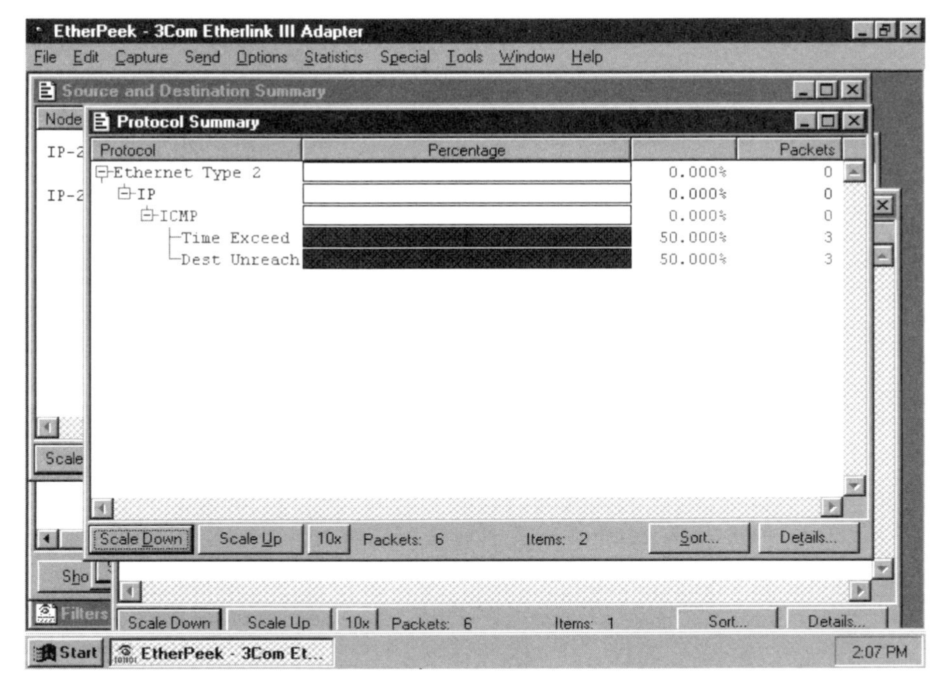

Figure 7.23 The EtherPeek Protocol Summary display

indicates common layer 2 errors. In our example no errors occurred during the sample period, resulting in an absence of erroneous activity.

Figure 7.23 illustrates the EtherPeek Protocol Summary display. Because we used a filter, this display summarizes the results of the filter, operation, indicating that half the ICMP packets were 'Time Exceeded' packets and half were 'Destination Unreachable' packets. Note that this screen also tells us that the ICMP packets were carried by IP and were delivered by Ethernet Type 2 frames. Thus, if you are monitoring a variety of packets, the Protocol Summary display can visually denote a number of significant items on one screen concerning packet delivery.

8

SNMP AND RMON

Any book covering the management of TCP/IP networks would be remiss without a detailed discussion of the role of the Simple Network Management Protocol (SNMP) and its Remote Monitoring (RMON) cousin, the latter making it practical to monitor the status of remote networks over relatively low speed wide area networking facilities. In this chapter we will first turn our attention to obtaining a detailed understanding of components of SNMP and RMON. Once this has been accomplished, we will focus our attention upon the SNMP protocol and the commands supported by different versions of the SNMP standard. The preceding information will form a foundation for examining the Management Information Base (MIB). In doing so, we will first examine the MIB in general terms and then focus our attention upon MIB II objects associated with the TCP/IP protocol suite.

8.1 SNMP AND RMON OVERVIEW

The development of SNMP to a large extent parallels the evolution of the Transmission Control Protocol/Internet Protocol (TCP/IP) protocol suite. A desire to monitor the performance of protocol gateways linking individual networks to the Internet resulted in the development of the Simple Gateway Monitoring Protocol (SGMP), which can be viewed as the predecessor of SNMP. The need for changes and improvements to SGMP resulted in the Internet Activities Board (IAB), which in 1992 was renamed the Internet Architecture Board, recommending the development of an expanded Internet network management standard in a Request for Comment (RFC). Under the auspices of the IAB, the Internet Engineering Task Force (IETF) became responsible for designing, testing, and implementing the new Internet network management standard. The result of the efforts of a group of IETF researchers and engineers was the publication of three RFCs in August, 1988, which formed the basis of SNMP.

Table 8.1 lists the initial RFCs that formed the basis for SNMP. It should be noted that RFCs are not static documents, as they go through several stages of review and refinement prior to being adopted as standards by the Internet community. Once standardized, over time they will commonly be superseded by another RFC.

Table 8.1 Original RFCs defining SNMP

RFC 1065	Structure and Identification of Management Information for TCP/IP-based internets.
RFC 1066	Management Information Base for Network Management of TCP/IP-based internets.
RFC 1067	A Simple Network Management Protocol

The original version of SNMP, which is usually described without reference to a version number, lacked authentication and security. These key limitations resulted in the establishment of an IETF working group that attempted to rectify many limitations of SNMP. The resulting standard, which is referred to as SNMPv2, added some additional commands beyond a core of five commands, supported by the first version of SNMP. However, arguments concerning the manner in which authentication and encryption should be added to the standard resulted in the failure to implement security in version 2. Recognizing the need for authentication and encryption resulted in the IETF forming another working group, which selected the best work of two competing groups for security and added security to the basic efforts of the SNMPv2 standard. The resulting work, which is referred to as SNMPv3, was a proposed standard when this book was being written, and will probably be standardized during late 1999. However, due to the failure of SNMPv2 to be adopted by more than a few vendors, the original version of SNMP will more than likely remain the most popular of this management protocol for the foreseeable future. In this chapter we will primarily focus our attention upon the original version of SNMP; however, when applicable, information concerning SNMPv2 and SNMPv3 will be discussed.

8.1.1 Basic architecture

An SNMP-based network management system consists of three components: a manager, agent, and a database referred to as a Management Information Base (MIB). Although SNMP is a protocol that governs the transfer of information between its three entities, it also defines a client–server relationship. Here the client program is the manager, while the agent that executes on a remote device can be considered to represent a server. Then, the database controlled by the SNMP agent represents the SNMP MIB. Figure 8.1 illustrates the general relationship between the three SNMP components.

Manager

The manager is a program that operates on one or more host computers. Depending upon its configuration, each manger can be used to manage a different subnet, or multiple managers can be used to manage the same subnet or a common network. The actual interaction between an end-user and the manager is obtained through the use of one or more application

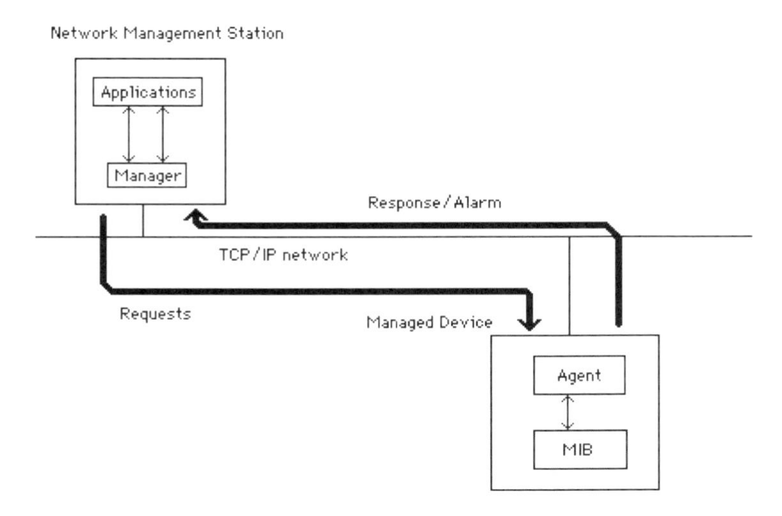

Figure 8.1 The general relationship between SNMP components

programs that, together with the manager, turn the hardware platform into a Network Management Station (NMS). Today, in the era of Graphic User Interface (GUI) programs, almost all application programs provide a point-and-click window environment that interoperates with the manager to generate graphs and charts providing visual summaries of network activities.

Through the manager, requests are transmitted to one or more managed devices. Originally SNMP was developed to be used on TCP/IP networks and those networks continue to provide the transport for the vast majority of SNMP-based network management products. However, SNMP can also be transported via NetWare IPX and other transport mechanisms.

Agents

Each managed device includes software or firmware in the form of code that interprets SNMP requests and responds to those requests. The software or firmware is referred to as an agent. Although a device must include an agent to be directly managed, non-SNMP compatible devices can also be managed if they support a proprietary management protocol. To accomplish this, you must obtain a proxy agent. The proxy agent can be viewed as a protocol converter, as it translates SNMP requests into the proprietary management protocol of the non-SNMP device.

Although SNMP is primarily a poll-response protocol, with requests generated by the manager resulting in agent responses, the agent also has the ability to initiate an 'unsolicited response.' That unsolicited response is an alarm condition resulting from the agent monitoring a predefined activity and noting that a predefined threshold was reached. Under SNMP, that alarm transmission is referred to as a trap.

Management Information Base

Each managed device can have a variety of configuration, status, and statistical information that defines its functionality and operational capability. This information can include hardware switch settings, variable values stored as data in-memory tables, records or fields in records stored in files, and similar variables or data elements. Collectively, those data elements are referred to as the Management Information Base (MIB) of the managed device. Individually, each variable data element is referred to as a managed object, and consists of a name, one or more attributes, and a set of operations that can be performed on the object. Thus, the MIB defines the type of information that can be retrieved from a managed device and the device settings you can control from a management system.

8.1.2 RMON

One of the problems associated with SNMP is the fact that its request–response (poll–select) operation, while having a relatively minor effect on the utilization of the bandwidth of a LAN, can result in the significant degradation of lower operating rate WAN bandwidth when monitoring geographically separated networks. Recognizing this problem, the Remote Network Monitoring Working Group of the IETF developed the Remote Monitoring (RMON) network management standard.

Probes and agents

RMON represents an extension of the network manager's operation to distant networks. At those networks intelligent devices known as probes or RMON agents monitor the data flowing on the remote network, organizing it into information the manager can easily access and interpret, with SNMP used as the transport mechanism between the manager and agent. Figure 8.2 illustrates the relationship between a network management station on one LAN used to manage a distant network through the use of an RMON agent or probe. Since the remote probe monitors and organizes data traffic occurring on the distant network, this considerably reduces the amount of information that would otherwise be required to be transmitted to the network management station for analysis. As the WAN circuit used to connect the geographically separated networks normally operates at a fraction of the operating rate of a LAN, the use of an RMON probe also reduces the potential for the saturation of the bandwidth of the WAN circuit that could occur if each device on the remote network had to be individually polled.

MIBs

Each RMON agent or probe includes an MIB that defines the attributes of the objects being monitored. The first RMON MIB, RFC 1271, was published in

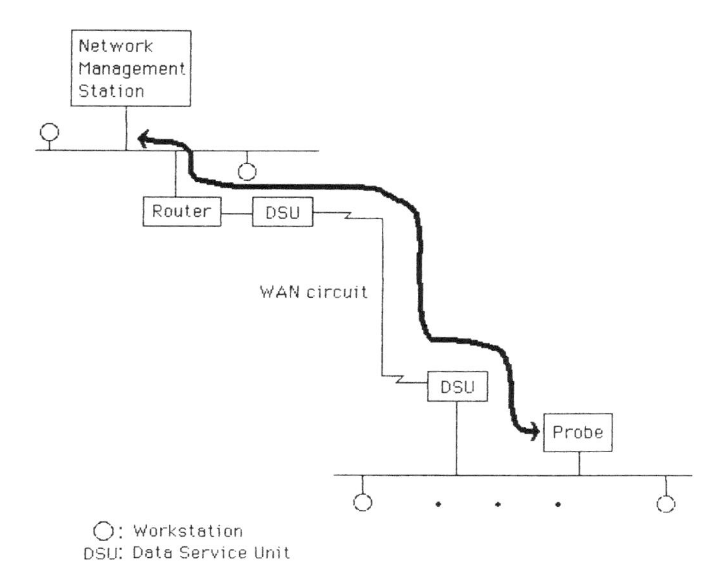

Figure 8.2 Extending network management to a remote location via an RMON probe

November 1991 and was limited to Ethernet LANs. RFC 1513, which was published in September 1993, extended RMON to Token-Ring networks. Extensions to other types of networks were developed after 1993.

Although SNMP and RMON MIBs define the attributes of objects that are monitored, the actual value and use of information is highly dependent upon the application program that controls the manager. Some application programs may support a full complement of the network management functional areas and tasks, while other applications may be limited to supporting a subset of those functional areas and tasks. Thus, it is extremely important to evaluate the capability and functionality of the application as well as its support of SNMP and RMON.

Operation

An example of the use of a GUI display supporting RMON operations is shown in Figure 8.3. This example of a Network General (acquired by Network Associates) Foundation Manager Cornerstone Agent monitoring a Token-Ring network. The Cornerstone Agent represents a RMON probe that can be connected to a network and accessed either directly or remotely via a management station located on the same network or a distant network. The Cornerstone Agent represents a software-based probe that is loaded and operates on a PC that is connected to a LAN and uses an adapter card that operates in promiscuous mode, meaning that it reads each frame flowing on the LAN in order to gather statistics that can be displayed directly by accessing the PC or remotely via a network management station. Other probes are firmware-based products with a network adapter that is

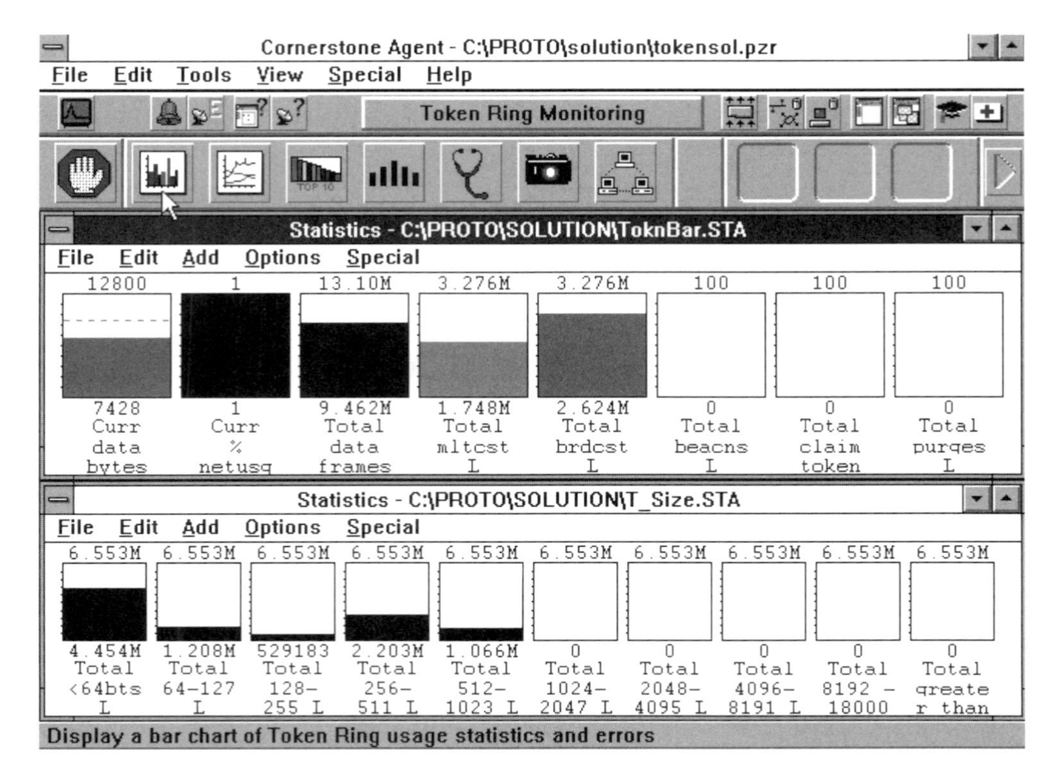

Figure 8.3 A Network General Foundation Manager Cornerstone Agent screen displaying statistical information about the operation of a Token-Ring network

connected to a LAN and can only be accessed through a network management station.

In examining Figure 8.3, note the display of two horizontal series of bar charts. Each series is displayed by clicking an icon, as the Cornerstone Agent provides a graphical shell which allows a series of operations to be performed based upon an icon selection. The top horizontal series of bar charts provides statistics concerning the percentage of Token-Ring network usage and traffic information in the form of 6 total data bytes, data frames, multicast frames, broadcast frames, beacons, claim tokens, and ring purges. The second series of statistics, which is displayed along the bottom of the screen, provides a distribution of traffic based upon frame length. Thus, simply clicking on icons on some GUI-based SNMP network managers can result in the display of information in a graphical format from which abnormalities may be easy to observe.

Evolution

Until the late 1990s RMON was restricted to layer 2 operations. This resulted in probes returning data that provided statistics concerning the flow of frames on LANs. Recognizing the necessity to provide information concerning the flow of data at higher layers in the protocol stack, several vendors introduced proprietary 'extended' RMON probes during the mid-to-late 1990s. Such probes looked further into each frame and provided statistics and other information covering the network layer, transport layer, and other higher layers in the protocol stack. By the late 1990s RMON version 2 was standardized, which extended remote monitoring above the data link layer. In addition to providing high layer information, RMON Version 2 probes do not have to be placed on each segment to provide information on the flow of traffic between networks. This is because monitoring at the network layer enables a probe to log packets flowing between segments. In comparison, under the original version of RMON, monitoring at the data link layer results in routers blocking layer 2 data at their boundaries and requires probes to be located on each segment. Thus, RMON version 2 probes provide a degree of monitoring flexibility beyond that obtainable from the earlier version of the remote monitoring standard.

8.2 THE SNMP PROTOCOL

In the previous section we have provided an overview of the basic components of SNMP and their relationship to one another. Although we noted that the manager and agent communicate with one another in essentially a client–server environment, we did not focus upon the specific mechanism by which they communicate. In this section we will do so, examining the commands supported by SNMP as well as the mechanism by which those commands are transported. In doing so, we will discuss the operation of the originally developed SNMP as well as SNMP versions 2 and 3, both of which were developed to address some of the inadequacies of their predecessors.

8.2.1 Basic SNMP commands

Under SNMP version 1, five types of commands or verbs, transported in a formatted message referred to as Protocol Data Units (PDUs), are defined. Table 8.2 lists each SNMP command as well as providing a short description of the operation of each command. Figure 8.4 illustrates the basic data flow of each command with respect to the manager and agent. From the

Table 8.2 SNMP version 1 commands

Command	Operational Result
GetRequest	Request the values of one or more Management Information Base (MIB) variables
GetNextRequest	Enables MIB variables to be read sequentially, one variable at a time
SetRequest	Permits one or more MIB values to be updated
GetResponse	Used to respond to a GetRequest, GetNext Request, or SetRequest
Trap	Indicates the occurrence of a predefined condition

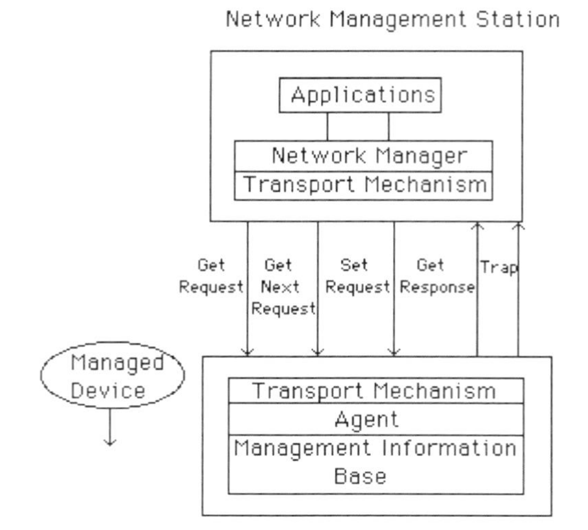

Figure 8.4 SNMP version 1 command flow

information provided in Table 8.2 and Figure 8.4, we can summarize the operation of each of the five initial SNMP commands as follows.

GetRequest

The GetRequest command is issued by the network manager to an agent. This command can be used to read a single MIB variable or a list of MIB variables from the destination agent. The GetRequest, as well as all other SNMP commands, requires the use of two addresses. The first address is that of a manager or agent. Since SNMP is transported by IP, this means that an IP address must be associated with each SNMP command. The second address represents the location of the MIB variable or object. As we will note later in this chapter, a standard tree structure defines the location of MIB variables.

GetNextRequest

The GetNextRequest command is similar to the GetRequest command; however, upon receipt by the agent the next entry in the MIB will be retrieved. Under SNMP, a variable stored at a device is technically referred to as an individual instance of a managed object. Thus, the GetNextRequest command results in the agent attempting to retrieve the next larger value than the requested managed object instance. However, in the remainder of this chapter we will simply reference entries in an MIB as an object, as the instance of an object, while technically correct, is a bit awkward to remember.

By issuing a sequence of GetNextRequest commands with incrementing values for the managed object, the network manager obtains the capability to 'walk' through an agent's MIB. One common use of the GetRequest command is to read the first value in a row, while a series of GetNextRequest commands are then used to read sequentially through the row, with the incrementation of the values for the requested managed object being used to control which entries in the MIB are retrieved.

SetRequest

The SetRequest command is similar to the GetRequest and GetNextRequest commands in that all three are issued by a network manager to a defined agent. Unlike the GetRequest and GetNextRequest commands, which seek to extract information from an MIB, the SetRequest command is used to request the agent to set the value of a managed object or objects contained in the command. Whether or not the command is successful depends upon several factors, including the existence of the managed object and whether or not it has an access mode of write-only or read–write. For example, a read-only mode would result in the failure of a SetRequest command, since the value of the managed object or objects could not be changed. Another fact that governs the ability of a management station to successfully issue a Set Request command involves the appropriate setting of a Community Name.

Under the original version of SNMP, an elementary form of security in the form of matching strings associated with Get, Set, and Trap Community Names occurs. Figure 8.5 illustrates an example of the Network General Cornerstone Agent RMON probe SNMP configuration screen. Note that the default's community name is 'public.' Because it is a relatively simple process to write a script to use each entry in an electronic dictionary against the configured community name associated with an RMON probe or SNMP management station, it is also relatively easy for a hacker to obtain a valid community name and use it with a SetRequest command to reset an organization's router, change a router's routing table, or perform another function that would curl the hair on the head of a network manager or LAN administrator. Due to this possibility, many SNMP-compliant products were designed either not to support SetRequest commands or by default disable the operation of this command.

GetResponse

As indicated in Figure 8.4, the flow of the GetResponse command is from the agent to the network manager. Thus, this command provides the mechanism by which an agent responds to GetRequest, GetNextRequest, and SetRequest commands.

Information returned by the GetResponse command as a GetResponse PDU consists of several fields that enable the response to be correlated to a previously received command, note whether or not the received command

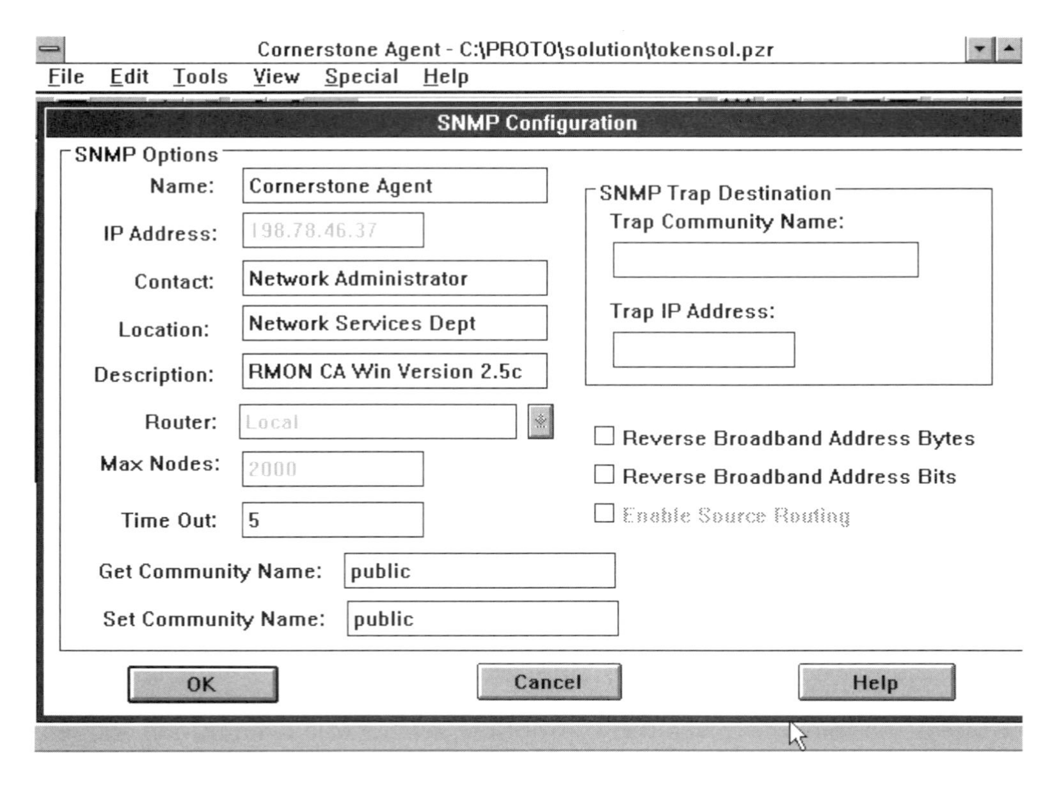

Figure 8.5 The original version of SNMP uses community name strings as a mechanism to vertify command requests

was successfully processed, and, if successfully processed, return a list of instances of the managed objects that were operated upon by the received command and their current values. Thus, the latter provides a mechanism to verify that a command issued to an agent was successfully executed as well as to determine the current value of one or more affected variables in the MIB.

Trap

Unlike the previously described commands, which are generated in response to a manager or agent request, the Trap command is unsolicited. Through the use of this command, an agent can take the initiative to inform the manager of the occurrence of a predefined condition, such as a Cold Start or Warm Start of equipment or a Link Down or Link Up condition.

8.2.2 SNMP version 2

One of the problems associated with SNMP version 1 was its lack of built-in security. This means that it could be relatively easy for the unintentional or intentional use of the SetRequest command to corrupt the configuration parameters of a managed device, which in turn could seriously impair network operations. As previously discussed, due to this problem some communications hardware and software developers elected to disable the ability of an agent within their SNMP implementation to process SetRequest commands. The original goal of SNMP version 2 (SNMPv2) was to alleviate this problem through the addition of encryption and authentication. Encryption would be used to protect the contents of messages, while authentication would be used to verify the originator of a message.

SNMPv2 dates to the fall of 1992, when the IETF formed two working groups to define enhancements to SNMPv1. One working group focused upon defining security functions, while the other focused its efforts upon defining enhancements to the SNMP protocol. Unfortunately, the working group tasked with developing security enhancements broke into separate camps with divergent views concerning the manner by which security should be implemented. Separate proposals for the implementation of encryption and authentication resulted in the emergence of two proposals referred to as SNMPv2μ and SNMPv2*. Due to a disagreement that appeared unresolvable at the time, SNMP version 2 proceeded to omit security enhancements and concentrated on improving several areas of weakness in the original version of SNMP. The resulting effort, referred to as SNMPv2, was issued in May, 1993. This version of SNMP is also referred to as SNMPv2c, where the 'c' references the fact that community names continues to represent the only security mechanism employed by this newer version of SNMP.

New features

SNMPv2 incorporates a manager-to-manager capability and two new PDUs. The manager-to-manager capability permits SNMP to support distributed network management in which one network management station can report management information to another management station. In comparison, SNMPv1 is restricted to supporting a manager-to-manager model.

To support effective manager-to-manager interactions, SNMPv2 added Alarm and Event groups to the SNMP MIB. The Alarm group permits thresholds to be established that, when crossed, will result in the initiation of alarm messages. The Event group specifies when a trap should be issued based upon one or more MIB element values.

The PDUs added to SNMPv2 are GetBulkRequest and InformRequest. Those PDUs supplement the five commands developed under SNMPv1, which were slightly improved with respect to their operations and operational capability under SNMPv2. In addition, the GetResponse command was renamed Response, which to many persons appears to be a much more appropriate name.

Two additional enhancements associated with SNMPv2 concern error processing and counter capability. Under SNMPv1, the attempt to retrieve data beyond the last element of a row would result in the termination of the retrieval request and the generation of an error message without any requested information being sent to the requester. Under SNMPv2, error processing was considerably enhanced, resulting in the return of all valid requested objects followed by an error message that would define the problem. This allows a management station to be programmed to adjust its retrieval method to the size of different tables that could represent new or proprietary additions to an MIB without having to simply stop its retrieval effort. Concerning counter capability, under SNMPv1, counters are 32 bits in width. While their counter width is sufficient to maintain statistics for a reasonable period of time for 10-Mbps Ethernet and 16-Mbps Token-Ring networks, the introduction of 100-Mbps Fast Ethernet, 100-Mbps Token-Ring, and Gigabit Ethernet LANs resulted in counters overflowing in a relatively short period of time. Recognizing this problem, SNMPv2 uses 64-bit counters to maintain statistics.

GetBulkRequest

The GetBulkRequest command functions similarly to the GetNextRequest command, with one key exception. Unlike the GetNextRequest command, which requires the amount of data to be retrieved to be specified, the GetBulkRequest tells the agent to return as much data as possible that can fit into a response message commencing with the next larger value than the requested managed object. Not only does the GetBulkRequest make more efficient use of the transmission facility but, in addition, its use can reduce or eliminate certain error conditions that can occur if a GetRequest command is used. For example, if a GetRequest command requests too much data, the GetResponse PDU will return a 'too big' error message without returning any data.

InformRequest

The InformRequest command provides SNMPv2 with the ability to support a hierarchy of network management stations. As previously discussed, the use of this command enables one management station to communicate with another management station, a feature conspicuous by its absence under SNMPv1.

Under SNMPv1, you can establish a series of management stations, with each station used to control one or more independently managed network segments. Since there is no standardized mechanism that enables each management station to communicate with other management stations, the construction of an Enterprise management station was based upon a proprietary design. The need for a proprietary system was eliminated under SNMPv2, as the InformRequest command standardizes communications between management stations.

Figure 8.6 illustrates the use of InformRequest to construct an Enterprise Management System. In this example, a two-level hierarchy of network management stations was created, with an Enterprise network manager at the top of the hierarchy. Through the use of the InformRequest command, the local network managers, which are local with respect to the networks being managed but which can be geographically distant with respect to the Enterprise Management system or other local network managers, can communicate with one another.

Similar to its predecessor, under SNMPv2 alarm conditions can be established when thresholds are reached which generate an event. However, under SNMPv2 the resulting event can be used to initiate the transmission of an InformRequest command which informs another management station of the occurrence of the predefined condition. Thus, through the establishment of a multi-level hierarchy of management stations, one or more stations can be informed of important conditions occurring at distant locations. In fact, with appropriate application software, information can be transferred on a timed basis so that the 'sun never sets' on the operational status of a truly global network. That is, one Enterprise network manager station could be continuously staffed while other network management stations are programmed to report information to the Enterprise manager only during those periods of time when they are unmanned.

8.2.3 SNMPv3

As previously discussed when we examined SNMPv2, there were two competing camps regarding security. SNMPv2μ supported the early

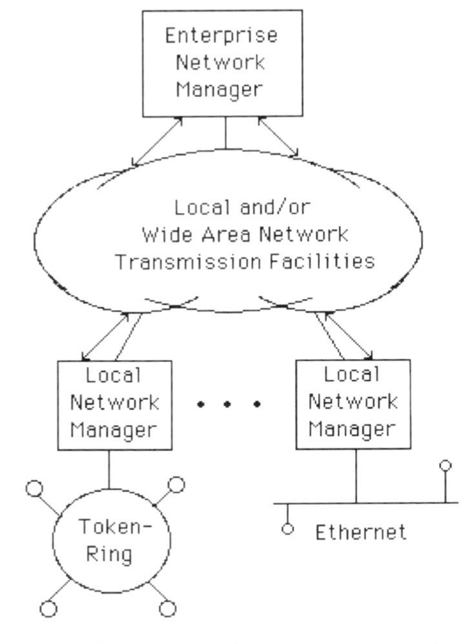

Figure 8.6 Through the use of the InformRequest command, communications between network managers becomes possible, permitting the construction of a multi-tier Enterprise network manager

standardization of security features and proposed a deferred standardization of features associated with the management of large networks to facilitate the rapid deployment of simple agents. In comparison, SNMPv2* supported the concurrent standardization of security and administrative features. Although the intentions of backers of both SNMPv2μ and SNMPv2* were admirable, their inability to compromise upon a single solution resulted in the formation of a Security and Administrative Framework Evolution Advisory Team in August, 1996 to recommend a single approach that would resolve the differences between the two versions. The team published its recommendations at an IETF meeting in December, 1996, which resulted in the chartering of the SNMPv3 working group in March, 1997. The goal of the SNMPv3 working group was to continue the effort of the disbanded SNMPv2 working group to define a standard for SNMP security and administration.

SNMPv3 specifications were published during January, 1998 as Proposed Standards in a set of five RFCs (RFC 2271 through RFC 2275). In October, 1998 SNMPv3 specifications were submitted as Draft Standard RFCs and should be finalized by the time this book is published. SNMPv3 uses almost all the prior efforts of SNMPv2 as is, primarily adding the ability to enhance security through a modular architecture that includes security and access control subsystems. Due to this, SNMPv3 is commonly referred to as SNMPv2 with security.

Architecture

SNMPv3 represents a more complex architecture than earlier versions, as it includes backward support for SNMPv1 and SNMPv2. In addition, the architecture of SNMPv3 is structured to support future unknown enhancements through the use of modular design. That modular design also enables software developers to design modules to support different levels of implementation commensurate with available resources. For example, a vendor could develop a minimal level of implementation to support environments that have resource constraints, while more robust implementation could be developed to support environments where resources are less constrained.

Figure 8.7 illustrates the modular architecture of SNMPv3. Under this modular architecture, there were a few changes in terminology to facilitate the application of the architecture to both managers and agents. One change in terminology is the use of the term ‘engine’ to reference the device that provides services for transmitting and receiving messages, authenticating and encrypting messages, and controlling access to managed objects. A second change in terminology involves the use of the term ‘SNMP entity’ to refer to both managers and stations. Although the modular architecture shown in Figure 8.7 is applicable to both managers and agents, under SNMPv3 the manager will contain a command generator and/or notification receiver application module, while the agent will contain command responder and notification originator application modules.

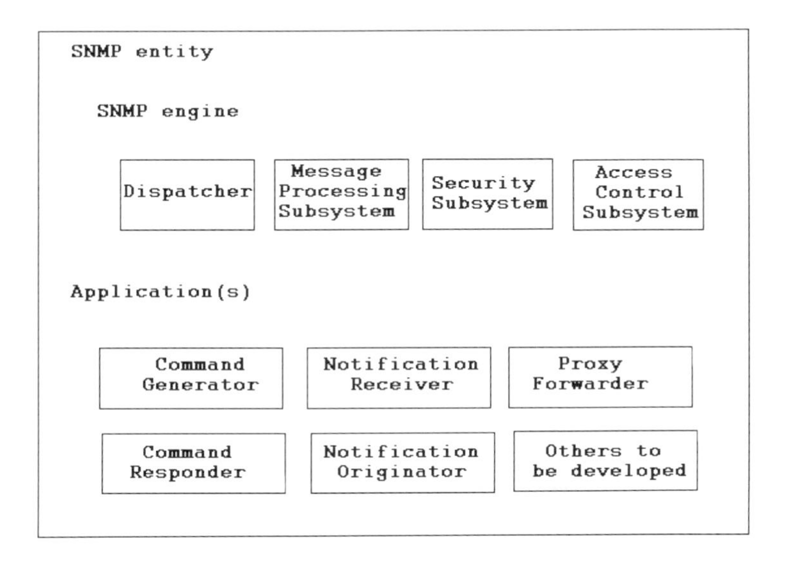

Figure 8.7 SNMPv3's modular architecture

SNMP engine modules

As indicated in Figure 8.7, there are four key SNMP engine modules. The Dispatcher is responsible for coordinating communications between subsystems and directing incoming messages to appropriate applications. The Message Processing Subsystem can include one or more message processing modules. Through the use of appropriate modules, multiple message formats can be supported for SNMP entities with sufficient resources. Thus, when resources are available, SNMPv3 can be backward-compatible with SNMPv1 and SNMPv2 message formats as well as forward compatible with future, to-be-developed message formats. The Security Subsystem can also consist of a variable number of modules included to match available resources and user requirements. One defined module is almost an exact duplicate of the User-based Security Module (USM) defined in SNMPv2μ and SNMPv2*. Through the ability to incorporate different security modules, it becomes possible to support SNMPv1 community names as well as the more robust authentication and encryption schemes supported by SNMPv3.

The fourth component of the SNMP engine is an Access Control Subsystem. This subsystem involves the security process applicable for access control, and is separate from the Security Subsystem, which involves the support of authentication and encryption for message verification and message security.

Application modules

Because the architecture of the SNMPv3 is modular, applications are similar to the basic engine, and consist of one or more modules. Applications use the services of engine modules for transmitting and receiving messages, authenticating and encrypting messages, and controlling access to managed devices. As illustrated in the lower portion of Figure 8.7, there are presently five defined SNMPv3 application modules.

These include command generators, command responders, notification originators, notification receivers, and proxy forwarders. Command generators monitor and manipulate management data, while command responders provide access to management data. Notification originators initiate asynchronous messages, while notification receivers process asynchronous messages. The fifth module, proxy forwarders, forward messages between entities.

Operation

Because SNMPv3 uses a modular architecture, its operation depends upon the modules incorporated into its management stations and agents. The key to SNMPv3's ability to interoperate with SNMPv1 and SNMPv2 devices is its Message Processing Subsystem, which was previously illustrated in Figure 8.7. That subsystem can include one or more modules, as illustrated in Figure 8.8. In examining the SNMPv3 message processing subsystem shown in Figure 8.8, note that the inclusion or omission of a particular module defines the compatibility of a particular vendor product with legacy as well as possible future to-be-developed versions of the protocol.

Similar to the message processing subsystem, the Security Subsystem and Access Control Subsystem are also modular in design. Thus, those subsystems can be configured to include one or more modules that will govern security and access control operations.

8.3 UNDERSTANDING THE MIB

The Management Information Base (MIB) can be viewed as a database whose structure and elements are both standardized. The process by which the MIB is standardized was originally defined in RFC 1065, which was published in August, 1988 and updated by a rerelease of that RFC as RFC 1155 in May, 1990. Although the official title of both RFCs is 'Structure and Identification of Management Information for TCP/IP-based internets,' they are commonly

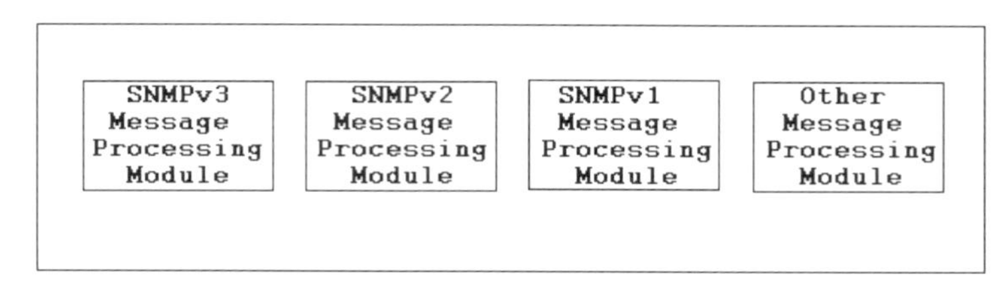

Figure 8.8 The SNMPv3 Message Processing Subsystem. Through the inclusion or omission of different message processing modules, SNMPv3 obtains its ability to support other versions of protocol

referenced as the Structure of Management Information (SMI). SMI denotes such information as how MIB variables in an MIB are related to one another, how variables are formatted, and similar information necessary for obtaining the standardization of the MIB.

8.3.1 The object identifier

As noted earlier in this chapter, the ability to retrieve an object from the MIB requires the specification of two addresses. One address represents the network address of the device that maintains the MIB data. In a TCP/IP environment that address is the IP address of the device. The second address represents the address of the object and is referred to as an object identifier.

The object identifier provides a mechanism for identifying a node within an inverted tree structure. It consists of a sequence of non-negative integers that identifies an object (node) within the inverted tree. That object can represent a registration authority, an algorithm, or another identifier based upon the entries in the naming tree. To illustrate the use of the object identifier requires a tree; thus, for illustrative purposes we need one to illustrate how to retrieve objects from an MIB. Fortunately, the International Organization for Standardization (ISO) and the International Telecommunications Union (ITU), Telecommunications section (ITU-T)—the latter until recently known as the Consultative Committee for International Telephone and Telegraph (CCITT)—coordinated the development of a global naming tree. This tree, which is partially illustrated in Figure 8.9, provides a mechanism for assigning an identifier to any object that requires naming.

The global naming tree illustrated in Figure 8.9 provides both a labeling mechanism and an identifier mechanism since each object is assigned to a specific node under the root of the tree. Then, describing an object identifier is accomplished by traversing the tree, starting at its root, until the intended object is reached. Several formats can be used to describe an object identifier, with integer values separated by dots most commonly used. That is, to identify the US Department of Defense (DOD), its object value in integer notation would become

1.3.6

The preceding integer string identifies the object DOD located by starting at the root of the tree and first moving to subordinate node 1, followed by moving to the subordinate node 3 under the first subordinate node. Next, moving to the sixth subordinate node under subordinate node 3 results in an integer string that uniquely identifies the US Department of Defense.

A second method used to identify an object in the global naming tree is accomplished through the use of a text string. This method results in the use of a series of text labels separated from one another by underscores. Thus, DOD would be identified by the following text string:

iso_identifier-organization_dod

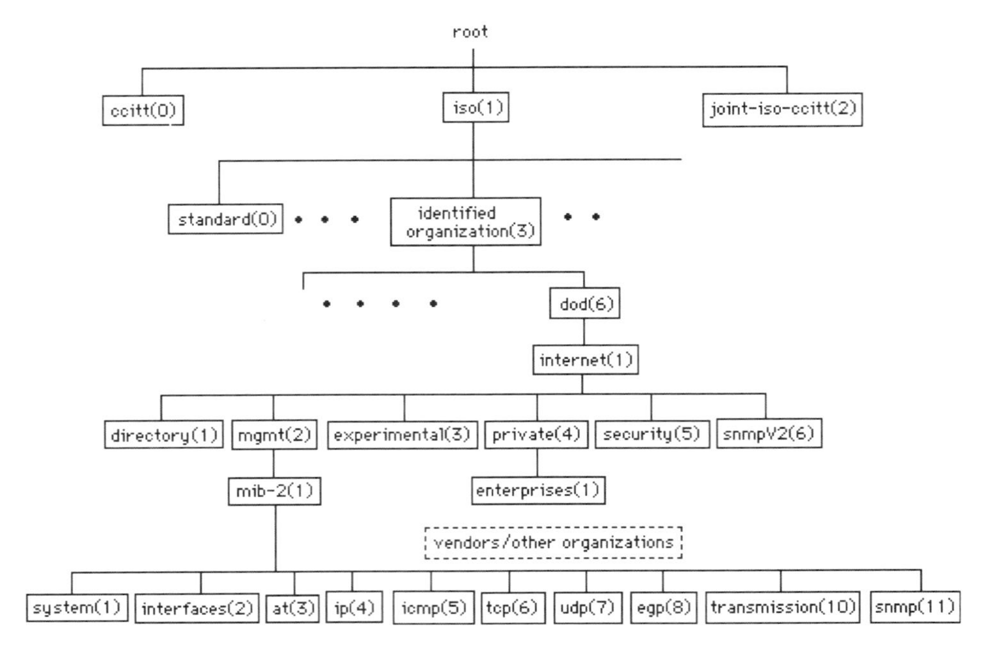

Figure 8.9 The ISO/ITU-T (CCITT) global naming tree

In examining the global naming tree illustrated in Figure 8.9, you will note that, as you might expect from a structure developed jointly by the ITU-T and ISO, two top level nodes are administered by each organization as a separate entity and one node as a jointly administered entity. Under the ISO-administered entity, that organization defined the org node as a mechanism to delegate authority to other organizations. One of those organizations just happens to be the US Department of Defense (DOD), which funded the development of ARPANet, the predecessor to the Internet. Today the Internet subtree under the DOD node is controlled by the Internet Architecture Board (IAB) and administered by the Internet Assigned Numbers Authority (IANA). As indicated in Figure 8.9, there are six nodes defined under the Internet subtree. Of particular interest is the mgmt (management) subtree, since that subtree is the placement area for approved standard network management variables. You should note that until a standard network management variable has been approved, it is placed under the experimental node. If, after a trial period, it is accepted as a standard, it is then moved to reside under the mgmt node.

8.3.2 Structure and identification of management information

As noted at the beginning of this section, the Structure and Identification of Management Information is the mechanism that defines the rules for describing management information. To accomplish this, RFC 1065, and its successor, RFC 1155, whose titles are similar to the heading of this section, provide formal descriptions of the defined management structure.

RFC 1065 and its successor, RFC 1155, are primarily focused upon two key areas. First, each RFC defines a naming structure for identifying managed objects. Secondly, each RFC defines a format to define managed objects. Although neither RFC defines objects in the MIB, they specify the format used by other RFCs to define network management objects. Due to this, those RFCs provide the framework for managing TCP/IP-based internets, commonly referred to as the Structure of Management Information (SMI). In this section we will review that portion of the global naming tree applicable to Internet network management.

8.3.3 Network management subtrees

In Figure 8.9 a portion of the global naming tree was illustrated. As noted, a sequence of integers that form the path from the root through one or more nodes to an object represents an object identifier. Currently there are six subtrees or nodes defined under the Internet node, of which three are most applicable for SNMP network management purposes. Those nodes or subtrees are mgmt, experimental, and private. Each of those nodes or subtrees are identified by specifying the path to the node or subtree. Thus, the directory node would be identified as

directory object identifier : := {1.3.6.1.1.}

As an alternative, since the directory node is the first node under the Internet node, it can be identified as follows using a combination of text and numeric identifiers:

directory object identifier : := {Internet 1}

The mgmt subtree

The mgmt subtree represents the location where objects defined in IAB approved documents are placed. That is, once an RFC that defines a new version of the Internet-standard MIB has been approved, it is assigned an object identifier under the mgmt subtree. Currently only one subtree, mib-2, is defined. That RFC uses the object identifier

{mgmt 1}

which can also be represented by the path 1.3.6.1.2.1.

The experimental subtree

The experimental subtree represents the location of objects used in Internet experiments. Newly defined objects are normally placed under the experimental subtree until they successfully pass a trial period during which any necessary revisions are performed. Once this has been accomplished, the objects are moved under the mgmt subtree.

Objects defined under the experimental subtree have the prefix {Internet 3}, which could also be specified as the path 1.3.6.1.3.

The private subtree

The private subtree represents a mechanism that enables a standardized expansion of SNMP to accommodate hardware and software developers, universities, and even government entities to define new MIB objects. To accomplish this, the subtree enterprises under the private node is used by the Assigned Numbers authority for the Internet as a mechanism for registering vendor- or organization-specific MIB objects. To accomplish this, the prefix {private 1} that represents the path 1.3.6.1.4.1 in the global naming tree forms the node under which a vendor's or organization's registered MIB objects are placed. To illustrate this concept, assume that the company Fudrucker manufactures bridges, routers, and gateways. Let us further assume that IANA assigned node 77 to Fudrucker and the vendor's subtree appears as shown in Figure 8.10. Then, the path to the firm's router MIB would become

1.3.6.1.4.1.77.2

Program utilization example

Through the use of certain SNMP programs, the structure and use of the global naming tree are facilitated by the generation of a graphic user interface. While such programs hide many of the 'complexities' of SNMP, as we will soon note, knowledge of certain aspects of the tree is required to efficiently operate such programs.

In this section we will illustrate the extraction of information from the private subtree of an RMON probe attached to a distant network. In doing so, we will use SimpleView, a Windows software product from Triticom of Eden Prairie, MN.

Figure 8.11 illustrates the SimpleView screen after an RMON probe at network address 198.78.46.37 was first selected and the GetNext entry from the Manage menu was next selected. In this example the term 'private' has been entered as the MIB variable, as we wish to examine the structure and values associated with the private subtree associated with the probe.

Although you can directly enter a subtree name, in the event you forget the name of a subtree or an entry under a subtree, SimpleView contains a most powerful MIB walk facility. This facility is initiated either from selecting MIB Browse from the dialog box previously shown in Figure 8.11 or from a View entry from the MIB Database menu. Either action results in the display of a window labeled MIB Walk, which is illustrated in Figure 8.12.

The SimpleView MIB Walk window allows a user to browse through MIB items defined in the MIB database. The program supports expandable/ collapsible trees and leaves that represent MIB objects. Thus, double clicking

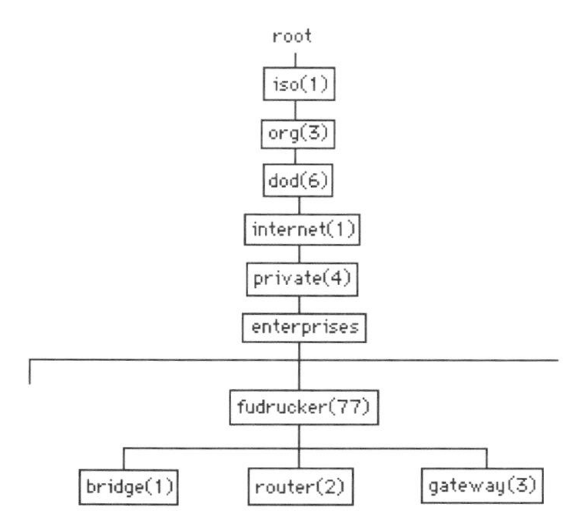

Figure 8.10 A hypothetical vendor subtree

on any entry in the window shown in Figure 8.12 would result in an expansion to display level 1 items or objects under the selected subtree. By highlighting an MIB object and clicking on the Select button, the selected object name is copied to the MIB field of GET, GET-NEXT, and SET dialog boxes, which are selected from the Manage menu.

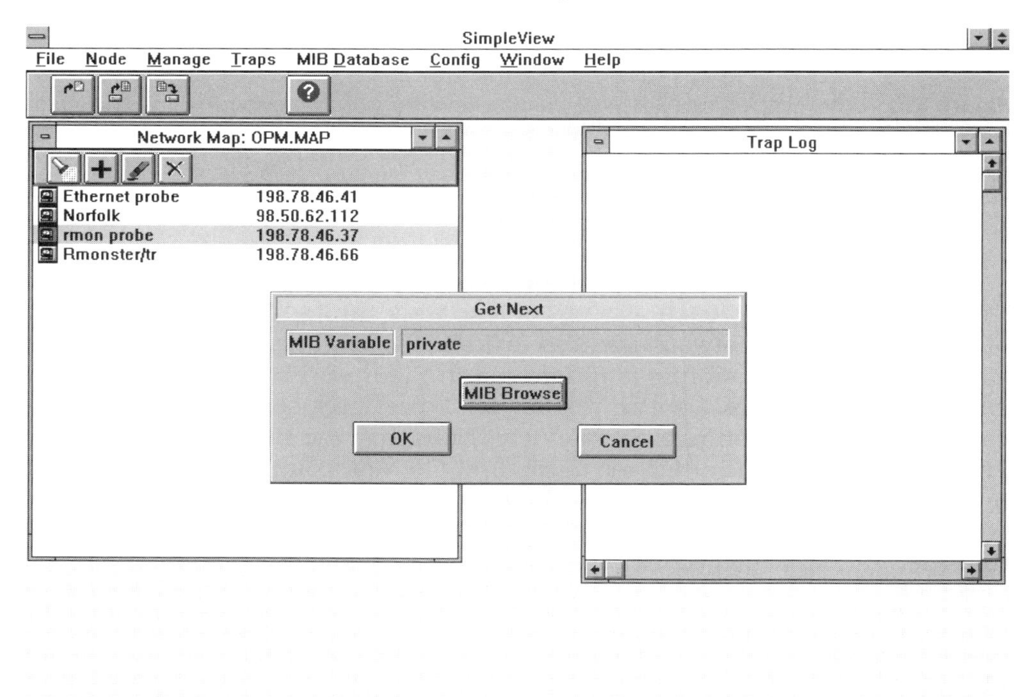

Figure 8.11 Using Triticom's SimpleView to display the structure and values of the private subtree of a device on the network

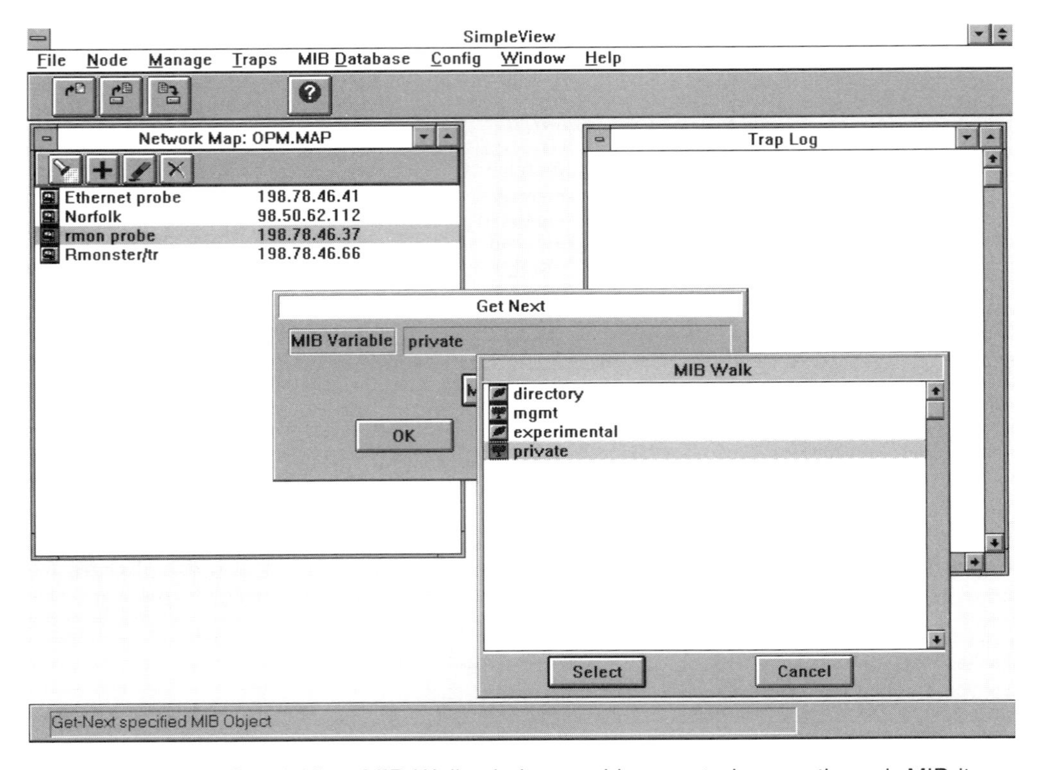

Figure 8.12 The SimpleView MIB Walk window enables you to browse through MIB items defined in the MIB as well as to copy a selected object for use by the GET, GET-NEXT, and SET dialog boxes

Through the use of a series of GetNext commands, of which the first used the MIB variable 'private', it was possible to walk through the private subtree of the probe whose network address is 198.78.46.37. A portion of this walk through the private subtree is illustrated in Figure 8.13, where the window labeled TrapLog displays the results of a previous sequence of GetNext commands issued through SimpleView. Note that the first two lines in the TrapLog represent the response to the first GetNext command, with the first line of the response listing the address that the command was sent to. The second line, which begins 'enterprises [209.1.1.01] = cornerstone,' tells us that the vendor of the product was registered or assigned node 209 by IANA. Subnode 1.1 under 209 returns the name of the product, which is Cornerstone, while subnode 1.2 under 209 returns the version of the product, which in this example is 1.53. Note that the terminating zero (0) for each displayed response is a mechanism to indicate that there is only one value associated with the object.

Since SimpleView was developed independently of the Cornerstone RMON agent, which is a product of Network General (now part of Network Associates), it was necessary to guess what the values returned from a

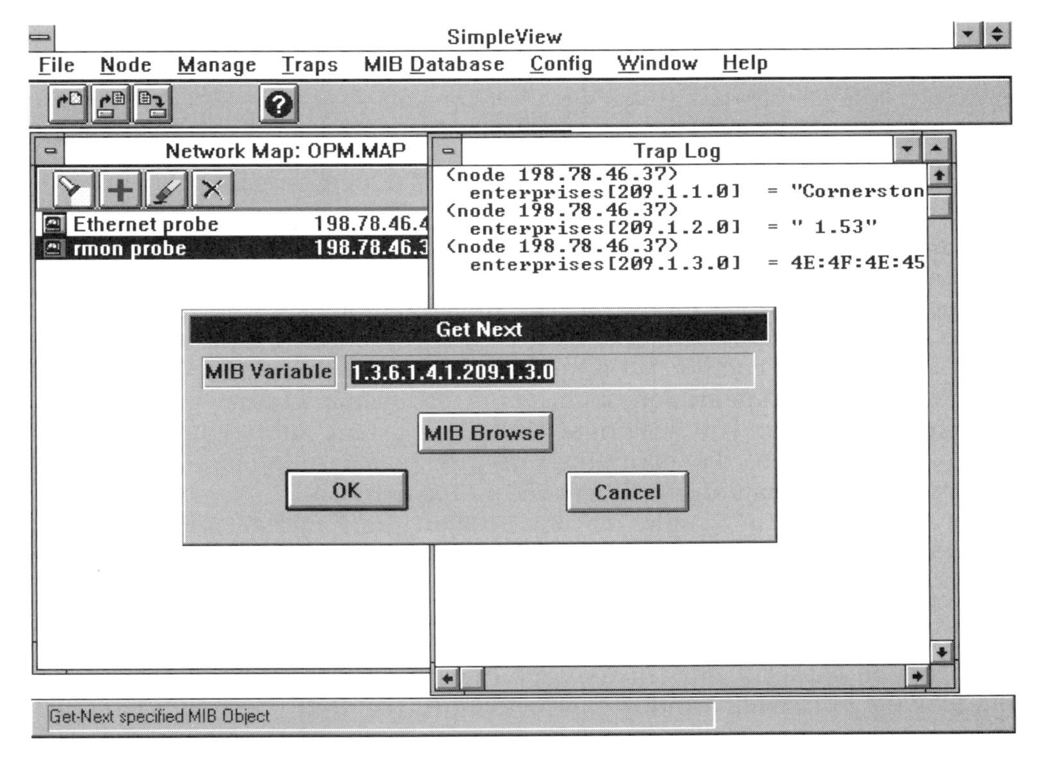

Figure 8.13 Walking through the private subtree of an RMON probe

walk through the private MIB represent or do the unthinkable and read the RMON agent manual, if one was available and described its private MIB. Since vendors can essentially do whatever they desire with respect to placing objects in a private MIB, coordination between vendors developing managers and vendors developing agents is best for the MIB-2 portion of the naming tree, and can represent the worst aspects of cooperation with respect to the private subtree. This is the price paid for a naming structure that provides a high degree of flexibility that enables vendors to improve their products beyond standardized functions.

8.3.4 MIB II objects

In concluding this section we will present a summary of the objects located in the current standard MIB, mib-2, whose location in the global naming tree is 1.3.6.1.2.1. As illustrated in Figure 8.9, located under that node are 10 MIB groups. After briefly reviewing the objects in the first few MIB groups, including a few examples showing the retrieval of objects from some groups, we will essentially provide a general overview of the remaining groups and focus attention upon the SNMP group. Although certain MIB groups are essentially skipped over while other groups are limited to a general explanation of their objects, a complete reference to MIB groups can be found in Appendix A.

The System Group

The System Group contains seven objects that are used to describe configuration information about the managed device. The object identifier for the system group is {mib-2 1}. The individual objects within the system group have an object identifier of {system *n*}, with *n* having a value of 1 to 7. A brief description of each object is contained in Table 8.3, while Appendix A.1 includes the formal definition of objects in the system group.

From Figure 8.9, you will note that the system subtree is located at 1.3.6.1.2.1.1. Thus, we can illustrate the object identifiers under that subtree in a tree diagram as illustrated in Figure 8.14.

Through the use of SimpleView or a similar SNMP management program, the global naming tree can, to a large degree, become hidden through the use of a program's graphic user interface. An example of this is shown in Figure 8.15, in which the SimpleView View option from the program's MIB Database menu was selected to obtain a view of the entries within the System group. As indicated earlier in this section, after selecting a desired object from the SimpleView MIB Walk window you could use the GET, GET-NEXT or SET entries from the program's Manage menu to perform a desired operation on a selected MIB object. By selecting the System Group entry and issuing a series of GET-NEXT commands, you can walk through the System Group of the selected device. The result of this operation, which in effect results in a dump of the objects of the System Group for the selected device at address 198.78.46.37, is shown in Figure 8.16. In this example the selected object at

Table 8.3 The System Group

Object Identifier	Access	Description
sysDescr (1.3.6.1.2.1.1.1)	r	A description of the device
sysObject ID (1.3.6.1.2.1.1.2)	r	An identification of an agent's hardware, software and/or resources
sysUpTime (1.3.6.1.2.1.1.3)	r	The length of the time since the agent was started or restarted
sysContact (1.3.6.1.2.1.1.4)	r–w	The name of the contact person responsible for the node
sysName (1.3.6.1.2.1.1.5)	r–w	The device name
sysLocation (1.3.6.1.2.1.1.6)	r–w	The physical location of the device
sysServices (1.3.6.1.2.1.1.7)	r	A coded number that identifies the set of services provided by the device

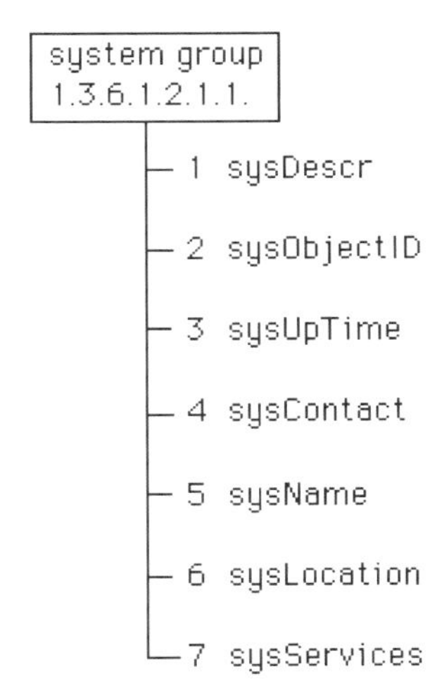

Figure 8.14 The System Group MIB objects

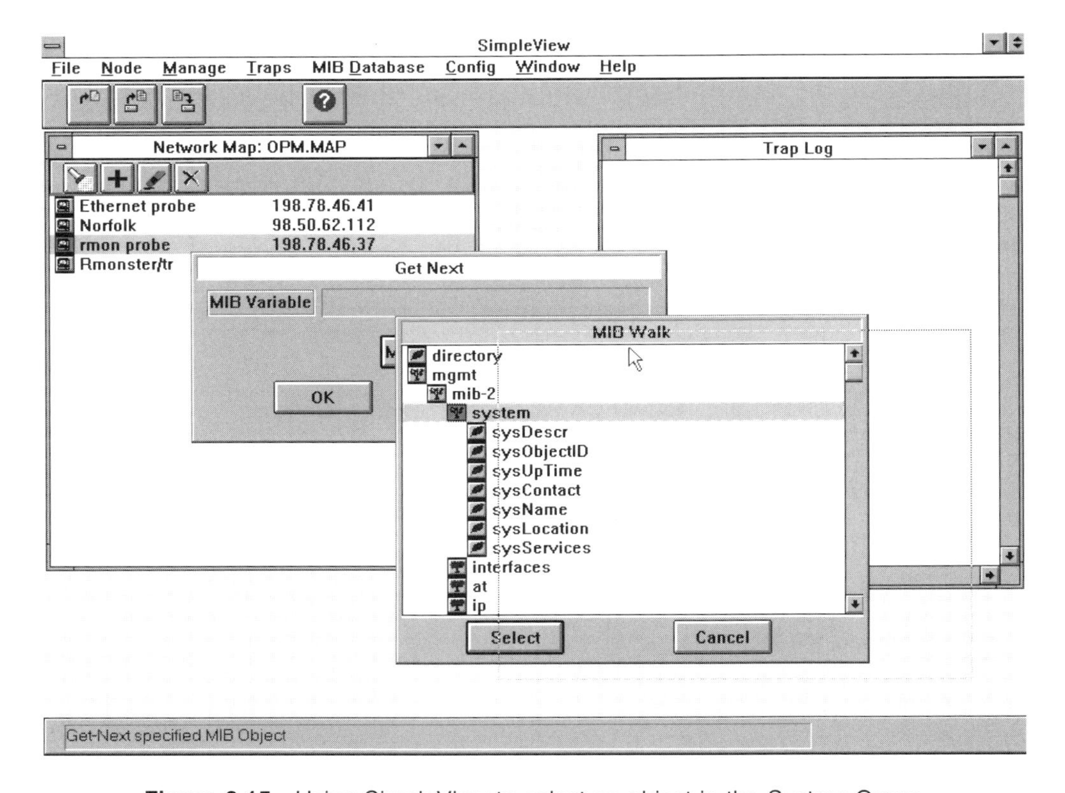

Figure 8.15 Using SimpleView to select an object in the System Group

address 198.78.46.37 is our familiar Network General RMON probe, more formally referred to as a Cornerstone Agent, which is its system name.

The Interfaces Group

The Interfaces Group consists of 23 object identifiers that provide generic performance, configuration, and status information for any type of interface. Although this generic information can provide an insight into the activity at an interface, by its nature generic information is not specific enough for many management applications. For example, although a count of inbound and outbound octets is relevant to both Ethernet and Token-Ring networks, you could not have a generic collision count, as it would not be applicable to Token-Ring networks. Recognizing this, the SNMP developers used the transmission subgroup as a location to define a set of interface-specific MIB groups.

Since a managed device can have more than one interface, a mechanism was required to identify the interface attachment to a device. Thus, the interfaces subtree is structured so that the first identifier (ifNumber) serves as a pointer to the interface. Information about a specific interface is included in a table consisting of 22 entries. Table 8.4 lists the object identifiers that provide information concerning the state of an interface as well as a brief description of each identifier. Note that the ifType object identifier is used to indicate the specific type of interface. Currently over 50 types of network interfaces are defined, ranging from a version of X.25 (interface type 4) to ATM Adaption Layer 5 (interface type 49).

To illustrate the retrieval of objects from the Interface group MIB, we will again turn to the use of SimpleView. Figure 8.17 illustrates the result obtained from 'exploding' the interfaces entry in the program's MIB Walk window. By selecting the interfaces entry and using a series of GET-NEXT requests, it was possible to walk through the Interfaces Group of the remote RMON probe. The results of this operation are shown in Figure 8.18.

```
(node 198.78.46.37)
    sysDescr[0] = "RMON CA Win Version 2.5c"
(node 198.78.46.37)
    sysObjectID[0] = enterprises[209.1.1.0] = 1.3.6.1.4.1.209.1.1.0
(node 198.78.46.37)
    sysUpTime[0] = 169830286 (19 days, 15:45:02)
(node 198.78.46.37)
    sysContact[0] = "Network Administrator"
(node 198.78.46.37)
    sysName[0] = "Cornerstone Agent"
(node 198.78.46.37)
    sysLocation[0] = "Network Services Dept"
(node 198.78.46.37)
    sysServices[0] = 72
```

Figure 8.16 The results obtained from using SimpleView to walk through the System Group of a selected device

Table 8.4 The Interface Group

Object Identifier	Access	Description
ifIndex (1.3.6.1.2.1.2.2.1)	r	The interface number
ifDeser (1.3.6.1.2.1.2.2.2)	r	Text describing the interface
ifType (1.3.6.1.2.1.2.2.3)	r	A numeric identifier that defines the type of interface
ifMtu (1.3.6.1.2.1.2.2.4)	r	The largest PDU in octets that can be sent or received
ifSpeed (1.3.6.1.2.1.2.2.5)	r	The transmission rate in bps
ifPhysAddress (1.3.6.1.2.1.2.2.6)	r	A media-specific address
ifAdminStatus (1.3.6.1.2.1.2.2.7)	r	The desired interface state
ifOperStatus (1.3.6.1.2.1.2.2.8)	r	The current interface state
ifLastChange (1.3.6.1.2.1.2.2.9)	r	The time when the interface state last changed
ifInOctets (1.3.6.1.2.1.2.2.10)	r	Total octets received, including framing
ifInUcastPkts (1.3.6.1.2.1.2.2.11)	r	The number of unicast packets delivered
ifInNUcastPkts (1.3.6.1.2.1.2.2.12)	r	The number of non-unicast packets delivered
ifInDiscards (1.3.6.1.2.1.2.2.13)	r	The number of inbound packets discarded due to resource limitations
ifInErrors (1.3.6.1.2.1.2.2.14)	r	The number of inbound packets discarded due to being in error
ifInUnknownProtos (1.3.6.1.2.1.2.2.15)	r	The number of inbound packets discarded as they were directed to an unknown protocol
ifOutOctets (1.3.6.1.2.1.2.2.16)	r	The total number of transmitted octets, including framing
ifOutUcastPkts (1.3.6.1.2.1.2.2.17)	r	The total number of unicast packets transmitted
ifOutNUcastPkts (1.3.6.1.2.1.2.2.18)	r	The total number of broadcast or multicast packets transmitted
ifOutDiscards (1.3.6.1.2.1.2.2.19)	r	The total number of outbound packets discarded due to resource limitations
ifOutErrors (1.3.6.1.2.1.2.2.20)	r	The number of outbound packets discarded due to error
ifOutQLen (1.3.6.1.2.1.2.2.21)	r	The number of packets in the output queue
ifSpecific (1.3.6.1.2.1.2.2.22)	r	An identifier for an MIB that contains additional information related to the interface type

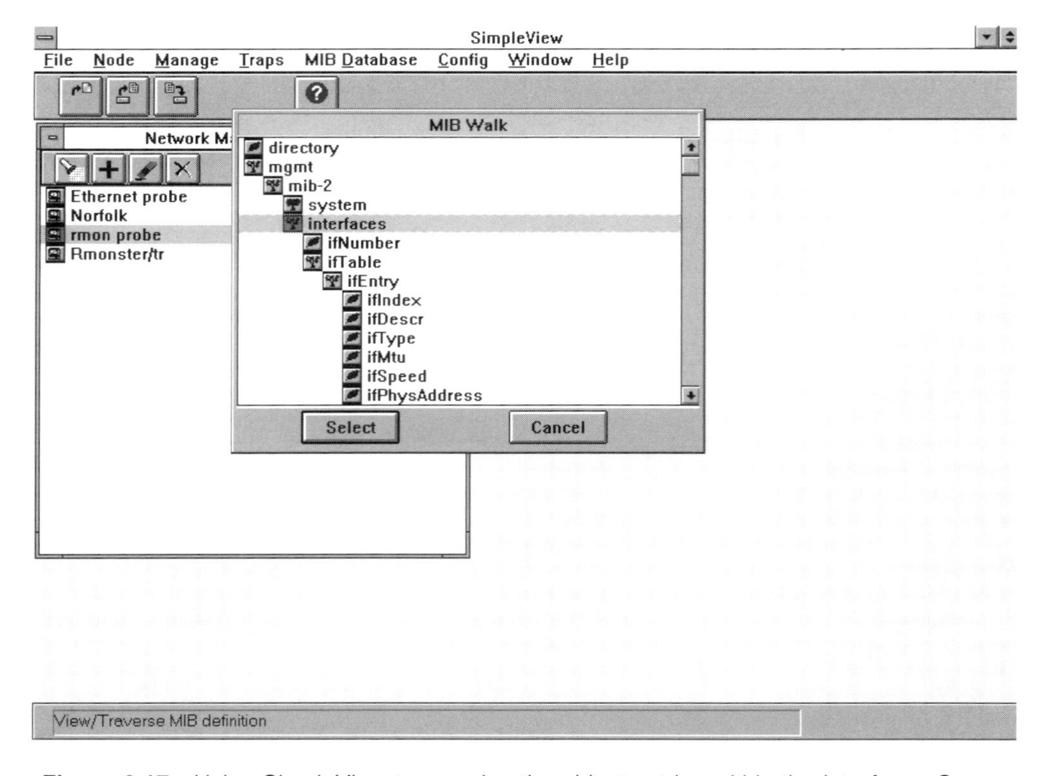

Figure 8.17 Using SimpleView to examine the object entries within the Interfaces Group

An examination of many of the object identifiers of the Interfaces Group can provide you with information that can be a valuable assistant in determining the health of a network component. From the results of the MIB walk shown in Figure 8.18, you will note that the probe is operational (ifOperStatus = 1) and no errors (ifInErrors = 0 and ifOutErrors = 0) were detected. Similarly, you can note that the values for ifInDiscards and ifOutDiscards, which denote the number of inbound and outbound packets discarded due to resource limitations, were both zero, indicating that all inbound and outbound packets were processed. Thus, at the time the objects from the Interfaces Group were retrieved, the monitored network did not exhibit any adverse resource-related problems. The preceding illustrates that a quick glance at a few of the values of objects in the Interfaces Group tells us that the RMON probe is operational and no input or output errors were encountered.

One of the more interesting aspects of the interfaces table is its last object identifier, ifSpecific. If additional variables for the interface exist, ifSpecific will function as a pointer to an MIB subtree where additional variables can be read. For example, if the ifType is Token-Ring, the value of ifSpecific could be set to identify the MIB subtree defining objects specific to Token-Ring. If there is no additional information for ifSpecific to point to, its value would be set to zero. Since the RMON probe just queried is attached to an Ethernet LAN, as you might expect, its ifSpecific value is zero. Readers are referred to Appendix

```
(node 198.78.46.37)
ifIndex[1] = 1
(node 198.78.46.37)
ifDescr[1] = "ProTools, Inc. Cornerstone Agent Serial No. NON Ver. 1.53
(node 198.78.46.37)
ifType[1] = 9
(node 198.78.46.37)
ifPcu[1] = 1500
(node 198.78.46.37)
ifSpeed[1] = 4194304
(node 198.78.46.37)
ifPhysAddress[1] = 00:00:F6:2B:5D:B7
(node 198.78.46.37)
ifAdminStatus[1] = 1
(node 198.78.46.37)
ifOperStatus[1] = 1
(node 198.78.46.37)
ifLastChange[1] = 1428 (00:00:14)
(node 198.78.46.37)
ifInOctets[1] = 1186429611
(node 198.78.46.37)
ifInUcastPkts[1] = 141222
(node 198.78.46.37)
ifInNUcastPkts[1] = 141216
(node 198.78.46.37)
ifInDiscards[1] = 0
(node 198.78.46.37)
ifInErrors[1] = 0
(node 198.78.46.37)
ifInUnknownProtos[1] = 2270730
(node 198.78.46.37)
ifOutOctets[1] = 350199
(node 198.78.46.37)
ifOutUcastPkets[1] = 1266
(node 198.78.46.37)
ifOutNUcastPkts[1] = 0
(node 198.78.46.37)
ifOutDiscards[1] = 0
(node 198.78.46.37)
ifOutErrors[1] = 0
(node 198.78.46.37)
ifOutQLEN[1] = 1
(node 198.78.46.37)
ifSpecific[1] = 0.0
```

Figure 8.18 The results obtained from a walk through the Interfaces Group of an RMON probe.

A.2 for detailed information concerning the object identifiers for the Interfaces Group.

The Address Translation Group

The Address Translation Group provides a mapping of the network layer address (i.e., its IP address) of a device to its physical or interface layer address. This group uses a single table for mapping, with each row containing three columns. Those columns contain instances for atIfIndex (the interface number), atPhysAddress (the media address), and atNetAddress (the IP address). In the wonderful world of TCP/IP the address translation process occurs through the use of the Address Resolution Protocol (ARP). Thus, the address translation tables contain the mapping or association of IP addresses and physical or MAC addresses. In actuality, the Address Translation Group maintains both physical address and network address tables, with equivalence between an index used to search through one table enabling a match to allow the retrieval of a desired object from the other table. By examining the entries in the address translation tables, you can note if ARP is performing correctly and if the size of the ARP cache should be increased to enhance performance of the address resolution process. Readers are referred to Appendix A.3 for detailed information concerning this MIB group.

The Internet Protocol Group

The Internet Protocol (IP) Group contains the managed objects that provide information about the IP subsystem of a managed node. Included in this group are 19 scalars and four tables whose object values provide such IP information as the state of routing tables and address conversion tables. The IP Group object identifier is {mib 4}, which results in a full naming tree address of 1.3.6.1.2.4. Table 8.5 lists the 19 scalar objects defined within the IP Group, including short descriptions of their uses.

The tables in the IP Group include address, route, and address translation. The IP address table is used to keep track of addresses associated with the managed node, while the address translation table tracks the mapping between IP and the physical address of the managed node. The routing table tracks IP routes associated with the managed node. Table 8.6 lists the object identifiers for the IP Group address and address translation tables.

By carefully examining the value of object identifiers in the IP group, you can determine if datagrams are being dropped due to errors in their IP headers, invalid IP addressing, unsupported protocols, lack of resources, and similar problems. Thus, retrieval of IP Group objects can provide you with an indication of the reason behind many IP-related network problems. Readers are referred to Appendix A.4 for detailed information concerning the object identifiers for the IP group.

The Internet Control Message Protocol Group

The Internet Control Message Protocol (ICMP) is responsible for handling error and control messages that are normally generated by gateways (routers) and hosts to report problems to the originators of datagrams. The ICMP Group provides statistics and error counts for the ICMP protocol. This group has the object identifier {mib 5}, while its numeric global naming tree address is 1.3.6.1.2.5.

Table 8.5 Internet Protocol Group scalar objects

Object Identifier	Access	Description
ipForwarding (1.3.6.1.2.1.4.1)	r–w	Whether the entry is acting as a gateway with respect to forwarding packets
iPDefault TTL (1.3.6.1.2.1.4.2)	r–w	The default Time-To-Live value of the IP header
ipInReceives (1.3.6.1.2.1.4.3)	r	The number of input datagrams received, including those in error
ipInHdrErrors (1.3.6.1.2.1.4.4)	r	The number of input datagrams discarded due to errors in their IP headers
ipInAddrErrors (1.3.6.1.2.1.4.5)	r	The number of input datagrams discarded due to IP address not being valid
ipForwDatagrams (1.3.6.1.2.1.4.6)	r	The number of forwarded datagrams
iPInUnknownProtos (1.3.61.2.1.4.7)	r	The number of datagrams discarded because of an unknown or unsupported protocol
ipInDiscards (1.3.6.1.2.1.4.8)	r	The number of datagrams discarded due to device resource limitations, i.e., buffer space
ipInDelivers (1.3.6.1.2.1.4.9)	r	The number of datagrams successfully delivered
ipOutRequests (1.3.6.1.2.1.4.10)	r	The number of IP datagrams that local IP user-protocols supplied to IP in request for transmission
ipOutDiscards (1.3.6.1.2.1.4.11)	r	The number of datagrams discarded due to lack of resources
ipOutNoRoutes (1.3.6.1.2.1.4.12)	r	The number of datagrams discarded due to no route
ipReasmTimeout (1.3.6.1.2.1.4.13)	r	The maximum number of seconds that received fragments held while awaiting reassembly
ipReasmReqds (1.3.6.1.2.1.4.14)	r	The number of IP fragments received that require reassembly
ipReasmOKs (1.3.6.1.2.1.4.15)	r	The number of IP datagrams successfully reassembled
ipReasmFails (1.3.6.1.2.1.4.16)	r	The number of failures detected by the IP reassembly algorithm
ipFragOKs (1.3.6.1.2.1.4.17)	r	The number of IP datagrams that were successfully fragmented
ipFragFails (1.3.6.1.2.1.4.18)	r	The number of IP datagrams needing fragmentation but that could not be fragmented
ipFragCreates (1.3.6.1.2.1.4.19)	r	The number of IP fragments generated

Table 8.6 Internet Protocol address and address translation table object identifiers

Address Table	Access	Object Identifier Description
ipAdEntAddr (1.3.6.1.2.1.4.20.1.1)	r	The IP address
ipAdEntIfIndex (1.3.6.1.2.1.4.20.1.2)	r	The interface number
ipAdEntNetMask (1.3.6.1.2.1.4.20.1.3)	r	The subnet-mask for the IP address
ipAdEntBeastAddr (1.3.6.1.2.1.4.20.1.4)	r	The least significant bit of the IP broadcast address
ipAdEntReasmMaxSize (1.3.6.1.2.1.4.20.1.5)	r	The largest IP datagram that can be reassembled

Address Translation Table	Access	Object Identifier Description
ipNetToMediaIfIndex (1.3.6.1.2.1.4.22.1.1)	r–w	The interface number
ipNetToMediaPhysAddress (1.3.6.1.2.1.4.22.1.2)	r–w	The media address of mapping
ipNetToMediaNetAddress (1.3.6.1.2.1.4.22.1.3)	r–w	The IP address of mapping
ipNetToMediaType (1.3.6.1.2.1.4.22.1.4)	r–w	The method by which mapping occurred

Because ICMP is the mechanism by which errors are reported, you can use ICMP object values to determine the type of errors users are experiencing. In addition, since ICMP is also used to transport echo requests and echo replies, you can also periodically retrieve or set thresholds for alerts to be issued when the number of echo messages exceeds the predefined level. The latter represents a technique by which you could use SNMP to alert you to the fact that your organization was under a so-called 'Ping of Death' attack in which one or more unscrupulous persons issue continuous pings to an address on your network. Readers are referred to Appendix A.5 for information concerning the object identifiers for the ICMP Group.

The Transmission Group

The Transmission Group is one of the more interesting MIB groups with respect to the fact that it is not actually a group. Instead, it represents a node in the global naming tree under which media-specific transmission groups are located.

The object identifier for the Transmission Group is {mib 10}, while its position in the global naming tree is 1.3.6.1.2.1.10. Directly under the Transmission Group node are a series of MIB groups. Figure 8.19 illustrates the relationship of a majority of presently defined MIB groups under the Transmission Group node to that node.

In examining the entries in Figure 8.19, you will note how this diagram illustrates the versatility of SNMP. In addition to providing the ability to work with such LANs as Ethernet (CSMA/CD) and Token-Ring, the MIB supports monitoring of T1/E1 and T3/E3 circuits through the use of SNMP-compliant Channel Service Units (CSUs) and Data Service Units (DSUs), as well as the monitoring of Frame Relay network operations.

The Transmission Control Protocol Group

The Transmission Control Protocol (TCP) Group provides statistics about the TCP connection, which represents the operation of the layer 4 protocol. The TCP Group has the object identifier {mib 6}, while its numeric global naming tree identifier is 1.3.6.1.2.6.

Through the retrieval of TCP MIB objects, you can determine such information as the minimum and maximum values permitted for retransmission time-outs, the limit on the total number of TCP connections a device can support, and the count of various TCP connection states, as well as statistics concerning segments received and segments in error. Thus, TCB MIB object values can facilitate understanding the reason behind certain types of transport layer problems and point you in the direction towards alleviating such problems. Similar to TCB MIB objects, UDP MIB objects provide a considerable amount of information about this second transport layer protocol

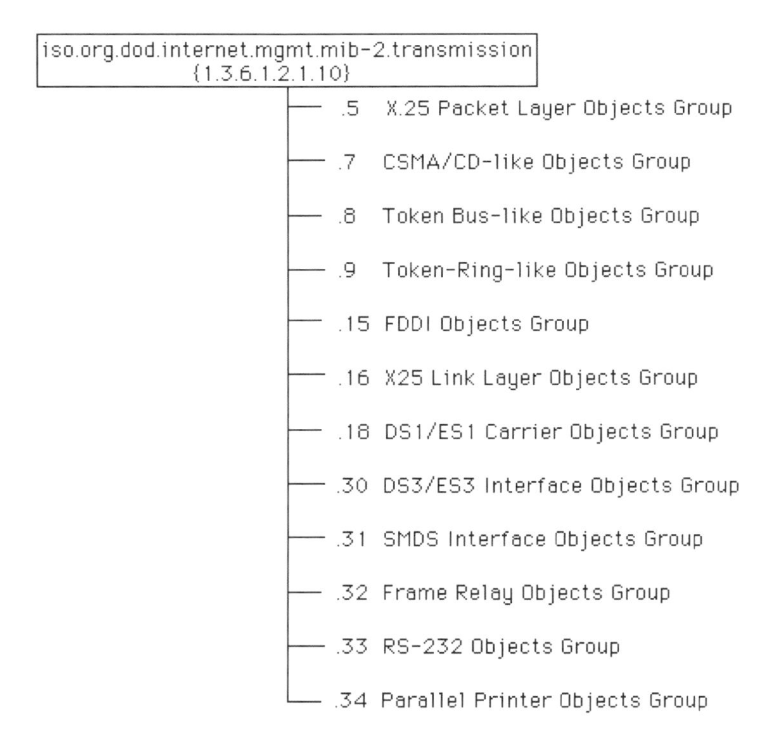

Figure 8.19 Currently defined MIB Transmission Groups

in the TCP/IP protocol suite. Readers are referred to Appendix A.6 for information concerning the object identifiers associated with the TCP Group.

The User Datagram Protocol Group

Statistics and information about the TCP/IP layer 4 protocol known as the User Datagram Protocol (UDP) are maintained in the object identifiers associated with this group. This group, which is a subtree under the MIB, has an object identifier of {mib 7}, while its global naming tree numeric address is 1.3.6.2.1.7. Through the retrieval of certain objects you can determine the total number of UDP datagrams that were delivered, the number of datagrams transmitted, and the reason why datagrams could not be delivered. Readers are referred to Appendix B.7 for a complete description of the object identifiers associated with this group.

The Exterior Gateway Protocol Group

The Exterior Gateway Protocol (EGP) represents a communications protocol that permits neighboring routers in different domains to exchange routing information. The protocol is defined in RFC 904, and the EGP Group's object identifiers provide statistical information about the EGP as well as a table of neighbor information.

The EGP Group has the object identifier {mib 8}, while its global naming tree numeric address is 1.3.6.2.1.8. Readers are referred to Appendix A.8 for a complete description of object identifiers associated with this group.

The SNMP Group

In concluding this section, we will examine the SNMP MIB Group, whose object identifier is {mib 11}, while its global naming tree address is 1.3.6.1.2.11. Although the SNMP Group is similar to other MIB groups covered in this chapter with respect to their object identifiers being fully described in Appendix A, due to the importance of those identifiers in examining the effectiveness of SNMP operations as well as the effect of SNMP on a LAN's non-management data flow, we will examine the use of the group's identifiers in some detail in this section.

If you examine the group's MIB definitions listed in Appendix A.10, you will note that this group has 28 identifiers, ranging from 1 to 30, as 7 and 23 are currently not used. Identifiers 1 and 3–19 perform traffic counts for incoming SNMP messages, while identifiers 2 and 20–29 provide traffic counts for outgoing SNMP messages. While all of the preceding identifiers are limited to read-only access, identifier 30, which has read–write access, governs the enabling or disabling of authentication traps.

Authentication traps When a station is properly configured, the enabling of the object identifier snmpEnableAuthenTraps will result in the transmission of a trap whenever an improperly authenticated message is received by the station. Although you will normally want to enable this identifier, upon occasion it can result in the generation of a large amount of repetitive traffic that can interfere with network operations.

To illustrate the preceding, consider the community naming process. For a manager to be able to correctly interoperate with each agent under SNMPv1 and SNMPv2, the community names, in the form of strings configured for each managed device, must exactly match the configured string of the manager, since the community name functions as a password. Figure 8.20 illustrates the community dialog box for the SimpleView manager. In this example, the SNMP GET, SET, and TRAP communities are each set to PUBLIC, which for many internal networks connected to public facilities can represent a poor choice. This is because PUBLIC is a commonly used default setting. Although you may not care who retrieves information from your agents, it is doubtful if you want any person with a bit of SNMP knowledge to be able to modify an agent via the use of the SET command. Thus, at the very least, you should consider modifying the manager and each agent's SET community name to an alphanumeric string that would be difficult to guess. Returning to our traffic generation problem, consider a management station inadvertently configured with one or more community names that do not match agent names. If the management station is configured to poll agents on a predefined basis, enabling snmpEnableAuthenTraps will result in the generation of a large number of authentication-failure traffic, which becomes more of a nuisance than being of assistance, especially if you have to manually reset each trap received by a management station. Another trap-associated problem can occur if you use two or more management platforms and have segmented your network with respect to the control of agents by each management platform. If you leave the string for the TRAP community name blank, many management platforms will display every trap received in their trap log which may not be your intention.

Now that we have covered the only SNMP object identifier that has read–write access capability, let us focus our attention upon incoming and outgoing SNMP traffic counts.

Incoming traffic counts Table 8.7 lists 17 SNMP object identifiers, as well as a brief description of those identifiers used to provide incoming traffic counts for a managed object. In examining the entries in Table 8.7, you will note that the 1st identifier indicates the total number of SNMP messages delivered to the managed agent while the 2nd through 10th identifiers represent error conditions. The 11th through 16th identifiers provide a summary of commands accepted and processed by the managed agent, while the 17th identifier indicates the number of traps accepted and processed. By examining the count of error condition identifiers and comparing that count with the number of SNMP messages tracked by snmpInPkts, you can obtain an appreciation of the performance of your network management platform, your communications infrastructure, and the operation of your managed agent.

Since SNMP is transported by UDP, you can reasonably expect the snmpInGenErrs count to reflect the error rate on your communications

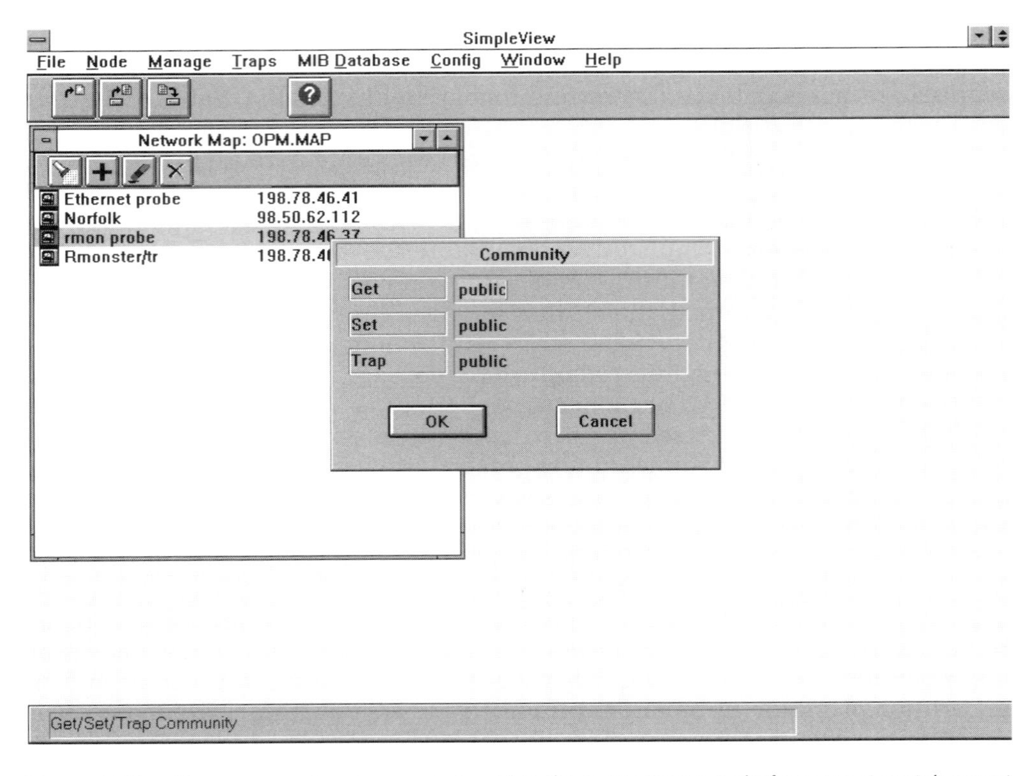

Figure 8.20 The community names assigned to the management platform must match agent community names for the manager and agent to interoperate with each other

infrastructure, which should be relatively low. Thus, a comparison of the values for the identifiers snmpInPkts and snmpInGenErrs permits you to determine if the error rate for other than defined SNMP errors is within an acceptable range. Concerning the defined SNMP errors listed in Table 8.7, by examining the count of different identifiers you can determine if the manager and agent are configured correctly, if a manager is performing correctly, and if the agent and manager provide support for the same version of SNMP. For example, a count for snmpInBadVersions indicates incompatibility between SNMP manager and agent support levels, while a count for snmpInBadCommunityNames indicates a configuration problem between the manager and agent.

Outgoing traffic counts Similar to the incoming SNMP traffic counts, the SNMP group has a number of identifiers used to track outgoing traffic. Table 8.8 lists the 10 identifiers used to track outgoing SNMP traffic to include a brief description of each identifier.

In examining the entries in Table 8.8, note that they are also subdivided into error counts and traffic counts. The first identifier listed in Table 8.8 provides a count of outbound messages from the agent, while the second

Table 8.7 Incoming SNMP traffic counts

Object Identifier	Access	Description
snmpInPkts (1.3.6.1.2.1.11.1)	r	The number of SNMP messages delivered by the transport service
snmpInBadVersions (1.3.6.1.2.1.11.3)	r	The number of SNMP messages delivered for an unsupported SNMP version
snmpInBadCommunityNames (1.3.6.1.2.1.11.4)	r	The number of SNMP messages that used an unknown community name
snmpInBadCommunityUses (1.3.6.1.2.1.11.5)	r	The number of SNMP messages requesting an unsupported operation by the community name
snmpInASNParseErrs (1.3.6.1.2.1.11.6)	r	The number of ASN.1 or BER errors encountered when decoding a message
snmpInTooBigs (1.3.6.1.2.1.11.8)	r	The number of SNMP message responses that could not fit in the largest message size supported by the manager and agent
snmpInNoSuchNames (1.3.6.1.2.1.11.9)	r	The number of messages that indicate a requested object is not supported by the agent
snmpInBadValues (1.3.6.1.2.1.11.10)	r	The number of SET-REQUEST commands that had an improper value
snmpInReadOnlys (1.3.6.1.2.1.11.11)	r	The number of messages indicating a local implementation error resulting from an incorrect SET-REQUEST command
snmpInGenErrs (1.3.6.1.2.1.11.12)	r	The number of messages with errors different from those listed above
snmpInTotalReqVars (1.3.6.1.2.1.11.13)	r	The number of MIB objects successfully retrieved in response to GET-REQUEST and GET-NEXT-REQUEST commands
snmpInTotalSetVars (1.3.6.1.2.1.11.14)	r	The number of MIB objects successfully updated as a result of SET-REQUESTs
snmpInGetRequests (1.3.6.1.2.1.11.15)	r	The number of GET commands accepted and processed
snmpInGetNexts (1.3.6.1.2.1.11.16)	r	The number of GET-NEXT commands accepted and processed
snmpInSetRequests (1.3.6.1.2.1.11.17)	r	The number of SET-REQUEST commands accepted and processed
snmpInGetResponses (1.3.6.1.2.1.11.18)	r	The number of GET-RESPONSE commands accepted and processed
snmpInTraps (1.3.6.1.2.1.11.19)	r	The number of traps accepted and processed

through fifth identifiers provide a traffic count for different types of error conditions. The last five identifiers listed in Table 8.8 provide a distribution of outgoing SNMP traffic counts based upon the type of outgoing message. Thus, an examination of the traffic counts listed in Table 8.7 provides you

Table 8.8 Outgoing SNMP traffic counts

Object Identifier	Access	Description
snmpOutPkts (1.3.6.1.2.1.11.2)	r	The number of messages passed to the transport service
snmpOutTooBigs (1.3.6.1.2.1.11.20)	r	The number of messages sent with the error status field set to 'tooBig,' indicating that the message could not fit in the largest message size supported by the manager and agent
snmpOutNoSuchNames (1.3.6.1.2.1.11.21)	r	The number of messages sent with the error status field set to 'noSuchName,' which indicates that a requested object is not supported by the agent
snmpOutBad (1.3.6.1.2.1.11.22)	r	The number of messages sent with the error status field set to 'badValue,' which indicates the number of SET REQUEST commands sent with an improper value
snmpOutGenErrs (1.3.6.1.2.1.11.24)	r	The number of messages sent with the error status field set to 'genErr,' indicating errors different from those listed above
snmpOutGetRequests (1.3.6.1.2.1.11.25)	r	The number of outgoing GET-REQUEST commands transmitted
snmpOutGetNexts (1.3.6.1.2.1.11.26)	r	The number of outgoing GET-NEXT-REQUEST commands transmitted
snmpOutSetRequests (1.3.6.1.2.1.11.27)	r	The number of SET-REQUEST commands transmitted
snmpOutGetResponses (1.3.6.1.2.1.11.28)	r	The number of GET-RESPONSE commands transmitted
snmpOutTraps (1.3.6.1.2.1.11.29)	r	The number of traps transmitted

with the ability to analyze communications between a manager and agent, while an analysis of the traffic count identifiers listed in Table 8.8 provides a mechanism for analyzing communications between the managed agent and the manager.

To illustrate the use of SNMP identifiers, we use SimpleView to walk through the SNMP group for an agent. The result of the MIB walk through the agent at address 198.78.46.37 is illustrated in Figure 8.21, which represents the listing of SimpleView's trap log after our MIB walk.

In examining the entries in Figure 8.21, note that the value of snmpInPkts was 1279 while the value of snmpOutPkts was 1272 when we started our walk through the SNMP group. Although you would expect the two to be equal, note that the value of snmpInBadCommunityNames was 7. Thus, out of the 1279 received packets only 1272 were valid and received responses. Another set of values in Figure 8.21 that may appear questionable and deserve a degree of explanation are the values for snmpInGetRequests and

```
(node 198.78.46.7)
snmpInPkts[0] = 1279
(node 198.78.46.7)
snmpOutPkts[0] = 1272
(node 198.78.46.7)
snmpInBadVersions[0] = 0
(node 198.78.46.7)
snmpInBadCommunity names[0] = 7
(node 198.78.46.7)
snmpInBadCommunityUses[0] = 0
(node 198.78.46.7)
snmpInASNParseErrs[0] = 0
(node 198.78.46.7)
snmpInTooBigs[0] = 0
(node 198.78.46.7)
snmpInNoSuchNames[0] = 0
(node 198.78.46.7)
snmpInBadValues[0] = 0
(node 198.78.46.7)
snmpInReadOnlys[0] = 0
(node 198.78.46.7)
snmpInGenErrs[0] = 0
(node 198.78.46.7)
snmpInTotalReqVars[0] = 0
(node 198.78.46.7)
snmpInTotalSetVars[0] = 0
(node 198.78.46.7)
snmpInGetRequests[0] = 40
(node 198.78.46.7)
snmpInGetNexts[0] = 1246
(node 198.78.46.7)
snmpInSetRequests[0] = 0
(node 198.78.46.7)
snmpInGetResponses[0] = 0
(node 198.78.46.7)
snmpInTraps[0] = 0
(node 198.78.46.7)
snmpOutTooBigs[0] = 0
(node 198.78.46.7)
snmpOutNoSuchNames[0] = 283
(node 198.78.46.7)
snmpOutBadValues[0] = 0
(node 198.78.46.7)
snmpOutGenErrs[0] = 0
(node 198.78.46.7)
snmpOutGetRequests[0] = 0
(node 198.78.46.7)
snmpOutGetNexts[0] = 0
(node 198.78.46.7)
snmpOutSetRequests[0] = 0
(node 198.78.46.7)
snmpOutGetResponses[0] = 1265
(node 198.78.46.7)
snmpOutTraps[0] = 0
(node 198.78.46.7)
snmpEnagleAuthenTraps[0] = 1
```

Figure 8.21 The SimpleView trap log obtained from a walk through an agent's SNMP group

snmpInGetNexts, which are shown as 40 and 1246, respectively. The total for those two identifiers is 1286; however, note that 14 operations were required between retrieving the value for snmpInPkts and retrieving the value of snmpInGetNexts during the walk through the SNMP group. Since seven of the 1279 input packets were bad, the 1279+14-7 equals 40+1246, which explains how the total for snmpInGetRequests and snmpInGetNexts became 1286. Although it is important to understand why the values retrieved from an SNMP group during an MIB walk may appear awkward when compared with other identifier values, the real value obtained from retrieving information about this group's values is in verifying configuration settings and focusing upon correcting errors. Thus, retrieving SNMP identifier values can provide you with information necessary to identify SNMP problems as well as an indication of the type of problem, which can be valuable when attempting to correct such problems.

9

MANAGEMENT BY UTILITY PROGRAM

In previous chapters we have examined the use of several comprehensive programs as well as SNMP and RMON as tools to facilitate the management of TCP/IP networks. In this chapter we will turn our attention to the use of several utility programs to test and troubleshoot TCP/IP networks. These programs also provide us with a mechanism to enhance TCP/IP performance. Utility programs examined in this chapter range in scope from Ping, Traceroute, and several statistical generation utility programs to a server performance program. Concerning the latter, although this book is focused upon TCP/IP networking, you cannot overlook the fact that a common source of perceived or actual adverse network performance is an adverse level of server performance. Rather than engage in finger-pointing, the ability to use a performance monitoring tool to ascertain the level of server performance can considerably facilitate isolating the true cause of many network-related problems. Thus, we will turn our attention to server performance in the second portion of this chapter by examining the operation and utilization of the Windows NT/Windows 2000 Performance Monitor Program.

9.1 NETWORK UTILITY PROGRAMS

In this section we will examine the use of a core set of network utility programs, several of which are built into many operating systems as well as being incorporated as menu options in several commercial communication monitoring and analysis programs. Utility programs that we will examine in this section include Ping, Traceroute, Nbstat, and Netstat.

9.1.1 Ping

One of the more common network-related tasks is to determine if a destination location is operational. Many times network users, including employees and customers, will call the network administrator or operations center to report their inability to access a particular destination address. In the wonderful world of Web browsing, users will commonly encounter the display of a dialog box that tells them the browser they are using cannot locate a server they are attempting to access. This dialog box can be generated by a number of causes, ranging from the user entering an incorrect address to the destination being down for maintenance, a network connection on the route to the destination becoming inoperative, or even traffic on the Internet resulting in an address resolution timeout. Although the typical solution to the inability of a browser to locate a destination address is to check the Uniform Resource Location (URL) used and try again, suppose that action fails. In this situation you can attempt to check the status of the destination through the use of the Ping utility program bundled with most TCP/IP software.

Overview

Ping, which stands for Packet Internetwork Groper, invokes a series of Internet Control Message Protocol (ICMP) Echo messages to determine if a remote device is powered on and operating a TCP/IP protocol stack. The remote device when Pinged will respond with an ICMP Echo Replay, enabling the round trip delay to be determined between the originator and the destination devices. Thus, the use of Ping can normally answer two common but very important questions: is the destination up, and if so, how long does it take to reach it? The answer to the first question can eliminate subsequent testing if a response is received. The second question can shed light upon why certain time-dependent applications, such as tunneling SNA within IP, the transmission of Voice over IP, or other delay-sensitive applications, fail to operate properly.

The implementation of Ping can vary between operating systems as well as in the manner in which the utility is incorporated into many communications programs. However, most implementations support a core set of options. A common form of the Ping command indicating some of the more popular options is as follows:

Ping [-q] [-v] [-t] [-m count] [-s size] Host

where

- -q Quiet input with no display except summary lines at startup and completion
- -v Verbose output lists ICMP packets received in addition to Echo responses
- -t Continuously Ping specified host until uninterrupted
- -m count Number of Echo Requests to send
- -s size Specifies the number of data bytes to be transmitted
- Host The host name or IP address of the destination system

From a security perspective, the -t option, when implemented, permits a continuous sequence of Pings to be transmitted to the destination. Get enough disgruntled persons to Ping a common destination using the -t option, and you have a situation referred to as the 'Ping of Death,' as the destination spends most of the time issuing Echo Replay messages instead of responding to legitimate queries. Due to this, many organizations may program router access lists and firewalls to prohibit Echo Requests from entering their network! When this occurs, you may have to use the telephone or email a query, as Ping will not illicit a response to the desired destination.

When a router or firewall is programmed to bar ICMP Echo Request message and you ping a site, your request will literally time-out. However, the lack of a response does not necessarily indicate that the destination is not operational. Thus, you should consider, as a minimum, the use of another utility referred to as Traceroute. Traceroute can be used to trace the route to the router or firewall to validate that the network connection to the distant site is operational.

Operation

Figure 9.1 illustrates the Microsoft Windows NT/Windows 2000 Ping help screen. Note that certain utility programs, including Ping, will display a help screen when the program name is entered without any parameters. Unfortunately, this action is not consistent, and other programs require the entry of 'help' after the command to display a help screen.

In examining Figure 9.1, you will note that in a Microsoft Windows environment most network utility programs operate in the command prompt window. Even if you enter the command name as a Windows Start/Run entry, the program will operate in the command prompt window. In comparison, if you are using a third party communications program that includes Ping or another communications utility program, most implementations provide a windows environment for executing the program.

```
Command Prompt
Microsoft(R) Windows NT(TM)
(C) Copyright 1985-1996 Microsoft Corp.

C:\>ping

Usage: ping [-t] [-a] [-n count] [-l size] [-f] [-i TTL] [-v TOS]
            [-r count] [-s count] [[-j host-list] | [-k host-list]]
            [-w timeout] destination-list

Options:
    -t             Ping the specifed host until interrupted.
    -a             Resolve addresses to hostnames.
    -n count       Number of echo requests to send.
    -l size        Send buffer size.
    -f             Set Don't Fragment flag in packet.
    -i TTL         Time To Live.
    -v TOS         Type Of Service.
    -r count       Record route for count hops.
    -s count       Timestamp for count hops.
    -j host-list   Loose source route along host-list.
    -k host-list   Strict source route along host-list.
    -w timeout     Timeout in milliseconds to wait for each reply.

C:\>_
```

Figure 9.1 The Microsoft Windows NT/Windows 2000 Ping help screen

Utilization

When issued behind a router or firewall, Ping represents a valuable mechanism to check the status of the TCP/IP protocol stack, network adapter, and cabling of the distant station to the network. If one or more of the previously mentioned entities are not correct, Ping will not produce an Echo Reply and will simply time-out.

When adding a new networking device to a network hub, you can easily verify its accessibility by Pinging it from another station on the network. If you do not receive a response from the newly installed device but can Ping other devices on the network, this situation will indicate that it is time to double-check the installation. Thus, you would want to determine if the protocol stack is up and running. Similarly, you would examine the station's communications software to determine if the stack is bound to the correct network adapter and the adapter is correctly cabled to a hub port.

Returning to the Ping options listed in the command format and in Figure 9.1, the -n option enables you to configure the program to transmit a specified number of Echo Requests. Some implementations of Ping use a default count of 3 or 4 Echo Requests, while other implementations use different default values. The -s option permits you to specify the number of data bytes to be transmitted during each Echo Request. Because the ICMP header adds 8 bytes, this results in an ICMP packet size being equal to the data byte count plus 8.

Another difference between Ping implementations concerns the presence or absence of summary statistics. Although many implementations of Ping provide a summary of Echo Request responses, some implementations lack this feature, requiring the user to take out their calculator and perform a few adds and a divide. A typical summary indicates the number of packets transmitted, packets received, percent packet loss, and the minimum average and the maximum round-trip delay. Concerning round-trip delay, this can be a bit deceptive if you enter a host name instead if an IP address. If the host name to IP address resolution was not previously performed and stored in cache on your local DNS, the first Echo Request will require a bit more time, since the IP address must be determined. Thus, if you need to determine latency through the use of Ping, it is advisable to issue the command several times and use the second and succeeding sets of results, which will not include address resolution time.

Operational example

Figure 9.2 illustrates the use of the Windows NT Ping utility program to ping the Web server at Yale University. Note that the -m option was used with a value of 5 to issue five Echo Request packets. Also note that after the command line entry the NT Ping utility program automatically displays the IP address of the Pinged host. Although we Pinged www.yale.edu, the application determined that the address is an alias and the host address elsinore.cis.yale.edu is associated with the Yale University Web server. Thus,

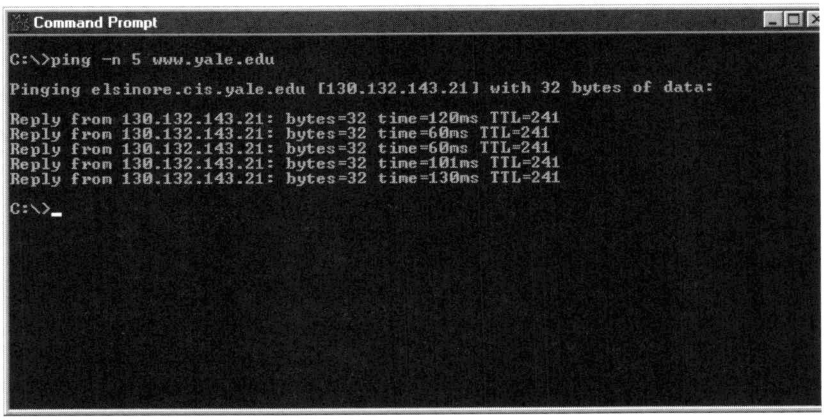

```
Command Prompt

C:\>ping -n 5 www.yale.edu

Pinging elsinore.cis.yale.edu [130.132.143.21] with 32 bytes of data:

Reply from 130.132.143.21: bytes=32 time=120ms TTL=241
Reply from 130.132.143.21: bytes=32 time=60ms TTL=241
Reply from 130.132.143.21: bytes=32 time=60ms TTL=241
Reply from 130.132.143.21: bytes=32 time=101ms TTL=241
Reply from 130.132.143.21: bytes=32 time=130ms TTL=241

C:\>_
```

Figure 9.2 Using Ping to generate a sequence of five EchoRequest packets

Ping can provide additional information beyond the round-trip delay time to the destination. Concerning that delay time, as expected, note that the first reply required more time than the second. However, the fifth reply was the longest. This could result from one or more routers on the path to the destination receiving a burst of traffic that could delay the Echo Request and/or the Echo Response, or activity at the destination resulting in a delay in issuing a response. This is why you should normally generate several sets of multiple Pings to obtain an average if you are concerned about round-trip delay or one-way latency.

Although Ping is a valuable utility, as previously discussed a firewall or router access list can bar its operation. Similarly, a problem on the route to the destination can cause Ping to time-out. When this situation arises, it is probably time to turn to the use of another popular network utility program: Traceroute.

9.1.2 Traceroute

The Traceroute utility program is similar to Ping with respect to its implementation under Microsoft Windows. That is, it is designed to operate in the command prompt window. As its name implies, Traceroute, which is named *tracert* under various versions of Microsoft Windows, provides information on the route packets take from the local host generating the command to the destination in the command.

Overview

The general format of the Traceroute command is as follows:

traceroute [-w timeout] [-h max-hops] host

where

-w Represents the amount of time in either seconds or milliseconds to wait for an answer from a router

-h Represents the maximum number of hops to search for the destination host

host Represents the host name or IP address of the destination

Figure 9.3 illustrates the Windows NT/Windows 2000 *tracert* help screen display. Note that the Windows NT/Windows 2000 version of Traceroute supports the previously mentioned -w and -h options as well as -d and -j options, whose operations are described in the figure.

Operation

Traceroute typically operates by transmitting a sequence of three UDP datagrams to the destination using an invalid destination port address and an initial Time-to-Live (TTL) field value of 1. The TTL value of 1 results in the datagram expiring as soon as it reaches the first router in the path to the destination. The router will respond to the sequence of three UDP datagrams with a sequence of three ICMP Time Exceeded Message's (TEMs), indicating that each datagram has expired. The typical default of three datagrams provides three round-trip delays to each router on the path to the destination. In response to the first sequence of three TEMs, another reference of three UDP messages will generated by the utility program, this time with a TTL value of 2. This action results in the second router in the path to the destination returning another sequence of three ICMP TEMs. This process continues until either the destination is reached or the default on the set maximum number of hops permitted by the program is reached. Once the destination host has been reached, the use of an invalid port address results

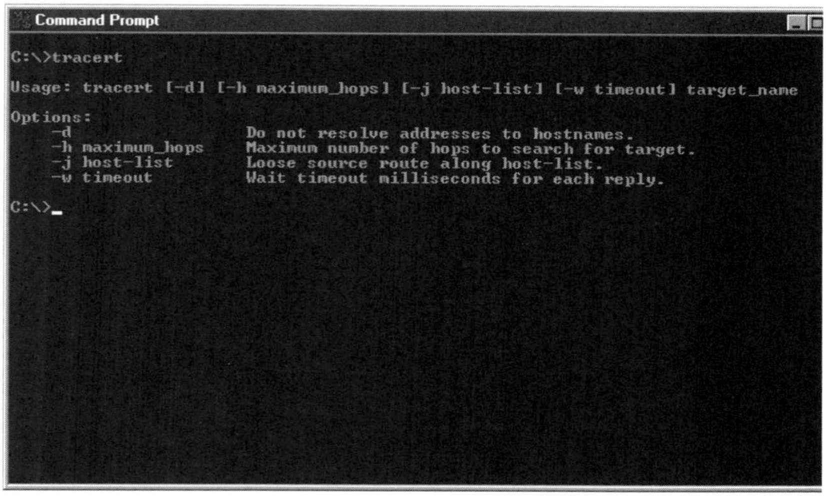

```
Command Prompt
C:\>tracert

Usage: tracert [-d] [-h maximum_hops] [-j host-list] [-w timeout] target_name

Options:
    -d                 Do not resolve addresses to hostnames.
    -h maximum_hops    Maximum number of hops to search for target.
    -j host-list       Loose source route along host-list.
    -w timeout         Wait timeout milliseconds for each reply.

C:\>_
```

Figure 9.3 The Windows NT/Windows 2000 *tracert* help screen

in the return of an ICMP Destination Unreachable Message. The return of this message informs Traceroute to terminate its operation.

Utilization

Similar to preventing Pings, some organizations configure their router access list or firewall to filter Traceroute ICMP messages. However, you can still use this utility program to check the status of the path to a destination router. This means that even when packet filtering precludes reaching a host, you can still use Traceroute to obtain valuable information about the path to a destination.

Operational example

Figure 9.4 illustrates the use of the Windows NT/Windows 2000 *tracert* utility program to trace the route to the Yale University Web server. By examining the response to *tracert*, you can obtain a significant amount of information concerning the structure of other networks linking your organization on the path to the destination host. For example, the first hop takes you on the path to your organization's router (205.131.176.1). The second hop takes you to a BBN Planet (now owned by GTE) router in Atlanta whose interface has the IP address 4.0.156.85. From Atlanta, another hop within the BBN Planet network remains in that city (IP address 4.0.5.114), but provides the route to a router located in Vienna, Virginia. By examining the routers in the path, you will note that the Vienna, VA BBN facility is connected to another BBN facility in New York City (NYC), where packets destined for Yale are apparently routed over an ATM connection to the university. Once the packet has reached the router at Yale's connection to the Internet, it requires three additional hops to reach that university's Web server. Note that from the last line in the *tracert* command shown in Figure 9.4, the host's real name is shown as elsimore.cis.yal.edu.

When used on a corporate internal IP network, the Traceroute utility program is ideal for determining the status of various components across your network. This is because an internal network normally would not bar the use of Traceroute by configuring router access lists or firewalls on the internal network. In addition, since you know the structure of your network, including possible routes from source to destination, Traceroute can be of considerable assistance in verifying the operational status of routers and paths between routers. Even when you are at the mercy of the Internet, Traceroute can assist you in immediately determining the status of your organization's connection to the Internet. If your organization has a leased line linking a LAN to the Internet, you can use either Ping or Traceroute to determine the status of the distant end of the connection and the transmission facility connecting your organization to an Internet Service Provider (ISP). If you previously noted the structure of your ISP's infrastructure, you can use Traceroute to verify the operational status of

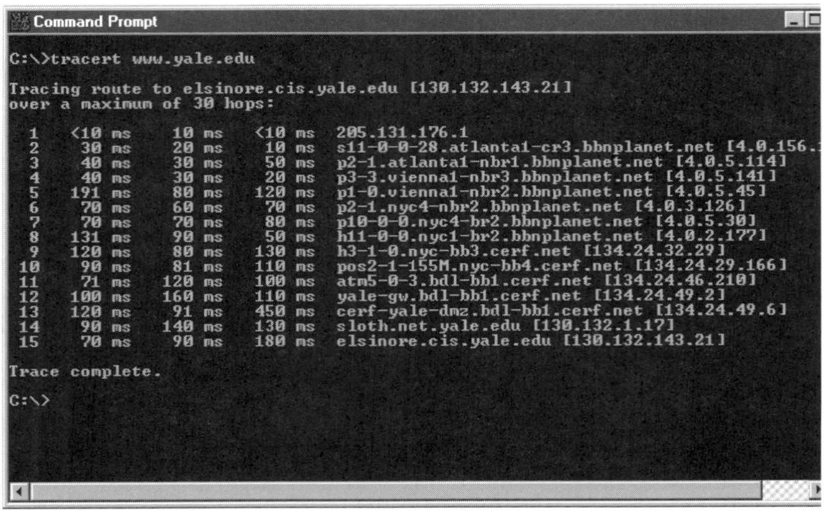

```
Command Prompt

C:\>tracert www.yale.edu

Tracing route to elsinore.cis.yale.edu [130.132.143.21]
over a maximum of 30 hops:

  1    <10 ms    10 ms   <10 ms  205.131.176.1
  2     30 ms    20 ms    10 ms  s11-0-0-28.atlanta1-cr3.bbnplanet.net [4.0.156.
  3     40 ms    30 ms    50 ms  p2-1.atlanta1-nbr1.bbnplanet.net [4.0.5.114]
  4     40 ms    30 ms    20 ms  p3-3.vienna1-nbr3.bbnplanet.net [4.0.5.141]
  5    191 ms    80 ms   120 ms  p1-0.vienna1-nbr2.bbnplanet.net [4.0.5.45]
  6     70 ms    60 ms    70 ms  p2-1.nyc4-nbr2.bbnplanet.net [4.0.3.126]
  7     70 ms    70 ms    80 ms  p10-0-0.nyc4-br2.bbnplanet.net [4.0.5.30]
  8    131 ms    90 ms    50 ms  h11-0-0.nyc1-br2.bbnplanet.net [4.0.2.177]
  9    120 ms    80 ms   130 ms  h3-1-0.nyc-bb3.cerf.net [134.24.32.29]
 10     90 ms    81 ms   110 ms  pos2-1-155M.nyc-bb4.cerf.net [134.24.29.166]
 11     71 ms   120 ms   100 ms  atm5-0-3.bdl-bb1.cerf.net [134.24.46.210]
 12    100 ms   160 ms   110 ms  yale-gw.bdl-bb1.cerf.net [134.24.49.2]
 13    120 ms    91 ms   450 ms  cerf-yale-dmz.bdl-bb1.cerf.net [134.24.49.6]
 14     90 ms   140 ms   130 ms  sloth.net.yale.edu [130.132.1.17]
 15     70 ms    90 ms   180 ms  elsinore.cis.yale.edu [130.132.143.21]

Trace complete.

C:\>
```

Figure 9.4 Using the Windows NT/Windows 2000 *tracert* utility to trace the path to the Yale University Web server

your ISP. In some situations, you can use this knowledge to separate the wheat from the chaff if your ISP claims the problem is at a 'higher level' on the Internet but the use of Traceroute indicates otherwise.

Another popular use of Traceroute is to determine the practicality of virtual networking via the Internet. For example, if your organization has several sites connected to the Internet, it is better to first investigate the route and potential delays between sites than to purchase Virtual Private Network (VPN) equipment and then determine that the routing infrastructure and resulting delays makes the interaction between sites resemble the Long Island Expressway on a Friday afternoon. Now that we have an appreciation for the use of the Traceroute utility program, let us turn our attention to two additional Windows NT/Windows 2000 utility programs that can provide additional information concerning the use of the network.

9.1.3 Nbtstat

Nbtstat is a Windows NT/Windows 2000 utility program that can be used to display protocol statistics as well as the state of TCP/IP connections using NetBIOS over TCP/IP. (NBT). Figure 9.5 illustrates the Windows NT/ Windows 2000 Nbtstat program's help screen. Note that you can specify an interval to effect a continuous display, with the interval value functioning as a pause between each succeeding redisplay. Also note that the Nbtstat program provides you with the ability to learn the MAC adapter address of remote stations via the use of the -a option. The Nbtstat program can also be used to ensure the appropriate use of a host's file for name resolution as well as a mechanism via the -c, -n, and -r options to view and reload different name tables. Thus, Nbtstat can provide a considerable amount of assistance

```
Command Prompt

C:\>nbtstat

Displays protocol statistics and current TCP/IP connections using NBT
(NetBIOS over TCP/IP).

NBTSTAT [-a RemoteName] [-A IP address] [-c] [-n]
        [-r] [-R] [-s] [-S] [interval] ]

  -a   (adapter status) Lists the remote machine's name table given its name
  -A   (Adapter status) Lists the remote machine's name table given its
                        IP address.
  -c   (cache)          Lists the remote name cache including the IP addresses
  -n   (names)          Lists local NetBIOS names.
  -r   (resolved)       Lists names resolved by broadcast and via WINS
  -R   (Reload)         Purges and reloads the remote cache name table
  -S   (Sessions)       Lists sessions table with the destination IP addresses
  -s   (sessions)       Lists sessions table converting destination IP
                        addresses to host names via the hosts file.

  RemoteName   Remote host machine name.
  IP address   Dotted decimal representation of the IP address.
  interval     Redisplays selected statistics, pausing interval seconds
               between each display. Press Ctrl+C to stop redisplaying
               statistics.

C:\>_
```

Figure 9.5 The Windows NT/Windows 2000 Nbtstat help screen

in tracking, testing, and troubleshooting NetBIOS over TCP/IP-related problems.

Operation

Figure 9.6 illustrates the use of the Nbtstat utility program. In this example the program's -a option was invoked to retrieve the MAC address of a distant computer as well as to display the NetBIOS name table of the remote

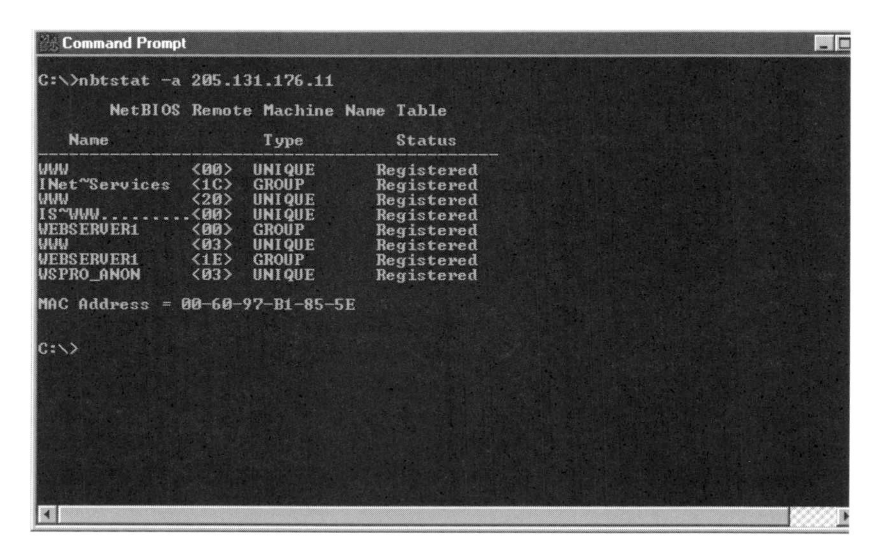

```
Command Prompt

C:\>nbtstat -a 205.131.176.11

       NetBIOS Remote Machine Name Table

   Name               Type         Status
---------------------------------------------
WWW            <00>  UNIQUE      Registered
INet~Services  <1C>  GROUP       Registered
WWW            <20>  UNIQUE      Registered
IS~WWW.........<00>  UNIQUE      Registered
WEBSERVER1     <00>  GROUP       Registered
WWW            <03>  UNIQUE      Registered
WEBSERVER1     <1E>  GROUP       Registered
WSPRO_ANON     <03>  UNIQUE      Registered

MAC Address = 00-60-97-B1-85-5E

C:\>
```

Figure 9.6 Using the Windows NT/Windows 2000 Nbstat program's -a option to learn the MAC address of a remote host as well as to display its machine name table

computer. Now that we have an appreciation for the operation and utilization of the Nbtstat program, let us turn our attention to another Windows program that can provide statistical information about the network. That program is the utility program.

9.1.4 Netstat

Netstat is a Windows NT/Windows 2000 utility program that you can use to display a variety of protocol statistics and current TCP/IP network connections. The general format of the Netstat command is

Netstat [-a] [-e] [-n] [-s] [-p protocol] [-4] [interval]

where

-a	Displays all connections and listening ports; server connections are normally not shown
-e	Displays Ethernet statistics. This may be combined with the -s option
-n	Displays address and port numbers in numerical form (rather than attempting name lookups)
-s	Displays per-protocol statistics. By default, statistics are shown for TCP, UDP, ICMP, and IP; the -p option may be used to specify a subset of the default.
-p protocol	Shows connections for the protocol specified by proto; proto may be TCP or UDP. If used with the -s option to display per-protocol statistics, proto may be TCP, UDP, ICMP, or IP
-r	Displays the contents of the routing table

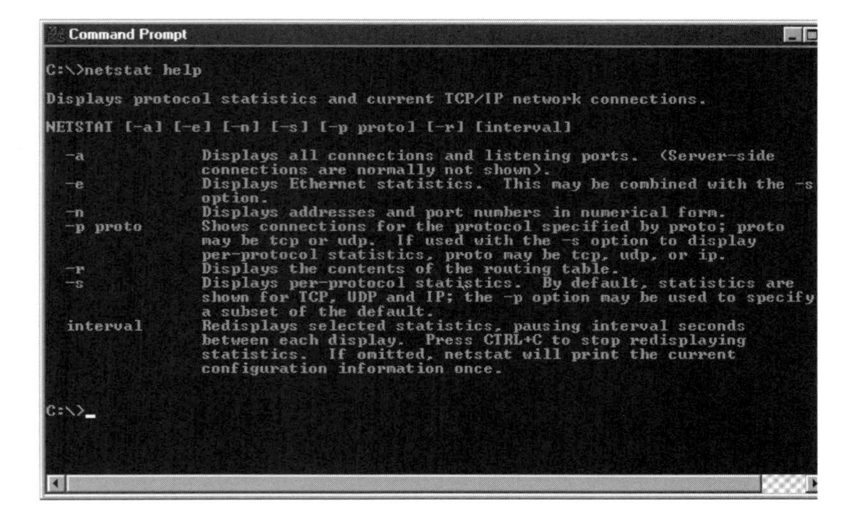

Figure 9.7 The display of the Windows NT/Windows 2000 Netstat help screen requires the entry of 'help' after the command

interval — Redisplays selected statistics, pausing interval seconds between each display. Press CTRL+C to stop redisplaying statistics. If this parameters is omitted, Netstat prints the current configuration information once

Figure 9.7 illustrate the Windows NT/Windows 2000 Netstat help screen. Note that, unlike other Windows commands, you must enter 'help' after the Netstat command name to display its help screen. Because the help screen provides the same information as previously listed, let us turn our attention to the use of the program.

Operation

Figure 9.8 illustrate the operation of the Netstat command using its -e and -p option parameters. Note that the -e option results in the display of Ethernet statistics, while the -p parameter that is followed by TCP results in the display of active TCP connections.

The top portion of Figure 9.8 illustrates the received and sent or transmitted data for the Ethernet network adapter installed in the station on which the utility program is operating. Note that there are no discards or errors. Because there are no discards, this indicates that the data flow to and from the station is within an acceptable level of processing by the station. If the station is a server, this metric would be very useful in determining if you need a more powerful server to keep up with the communications workload. In the lower portion of Figure 9.8 note the active TCP connections are displayed. If you reenter the command using the parameters *-p udp*, you would display active UDP connections. Here the ability to display TCP and UDP connections can be of considerable assistance when attempting to isolate a network problem.

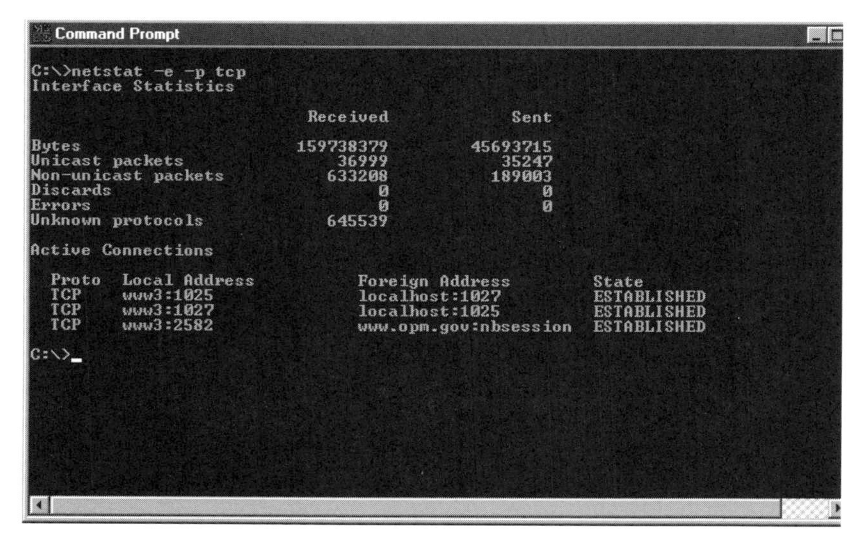

Figure 9.8 Using the Windows NT/Windows 2000 Netstat command with its -e and -p options to display Ethernet statistics and active TCP connections

9.2 MONITORING SERVER PERFORMANCE

As discussed at the beginning of this chapter, many times a perceived network performance problem may actually represent a server performance problem. This means that network managers and LAN administrators who understand this fact as well as having knowledge concerning the ability to monitor server performance may be better positioned to isolate network problems than persons with a more parochial view of networks solely representing communications.

9.2.1 Using Windows NT/2000 Performance Monitor

One of the more valuable features included in Windows NT/Windows 2000 is its Performance Monitor, a graphical tool that can be used for measuring the performance of a Windows/NT/Windows 2000-based computer, other computers on a network, or metrics associated with the performance of different transmission protocols used on a network. By understanding how to use the built-in Performance Monitor as well as its alert capability network, managers and administrators obtain a window that provides a view of numerous computer- and network-related parameters. Those parameters in turn provide an understanding of the use of different computer components and network protocols, and can provide information necessary to upgrade facilities prior to the occurrence of bottlenecks adversely affecting network performance, device utilization information necessary to optimize the use of computer resources, and similar information necessary to provide a high level of client-server performance, which in turn enhances employee productivity.

Overview

Performance Monitor is a graphical tool built into both Windows NT/2000 Workstation and Windows NT/2000 Server. This utility program enables you to view the behavior of a variety of computer- and network-related objects. Each of these objects has an associated set of counters that will provide metrics concerning the use of selected objects. Included in Performance Monitor are charting, alerting, and reporting capabilities that enable the utility to track selected objects over a period of time.

The charting capability included in Performance Monitor permits you to view counter values that are updated at a user-defined frequency. You can display multiple counter values on a single chart that can represent metrics associated with different computers.

The alert capability included in Performance Monitor enables you to specify thresholds for different counters that, when reached, are listed in the Alert Log or are used to notify the computer operator by displaying the alert on the computer's display. You can set several types of thresholds, which can be valuable when attempting to determine if a computer can handle a given traffic load arriving via the network. You can also use thresholds to alert you to the computer reaching disk or memory utilization levels that warrant upgrades, as well as obtaining an understanding of the use of other computer- and network-related objects that might warrant modification to enhance computer or network performance prior to the occurrence of performance-related problems. Since the best way to understand the use of Performance Monitor is by example, let us view its use.

Utilization

Performance Monitor is bundled with Windows NT/2000. The program is accessed from the Administrative Tools menu. Figure 9.9 illustrates the method used to invoke Performance Monitor.

The initial Performance Monitor window contains a blank or empty chart display area and familiar menu entries File, Edit, View, Options, and Help as illustrated in Figure 9.10. The icons displayed under the menu bar are used to invoke predefined functions, with the plus (+) sign used to add a counter for display. As you move the mouse pointer over an icon, a short description of its use is displayed below and to the right of the icon. This is indicated in Exhibit 2, in which the message label 'Add counter' was displayed after this author placed the pointer on the plus (+) icon. By clicking on the plus (+) icon, you obtain the capability to select one or more objects, as well as different counters associated with different objects that will be charted.

Figure 9.11 illustrates the dialog box labeled 'Add to Chart' which is displayed after you select the plus (+) icon from the Performance Monitor

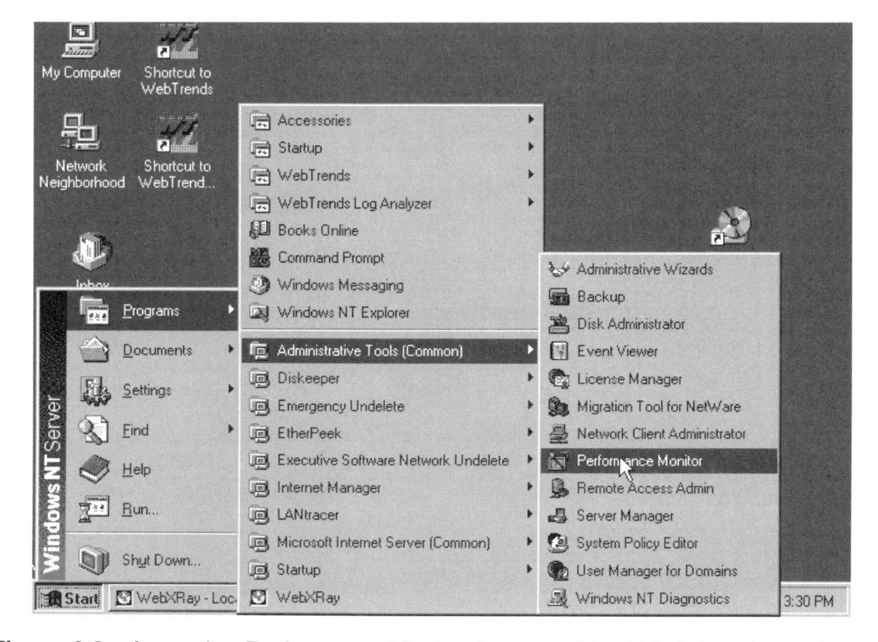

Figure 9.9 Accessing Performance Monitor from the Start/Administrative tools menu

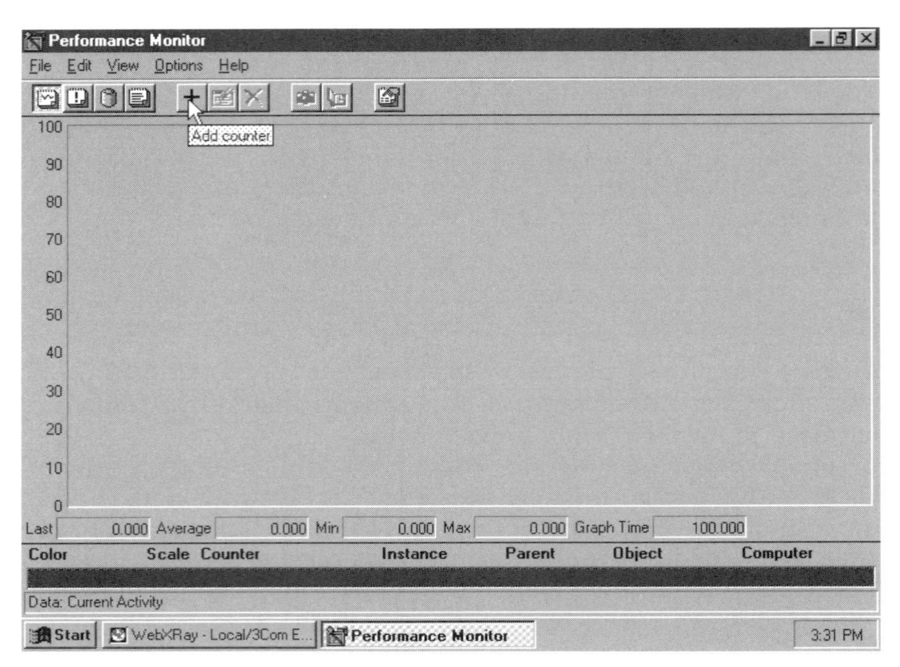

Figure 9.10 Invoking Performance Monitor results in the initial display of an empty chart display area

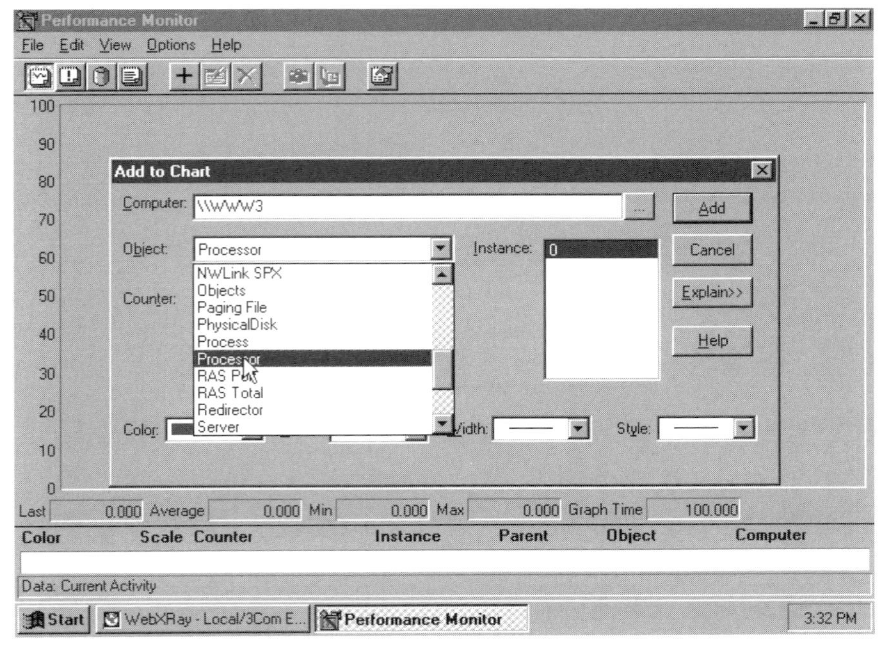

Figure 9.11 Using the Performance Monitor 'Add to Chart' dialog box to select the Processor object

main window. This dialog box contains four main selectable items: Computer, Object, Counter, and Instance.

The Computer item enables you to select a specific computer in a Windows NT/2000 network for which an object, counter, and instance will be selected. The Object represents a standard mechanism used for identifying and using a system resource. Objects represent the processor, memory, cache, hard disk, different network protocols, and other entities for which it is important to track statistical information. Certain types of objects, such as processor, memory and cache, and their respective counters are present on all computers. Other objects, such as different network protocols, are only applicable to computers that are configured to use the appropriate protocol stack.

The Counter item represents statistical information tracked for a defined object. Most objects have a number of counters that provide you with the ability to track different metrics associated with a selected object.

Each object type can have several instances, with the term 'instance' used by Microsoft to identify multiple objects within the same object type. For example, the Processor object type will have multiple instances if a computer has multiple processors. Similarly, the Physical Disk object type will have multiple instances if the computer system has two or more physical hard disks. When an object type has multiple instances, each instance will produce the same set of statistics, as they support the same counter values.

In Figure 9.11, Processor was selected as the Object. Once you select an appropriate object, you can then select one or more counters associated with that object. Figure 9.12 illustrates the selection of the '% Processor Time' counter for the previously selected Processor object. The % Processor Time

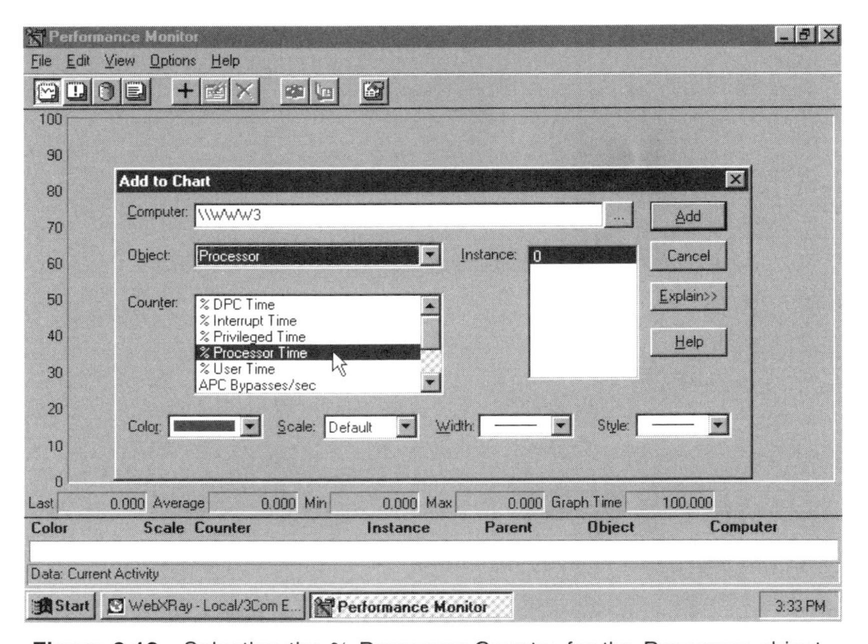

Figure 9.12 Selecting the % Processor Counter for the Processor object

counter tracks the percentage of time the processor is busy, and represents a valuable metric to track server performance. That is, if the %Processor Time counter stays relatively high during the busy hour, this could indicate that complaints of users accessing the server via the communications network will more than likely be alleviated by a processor upgrade or, if the server supports multiple processors, the addition of one or more processors. Returning to Figure 9.11, in this display the Explain button located on the right side of the dialog box was clicked on. If you click on this button, it will result in the display of the definition of the selected counter appearing at the bottom of the screen. The four selectable bars above the counter definition section enable you to set the color, scale, width, and style of the graph used to chart each selected counter. Those features can be extremely valuable when you want to chart multiple counters.

Observing processor performance

Once you have selected one or more counters, you can view the change in the value of those counters over a period of time through the use of the Performance Monitor charting feature. Figure 9.13 illustrates the display of the previously selected % Processor Time counter in the form of a time chart. In this example, the heavy line near the center of the display represents the present monitoring time and moves from left to right across the display. The display shows that processor activity varied from near zero to a high of

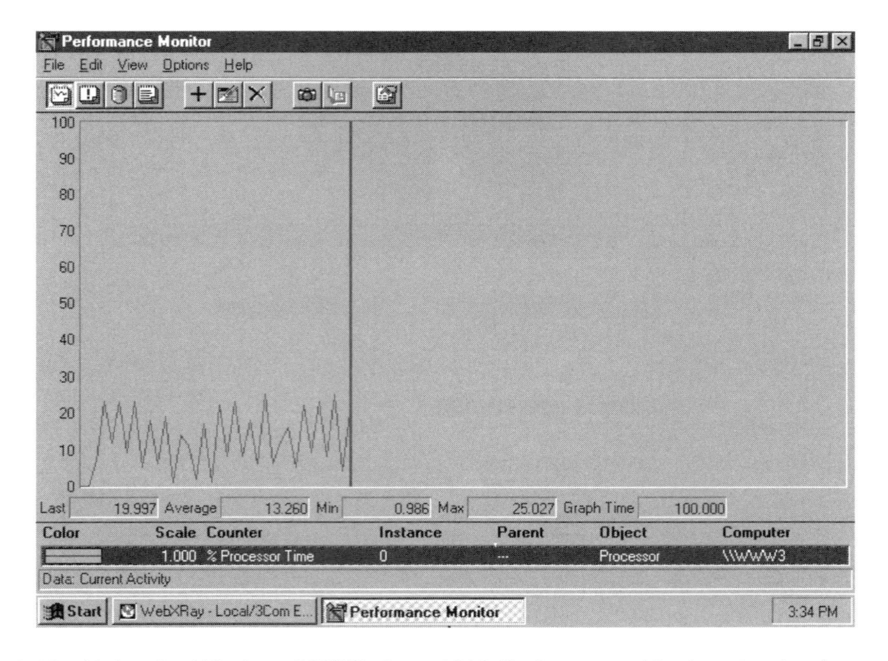

Figure 9.13 Using the Windows NT/Windows 2000 Performance Monitor charting feature to observe the percentage of processor utilization

approximately 25%, but that there was no continuous high level of processor activity that might warrant a processor upgrade. Since the monitoring of just one counter might not indicate other server-related problems that can occur without a high level of network utilization being reached, let us return to the 'Add to Chart' dialog box so we can note how the values for multiple counters can be simultaneously displayed.

Figure 9.14 illustrates the plot of two counters after the Committed Bytes counter from the memory object was added to the Performance Monitor chart. Note that different counters can be displayed in different colors using different widths or shapes that can make them easier to view. In addition, you can display multiple objects from the same or a different computer, which provides you with the ability to examine the performance metrics associated with a variety of servers operating Windows NT/Windows 2000.

9.2.2 Working with alerts

In addition to charting counters, Performance Monitor includes the ability to generate alerts. Alerts enable you to continue working, while Performance Monitor tracks predefined events and notifies you when an event threshold is reached.

In a manner similar to charting, you select alerts. First you select an icon or use the View menu 'Add to Alert' menu entry to display a dialog box with that label. Figure 9.15 illustrates the Add to Alert dialog box, for which the % Processor Time counter was selected for the Processor object. Note that this dialog box enables you to specify a threshold for the alert as well as if the alert

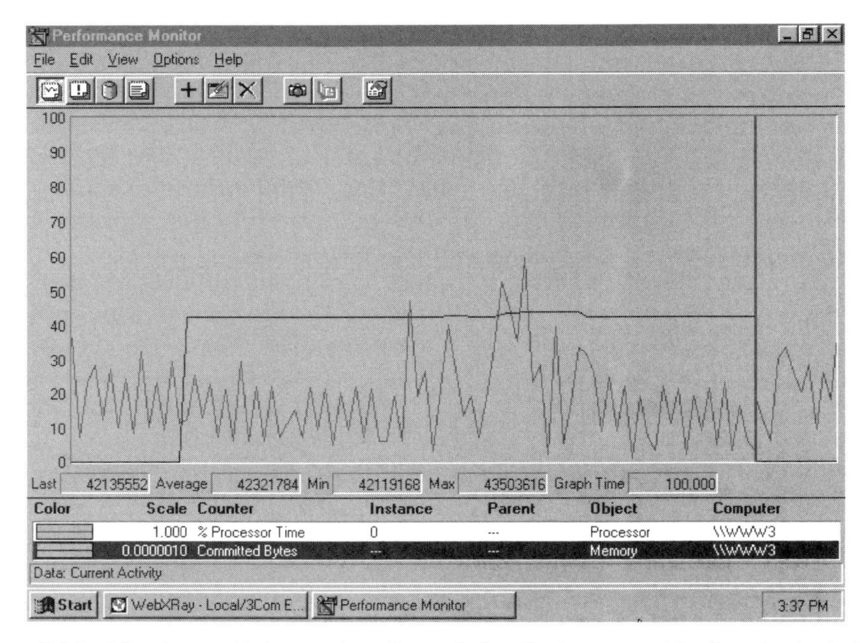

Figure 9.14 Viewing multiple counters through the Performance Monitor chart window

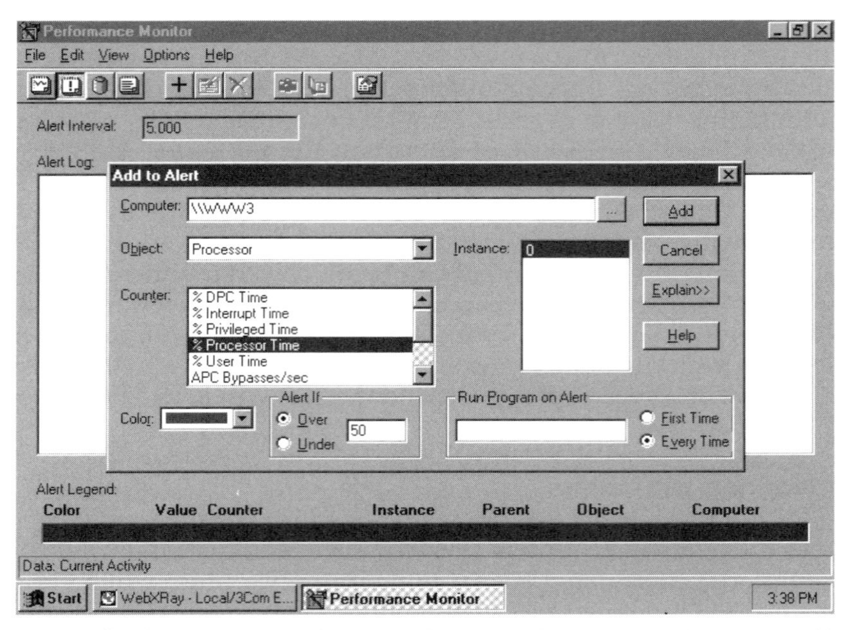

Figure 9.15 Configuring an alert for the % Processor Time counter Processor object

should be generated when the value of the selected counter is over or under the threshold.

In the example shown in Figure 9.15 the alert threshold is shown initially set to 50%. Also note that you can specify a program or macro that you want to run whenever the specified alert occurs. If you do not specify a program or macro, you can observe the Performance Monitor Alert log to determine if any alerts occurred and if so, when they occurred, as well as the values of counters being tracked when the alert occurred.

After selecting the % Processor time counter to generate an alert when the percentage of processor use exceeded 50%, this author executed several compute intensive applications on the computer being monitored in an attempt to generate several processor-related alerts.

Unfortunately, the 450-MHz processor was a bit difficult to overload, let alone exceed 50% processor time utilization. Due to this, it was not possible to generate a 50% level of processor time utilization. Thus, the resulting alert log shown in Figure 9.16 does not contain any entries. This resulted in this author revising the threshold to 30% to obtain a log entry that illustrates information provided by the Performance Monitor alert log.

Figure 9.17 illustrates the alert log after an entry occurred due to processor time utilization exceeding 30%. Note that the log records the date and time of the event, the actual value of the counter, the threshold value of the counter, and the counter, object, and computer for which the alert occurred.

In addition to charting counter values and generating alerts, Performance Monitor can be used to generate a variety of performance-related

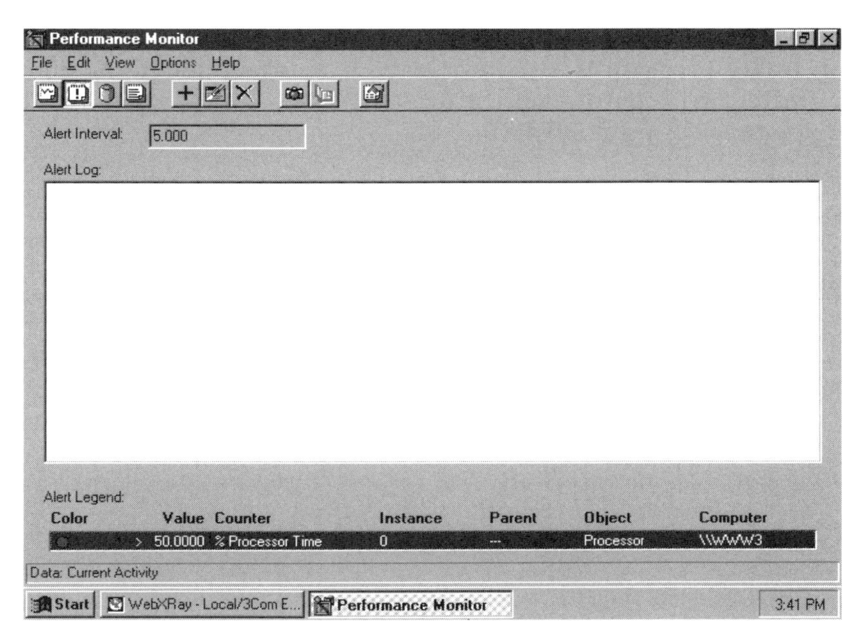

Figure 9.16 Examining the Performance Monitor alert log when a threshold of 50% for processor time was selected for an alert

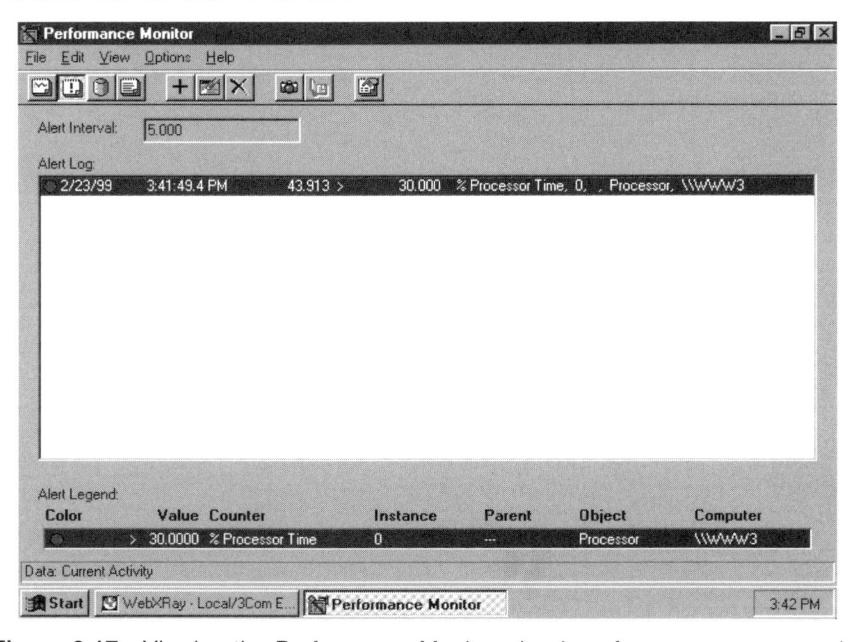

Figure 9.17 Viewing the Performance Monitor alert log after an entry occurred

reports. Due to this capability, Performance Monitor provides network administrators and managers with a significant capability to view and track a variety of computer- and network-related metrics that are extremely valuable when attempting to support a modern client-server environment. By periodically using Performance Monitor, you can use this built-in utility program to ensure that your network, including your client-server computing environment, provides the level of support necessary to enhance employee productivity.

10

SECURITY

No discussion of the TCP/IP protocol suite or of its management would be complete without focusing attention upon security. When the TCP/IP protocol suite was being developed, its user community was restricted to government research laboratories and a small number of academic institutions. This made security problems relatively easy to trace, and limited the need to develop security products and techniques that today are considered indispensable. As ARPAnet evolved into the Internet, its user community expanded by multiple orders of magnitude, with any individual with a credit card able to obtain a dial-up modem Internet access account. All of a sudden a person could easily obtain or attempt to obtain access to network devices located at the other side of the globe. If the old adage that there are only a few bad apples in a bushel is true, then the rapid expansion of the Internet can be considered as equivalent to providing many bushels of bad apples with the ability to harm the computer resources of others. A few well-publicized break-ins into corporate and government Web sits resulted in the development and enhancement of a variety of security-related tools and techniques that are the focus of this chapter.

The term 'security' can have many meanings, ranging from locked doors and guarded buildings that depend upon physical barriers to a variety of electronic-based methods concerned with computer and file access. In this chapter we will ignore the physical aspects of security and focus our attention upon electronic-based methods. In doing so, we will turn our attention to the use of two key hardware products that facilitate the security of a network: routers and firewalls. Because a router provides the network connection to the Internet, we will examine this device with respect to other persons possibly accessing its configuration capability as well as the use of this capability to block the flow of packets between networks. In the first section of this chapter we will turn our attention to router security; however, we will do so with respect to electronic access security and avoid a discussion of physical security. In the second section of this chapter we will turn our attention to the creation and utilization of router access-lists as a mechanism to inhibit the flow of packets between networks. Because Cisco Systems currently has over 70% of the market for router products, the first two sections in this chapter will use Cisco Systems router examples; however, the concepts presented will be applicable to other vendor products.

Because router access-lists represent only a partial solution to network security, in the third section of this chapter we will turn our attention to proxy services and their incorporation into a category of products referred to as firewalls to enhance the level of network security. This overview of firewall proxy services will be followed by a concluding section covering network address translation, which represents a technique that can be used both to save scarce IPv4 addresses and to hide the IP addresses of organizational computers, making it much more difficult for hackers to attempt to directly attack hosts behind routers and/or firewalls that perform this function.

10.1 ROUTER SECURITY

A router represents an entry point into most networks as well as the primary communications device used to move data between networks. As such, it represents a very strategic communications networking device, since an inadvertent or intentional change in its configuration could have a major bearing upon the ability of an organization to maintain its network in a desired state of operation. In addition, if routing tables or other parameters are altered, it becomes possible for organizational data to be directed to locations where such information could be recorded and read by third parties. Thus, it is important to understand how persons can access and take control of a router and steps you can employ to secure this communications networking device.

In this section we will examine and discuss methods of router access in both general and specific terms. Our discussion of router access in general terms will be applicable to products manufactured by different vendors. However, when we turn our attention to specific methods of access and methods we can use to secure access to a router, we will supplement generalizations with specific details applicable to routers manufactured by Cisco Systems. Although specific examples of methods to protect access to routers in this section will be oriented towards Cisco routers, most routers manufactured by other vendors include similar capabilities. If your organization uses routers manufactured by another vendor, you can check the access security functionality of that router and specific commands supported by the router to facilitate one or more access security features for enabling, disabling, and protecting access to the device by referring to the specific vendor router manual.

10.1.1 Need for access security

When they consider router security, most persons automatically think of router access-lists. These are used to establish restrictions on the transfer of data through router ports and are considered by many persons to represent the first line of defense of a network. Although router access-lists are an extremely important aspect of network security, this author considers them to actually represent the second line of defense of a network. This is because the ability to access and configure a router represents the first line of defense of a network. If other than designated personnel obtain the ability to access and change the configuration of organizational routers, this means that any previously developed access-lists can be altered or removed — in effect stripping away any previously developed network protection. Similar to the farmer who constructs a solid hen house but inadvertently goes home at the end of the day and leaves the door ajar, failure to secure access to organizational routers permits predators of the two-legged variety to gain access to valuable resources. This also explains why we will discuss router access security in this section prior to discussing router access-lists in the second section.

As we probe deeper into router access, we will note several methods to bar the proverbial door to this communications device. In fact, one method we will discuss involved the use of an access-list as a mechanism to control access to the router to certain predefined IP addresses. However, prior to doing so, we must lock the door, which should be accomplished prior to the use of the access-list capability of the router. Thus, the use of access-lists should be viewed as a second line of defense.

10.1.2 Router access

For the purposes of this discussion, the term router access represents the ability of a person to connect to a router and gain access to its operating system. Most routers include one or more serial ports built into the device that permit terminals or personal computers operating a specific type of terminal emulator to gain access to the router. This terminal access can occur directly via a direct cable connection or remotely via a communications path. The latter is accomplished through the use of a modem or DSU connected to a router's serial port. Although the use of a serial port connection is the primary method used by most organizations to provide access to a router's operating system to enable the device to be configured, it is not the only method of access. Additional methods supported by many routers include Telnet access and the use of the Trial File Transfer Protocol (HTTP) to store and transmit system images and configurations files to and from routers and workstations.

10.1.3 Telnet access

Telnet provides the ability to access a remote device including a router as if the terminal device operating a Telnet client program was directly connected to the remote device. Telnet access to a router can occur from 'in front' or 'behind' a router, with the term 'in front' used to refer to access to a router via a wide area network connection from a station located on another network not directly connected to a specific router, while the term 'behind' refers to a station located on a network directly connected to a router's local area network port. This means that Telnet access to a router can occur from a device located on a local organizational network or, if the router is connected to the Internet, from virtually any terminal device in the world that has Internet access. This also means that, regardless of the location of the client operating the Telnet program, the operator of the program only needs to know the IP address of the network interface of the router to attempt to initiate a Telnet session to the router and gain access to the device. If the operator of the Telnet client makes a connection to the router the operator will receive a prompt, such as

routername > or

User Access Verification

Password:

where routername represents the name an organization assigned to the router, while Password represents a prompt for entering the appropriate password to access the router.

Figure 10.1 illustrates the use of a Telnet client readily available under Windows 95 and Windows 98 to millions of persons to attempt to access a router whose IP address is 205.131.176.1. In the example illustrated in Figure 10.1 the router breaks the connection after three bad logon attempts. However, the hacker can immediately try again three more combinations. With the use of a script file and an electronic dictionary, it becomes a relatively simple task for a person to create a password cracking Telnet program in an attempt to gain access to a router's configuration capability. Thus, it is important to select passwords that are not only not in a dictionary but do not represent minor additions to dictionary words, such as dog7, since a hacker might program iterations of words for a successful attack over a long weekend.

It should be noted that many organizations have an IP addressing policy where they assign low addresses to router interfaces. For example, if an organization's Class C IP network address was 205.123.124.0, they might assign 205.123.456.1 as the address of the interface from the 205 network to the router. Due to this common scheme of addressing used by many organizations, it is often easy to determine the address of a router for subsequent Telnetting attempts. At this point in time a Telnet client operator may be able to directly access all of the router's configuration capabilities and in effect take over control of the router. As an alternative to developing a sophisticated script with the use of an electronic dictionary, many hackers know that many routers are configured at the factory to have a default password for Telnet access. Unfortunately for many organizations that should know better, you should never use a default password. This is because such passwords are listed in the vendor's router manual, which may be available for purchase for $29.95 or available for access via the World Wide Web for free. This means that a virtually unlimited number of persons have the ability to discover the default password needed to access a router via a Telnet connection. If the router administrator failed to change the default Telnet password or did not place any additional restrictions upon Telnet access, anyone with knowledge of the IP address of the router interface could gain access to the device.

10.1.4 TFTP access

Most routers have two types of memory: conventional random access memory (RAM) and non-volatile memory. Unlike conventional RAM, whose contents are erased upon the removal of power, the contents of non-volatile memory remain in place. When configuring, a router non-volatile memory is commonly used to store an image of router memory as well as backup or alternative router configurations. Because routers do not contain diskettes or have hard drives, their ability to store more than one or perhaps a few alternative configurations is severely limited. This means that administrators that require the ability to store backup or alternative router configurations beyond the capacity of the limited amount of router non-volatile memory typically do so on a workstation and use the Trivial File Transfer Program (TFTP) to load and save router system images and configuration files. This also means that if TFTP access is enabled, depending upon how the router supports TFTP access, it may be possible for unauthorized persons to create configuration data that when used by the router results in a breach of security or an unintended operational environment.

Now that we have an appreciation of the main methods that can be used to gain access to a router, let us turn our attention to the methods that can be used to either protect such access or lock the door upon the access method. This will provide us with the ability to make it extremely difficult for non-authorized persons to gain access to a router and obtain the ability to view and possibly change the configuration of the device. In doing so, we will also, when applicable, discuss certain Cisco Systems router commands.

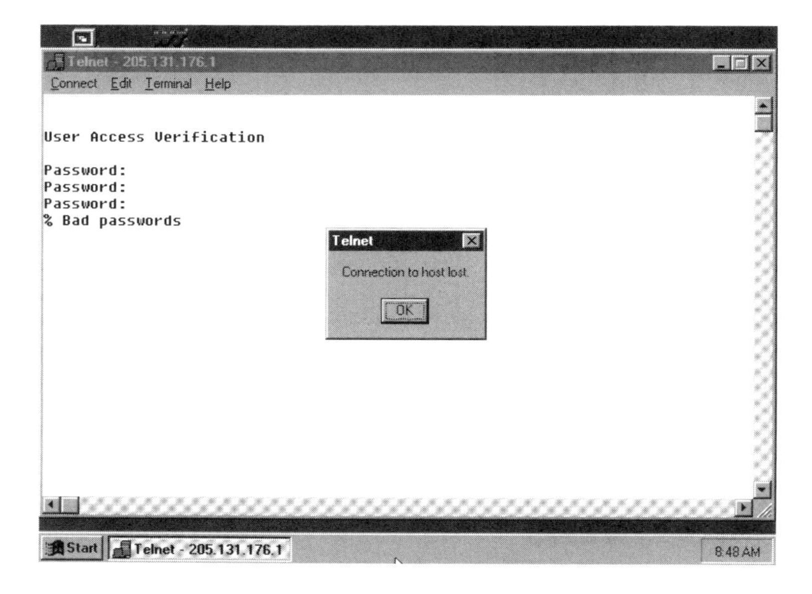

Figure 10.1 Via the use of a Telnet client included in millions of copies of Windows 95 and Windows 98, hackers can attempt to gain access to the configuration capabilities of routers connected to the Internet

10.1.5 Securing console and virtual terminals

After you unpack a router and initiate its installation process, it is extremely important to consider the manner in which access to configuring the device will occur. If you only plan to enable configuration changes to occur from a directly connected terminal device, you should ensure that Telnet and TFTP access is disabled. In a Cisco router environment you can configure access from the console and virtual terminals via the use of the 'line' command. This command has the following format:

line [type-keyword] first-line [last-line]

where information in brackets represents options.

The type-keyword entry can be entered as either 'console', 'aux' or 'vty.' The console entry is used to represent a console terminal line, representing a device directly cabled to a port on the router. In comparison, aux is used to indicate an auxiliary line, and allows you to specify access via a port on the router connected to a CSU, DSU, or modem, permitting serial communications from afar. The third option, vty, represents a virtual terminal connection for remote console access. Note that when entering the line command, the first-line and last-line represent a number of contiguous entries that are applicable to a specific device and can be represented and associated with a line number.

When configuring access through the use of the line command it is also important to consider associating a password with the device you enable for access. Even if you only plan to allow access to a router via a directly cabled terminal device located in a secure technical control center, every once in a while a situation can occur that would justify the password protection. In one event with which this author is familiar, a tour of the technical control center of a government network by a group of Boy Scouts resulted in one extremely inquisitive individual inadvertently causing a bit of havoc. As the rest of the group moved to an area of the technical control center to view a graphical display of the status of the network, this inquisitive individual started playing with a terminal directly cabled to a router and which functioned as the router console. Not knowing what to enter, the Boy Scout entered a question mark (?), which resulted in the display of router commands. Within a short period of time, this Boy Scout managed to mis-configure the router while the rest of the group was on the other side of the center listening to a briefing given by the manager of the center. Needless to say, if a password had been previously associated with terminal access, the unintentional mis-configuration of the router and the resulting havoc it created would not have been possible.

In a Cisco router environment you can associate a password with a remote access method. To do so, you would use the password command. For example,

```
line console
password bugs4bny
```

would block console access until the console operator responded with the password bugs4bny to a prompt generated by the router for a password.

The password associated with the Cisco password command can be up to 80 characters in length. The password is case-sensitive and can contain any combination of alphanumeric characters, including spaces. While this capability provides the router administrator with the ability to be innovative, it also provides the ability to make it extremely difficult for authorized users to gain access to the router. This is because selecting a password based upon a large number of varying upper and lower case letters mixed with numerics makes it subject to erroneous entry. While this type of password will certainly be difficult to guess and should avoid the possibility of a successful dictionary attack, it is also easy for an authorized router administrator to enter incorrectly. If incorrectly entered three times, a Cisco router will return the terminal attempting access to the idle state of operation. Thus, when selecting a password, it is important to remember several password principles. First, use a mixture of alphabetic and numeric characters to alleviate the potential of a dictionary attack being successful. Secondly, as you structure a password remember that as you extend the length of the password you also increase the possibility of password entry error. In general, passwords that are between 10 and 15 characters in length should be sufficient if they are structured to join a few abbreviated words with a sequence of numerics.

10.1.6 File transfer

Previously we noted that the Trivial File Transfer Protocol is commonly supported by routers as a mechanism to permit system image and configuration files to be stored on workstations. In a Cisco router environment, to enable the loading of network configuration files at router reboot time, you must specify the 'service config' command, as the default is the disabling of this capability. If this capability is enabled, the router will broadcast a TFTP read request message, and the first station to respond will have the file with a specific name based upon the router's configuration loaded into the router across the network. Because a standardized file naming scheme is used, this author believes it is best to consider leaving this feature in its disabled state instead of opening the ability for an inquisitive employee with a bit of knowledge to 'see what would happen' if they created a configuration file.

10.1.7 Internal router security

Once access has been gained into a router, the operating system of the device may provide a further level of protection capability you can use for additional router access security. In a Cisco router environment the command interpreter included in the operating system is referred to as the EXEC. The EXEC has two levels of access: user and privileged.

The user level of access allows a person to use a small subset of all router commands, such as commands that enable the listing of open router connections, and commands for providing a name to a logical connection and displaying certain statistic concerning router operations. In comparison, the privileged level of access includes all user access commands as well as commands that govern the operation of the router, such as the configure command, which allows a router administrator to configure the router, the reload command, which halts the operation of the device and reloads its configuration, and similar commands that have an active effect upon the operational state of the device.

Due to the ability of a person gaining access to the privileged mode of operation of a Cisco router obtaining the ability to directly control the operation of the router, this level of access can also be password-protected. Thus, when installing a Cisco router, it is important to use the enable-password configuration command to protect access to the privileged level of router access. For example, to assign the password power4you for the privileged command level, you would use the enable password command as follows:

```
enable password power4you
```

Similar to the password associated with a serial terminal line, the password assigned to the privileged command level is case-sensitive, can contain any mixture of alphanumeric characters, including spaces, and can consist of up to 80 characters. Thus, by placing a password on the serial port or on any allowed virtual terminal connections as well as on the privileged command level of the router, you protect both access into the router as well as the use of privileged commands once access has been obtained.

Figure 10.2 illustrates the initial router configuration process and the assignment of passwords. Note that after router interfaces have been displayed and a name (BigMac) has been entered for the router, you are prompted to enter three passwords. The first password, referred to as 'enable secret,' is a one-way cryptographic secret that is used instead of the enable password. The second is the enable password, which will be used when there is no enable secret and when using older software and some boot images. The third password is the virtual terminal password. Once those passwords have been entered, the router prompts you to enter specific configuration data, only a portion of which is shown in the lower part of Figure 10.2. The passwords shown entered in Figure 10.2 are for illustrative purposes only and violate the description of password compositions mentioned previously in this section.

10.1.8 Additional protective measures

If you need to provide one or more persons on a network with the ability to configure one or more routers, you can add an additional layer of protection beyond passwords. To do so, you can program one or more router access-lists. Although the use of router access-lists is covered in the next section, we can briefly note that they represent a sequential collection of permit and deny conditions that can be applied to Internet addresses. This means that if you can determine the IP addresses of stations that will require the ability to have

First, would you like to see the current interface summary? [yes]: y

Any interface listed with OK? value "NO" does not have a valid configuration

Interface	IP-Address	OK?	Method	Status	Protocol
Ethernet0	unassigned	NO	not set	up	down
Serial0	unassigned	NO	not set	down	down
Serial1	unassigned	NO	not set	down	down
TokenRing0	unassigned	NO	not set	reset	down

Configuring global parameters:

Enter host name [Router]: BigMac

The enable secret is a one-way cryptographic secret used
instead of the enable password when it exists.

Enter enable secret: gizmo

The enable password is used when there is no enable secret
and when using older software and some boot images.

Enter enable password: beverly
Enter virtual terminal password: spring
Configure SNMP Network Management? [yes]: y
Community string [public]:
Configure IPX? [no]:
Configure bridging? [no]:
Configure IP? [yes]:
Configure IGRP routing? [yes]:

Figure 10.2 Configuring a Cisco router allows three passwords to be specified

operators configure one or more routers via a network connection, you can use the access-list capability of each router to restrict Telnet access to each router to one or more specific IP addresses. This means that not only does the terminal operator need to know the correct passwords to gain access to an appropriate router, but, in addition, they can only perform such access from predefined locations. By combining password protection into a router with password protection to its privileged mode of operation and restricting configuration access to predefined locations via the use of one or more access-lists, you can in effect close the proverbial door to the router.

10.2 ROUTER ACCESS-LISTS

In the first section of this chapter we focused our attention upon securing access to a router's configuration capability. In that section it was noted that one method to secure access to a router was through the use of an appropriate access-list. It was also mentioned that to facilitate access control, router access-lists could control such access; however, it was left until this section to probe deeper into this security capability.

In this section we will examine the operation, utilization, and limitations of access-lists. Although we will discuss access-lists in general terms applicable to products manufactured by different vendors, we will also discuss and describe specific types of access-lists and give several examples that illustrate how they operate. In doing so, we will discuss and describe access-lists supported by routers manufactured by Cisco Systems, as this vendor currently represents approximately 70% of the market for this popular communications device. Although specific examples of the use of access-lists in this section are oriented towards Cisco Systems products, it should be noted that other router manufactures offer an equivalent capability. This means that the examples presented in this section are relevant to other vendor products—often with little or no modification required. It should also be noted that because there are many versions of the Cisco Systems router operating system referred to as IOS, the actual capability and coding of an access-list depends upon the version of the IOS used. In this discussion we will focus our attention upon a common core set of access-list parameters supported by the vast majority of different versions of Cisco Systems' IOS released over the past few years.

10.2.1 Overview

An access-list represents a sequential collection of permit and deny conditions that are applied to certain field values in packets that attempt to flow through a router interface. Once an access-list has been configured, it is applied to one or more router interfaces, resulting in the implementation of a security policy. As packets attempt to flow through a router's interface, the device compares data in one or more fields in the packet with the statements in the access-list associated with the interface. Data in selected fields in the packet are compared against each statement in the access-list in the order in which the statements were entered to form the list. The first match between the contents or conditions of a statement in the access-list and one or more data elements in specific fields in each packet determine whether or not the router permits the packet to flow through the interface. If the condition for packet flowing through the router is not met then the router sends the packet to the great bit bucket in the sky via a filtering operation.

At a minimum, router access-lists control the flow of data at the network layer. Because there are numerous types of network layer protocols, there are also numerous types of access-lists, such as Novell NetWare IPX access-lists, Internet Protocol IP access-lists, and Decnet access-lists. Because of the focus of this book on the management of the Transmission Control Protocol/Internet Protocol (TCP/IP) and the important role of the Internet Protocol in accessing them we will narrow our examination of access-lists to those that support the TCP/IP protocol suite.

10.2.2 TCP/IP protocol suite review

To obtain an appreciation of the manner in which IP access-lists operate, a brief review of a portion of the TCP/IP protocol suite is in order. At the application layer the contents of a data stream representing a particular application in the protocol suite, such as a file transfer, remote terminal session, or an electronic mail message, is passed to one of two transport layer protocols supported by the TCP/IP protocol suite: the Transmission Control Protocol (TCP) and the User Datagram Protocol (UDP).

Both TCP and UDP are layer 4 protocols that operate at the transport layer of the International Organization for Standardization (ISO) Open Systems Interconnection (OSI) Reference Model. Because a host computer operating the TCP/IP protocol stack can support the operation of multiple concurrent applications, a mechanism is required to distinguish one application from another as application data is formed into either TCP or UDP datagrams. The mechanism used to distinguish one application from another is the port number, with each application supported by the TCP/IP protocol suite having an associated numeric port number. For example, a host might transmit a packet containing an email message followed by a packet containing a portion of a file transfer, with different port numbers in each packet identifying the type of data contained in each packet. Through the use of port numbers, different applications can be transmitted to a common address, with the destination address using the port numbers in each packet as a mechanism to demultiplex one application from another in a data stream received from a common source address. Port numbers are assigned by the Internet Assigned Numbers Authority (IANA). IANA maintains a list of assigned port numbers that anyone with access to the Internet can access.

TCP is a connection-oriented protocol that provides a guaranteed delivery mechanism. Because a short period of time is required to establish a TCP connection prior to obtaining the ability to exchange data, it is not extremely efficient for transporting applications that only require small quantities of data to be exchanged, such as a management query that might simply retrieve a parameter stored in a remote probe. Recognizing that this type of networking situation required a speedier transmission method resulted in the development of UDP. UDP was developed as a connectionless, best effort delivery mechanism. This means that when a UDP session is initiated, data transmission begins immediately instead of having to wait until a session connection is established. This also means that the upper layer application becomes responsible for having to set a timer to permit a period of time to expire without the receipt of a reply to determine that a connection either was not established or was lost.

Although both TCP and UDP differentiate one application from another by the use of numeric port values, actual device addressing is the responsibility of IP, a network layer protocol that operates at layer 3 of the ISO OSI's Reference Model. As application data flows down the TCP/IP protocol stack, either a TCP or a UDP header is added to the data, with the resulting segment of data containing an appropriate port number that identifies the application being transported. Next, as data flows down the protocol stack, layer 3 operations result in an IP header being prefixed to the TCP or the UDP header. The IP header contains the IP destination and IP source addresses as 32-bit numbers under IPv4. We frequently code the IP address when configuring the protocol stack as four numerics separated by decimal points, referred to as dotted decimal notation. Based upon the preceding, there are three addresses that can be used in an IP access-list for enabling or disabling the flow of packets through a router's interface: the source IP address, the destination IP address, and the port number that identifies the application data in the packet. In actuality Cisco Systems and other router manufacturers also support other IP-related protocols, such as the Internet Control Message Protocol (ICMP) and the Open Shortest Patch First (OSPF) protocol as a mechanism to enable or disable the flow of predefined types of error messages and queries, with an example of the latter being an ICMP echo packet request and echo packet response.

10.2.3 Using access-lists

In a Cisco router environment there are two types of IP access-lists that you can configure: standard or basic access-lists and extended access-lists.

A standard access-list permits filtering by source address only. This means that you can only permit or deny the flow of packets through an interface based upon the source IP address in the packet. Thus, this type of access-list is limited in its functionality. In comparison, an extended access-list permits filtering by source address, destination address and various parameters associated with upper layers in the protocol stack, such as TCP and UDP port numbers.

Configuration principles

When developing a Cisco router access-list, there are several important principles to note. First, Cisco access-lists are evaluated in a sequential manner beginning with the first entry in the list. Once a match occurs, access-list processing terminates and no further comparisons occur. Thus, its important to place more specific entries towards the top of your access-list.

A second important access-list development principle to note is the fact that there is always an implicit deny at the end of the access-list. This means that the contents of a packet that does not explicitly match one of the access-list entries will automatically be denied. You can override the implicit deny by placing an explicit permit 'all' as the last entry in your list.

A third principle concerning the configuration of access-lists concerns additions to the list. Any new access-list entries are automatically added to the bottom of the list. This fact is important to note, especially when attempting to make one or more modifications to an existing access-list. This is because the addition of statements to the bottom of an access-list may not result in the list being able to satisfy organizational requirements. Many times it may be necessary to delete and recreate an access-list instead of adding entries to the bottom of the list.

A fourth principle concerning access-lists is that they must be applied to an interface. One common mistake some persons make is to create an appropriate access-list and forget to apply it to an interface. In such situations the access-list simply resides in the router's configuration memory area but will not be used to check the flow of data packets through the router, in effect similar to leaving the barn door ajar after spending time constructing a fine structure. Now that we have an appreciation of key access-list configuration principles, let us turn our attention to the creation of standard and extended Cisco router access-lists.

Standard access-lists

The basic format of a standard access list is as follows:

access-list number {permit|deny} [ip address] [mask]

Each access-list is assigned a unique number that identifies the specific list as well as informing the router's operating system of the type of the access-list. Standard Cisco IP access-lists are assigned an integer number between 1 and 99.

A new release of Cisco's router operating system permits access-list names to be defined. However, because named access-lists are not backward-compatible with earlier versions of router operating system, we will use numbered lists in the examples presented in this section.

Because standard access-lists only support filtering by source address, the IP address in the above access-list format is restricted to representing the originator of the packet. The mask that follows the IP address is specified in a manner similar to the way in which a network mask is specified when subnetting an IP address. However, when used in an access-list, the binary 0 in the mask is used as a 'compare' while a binary 1 is used as an 'unconditional' match. This is exactly the opposite of the use of binary 1s and binary 0s in a network mask to subnet an IP address. Another difference is that in Cisco router terminology the mask used with an access-list is referred to as a wildcard mask and not as a network mask or a subnet mask.

To illustrate the use of a Cisco router wildcard mask, let us assume that your organization's router is to be connected to the Internet, and that your network configuration is as illustrated in Figure 10.3, with a World Wide Web server located behind the router. Let us further assume that you want to allow all hosts on the Class C network at another location whose IP address is 205.131.176.0 access to the server. If you used a traditional network mask its composition would be 255.255.255.0. Writing the network and mask in binary would result in the following, where the letter x represents a 'don't care' condition where either a binary 1 or binary 0 can occur in the appropriate bit position:

network address 205.131.176.0 =
11001101.10000011.10100110.00000000

network mask 255.255.255.0 =
11111111.11111111.11111111.00000000
- -
resulting address match =
11001101.10000011.10100110.xxxxxxxx

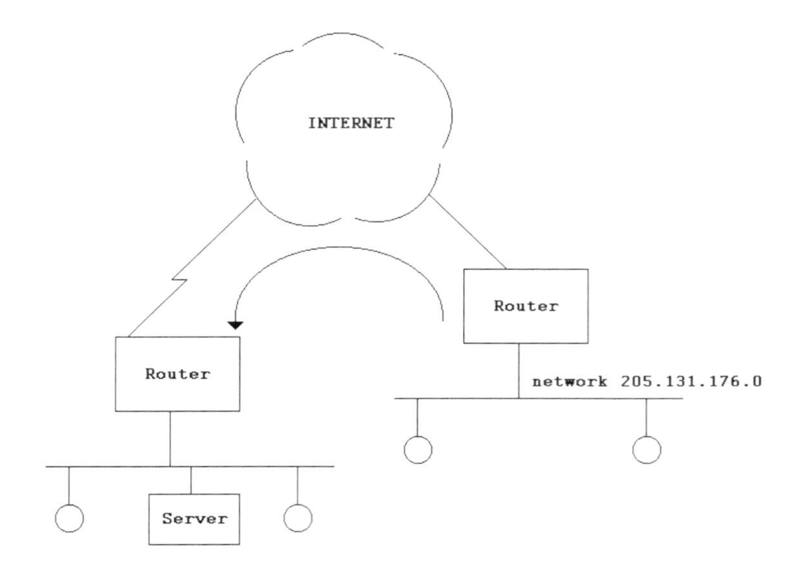

Figure 10.3 Sample network configuration for illustrating access-list creation and the use of Cisco wildcard masks

Note that a binary 1 in the network mask represents a compare, while a binary 0 represents an unconditional match. When working with Cisco access-lists, the use of binary 1s and 0s in the wildcard mask is reversed. That is, a binary 1 specifies an unconditional match while a binary 0 specifies a compare condition. However, if you attempt to use the same mask composition instead of reversing its composition, you will more than likely obtain a result that does not meet your operational requirements. This is illustrated by the following example, where a wildcard mask is used instead of a network mask:

network address 205.131.176.0 =
11001101.10000011.10100110.00000000

wildcard mask 255.255.255.0 =
11111111.11111111.11111111.00000000
- -

resulting address match =
xxxxxxxx.xxxxxxxx.xxxxxxxx.00000000

In the above example any values in the first three octet positions are allowed as long as the value in the last octet was all 0s. This is obviously not a satisfactory solution to our previously assumed Web server requirement. However, if we place 0s in the wildcard mask where we normally place binary 1s in the network mask and vice versa, we will properly define the wildcard mask. Modifying the masking operation one more time, we obtain the following:

network address 205.131.176.0 =
11001101.10000011.10100110.00000000

wildcard mask 0.0.0.255 =
00000000.00000000.00000000.11111111
- -

resulting address match =
11001101.10000011.10100110.xxxxxxxx

Note that the creation of the above mask results in specifying any host on the 205.131.176.0 network, which is the requirement that we were attempting to satisfy. Although the use of Cisco wildcard masks can be a bit confusing at first—especially if you have a considerable amount of experience in using subnet masks—once the concept has been grasped, it is as easy to apply to an access-list as a subnet mask to a network address. However, it is extremely important to remember that the wildcard mask is a reverse of the network mask, including the function of binary 1s and 0s and their positioning in the mask, and to apply it accordingly. Now that we have an understanding of the creation and use of Cisco wildcard masks, let us return to our example and complete the creation of a standard access-list. That access-list would be constructed as follows:

access-list 77 permit 205.131.176.0 0.0.0.255

In this example we have used a list number of 77, which, being between 1 and 99, identifies the access-list as a standard access-list to the router's operating system. Also note that the network address of 205.131.176.0 and a wildcard mask of 0.0.0.255 results in a don't care condition for any value in the last octet of the network address, permitting any host on the 205.131.176.0 network to have its packets flow through the router without being filtered.

A few additional items concerning access-lists warrant attention. First, if you omit a mask from an associated IP address, an implicit mask of 0.0.0.0 is assumed, which then requires an exact match between the specified IP address in the access-list and the packet to occur for the permit or deny condition in the access-list statement to take effect. Secondly, as previously mentioned, an access-list implicitly denies all other accesses. This is equivalent to terminating an access-list with the following statement:

access-list 77 deny 0.0.0.0 255.255.255.255

To provide another example of the use of a standard access-list, let us assume that your organization uses a router to connect two Ethernet segments together as illustrated in Figure 10.4.

In examining the use of the router illustrated in Figure 10.4, let us assume that segment 1 has the network address 198.78.46.0 and you want to enable clients with the host addresses .16 and .18 on segment 1 to access any server located on segment 2. To do so, your initial router configuration, including applying the access-list to the outgoing interface on Ethernet 1 (E1), would consist of the following statements:

```
interface ethernet 1
access-group 23 out
access-list 23 permit 198.78.46.160.0.0.0
access-list 23 permit 198.78.46.180.0.0.0
```

In the preceding example note that the access-group statement is used to define the data flow direction that is associated with an access-list. Also note that the access-list was applied to the outgoing interface on Ethernet 1 instead of to the inbound interface on Ethernet 0 (E0) towards the router from segment 1 as an inbound access-list. While either method would work, the latter method would have the potentially undesirable effect of blocking all other traffic from leaving segment 1. Thus, in this example we have elected to apply the access-list to the outgoing interface on E1. Now that we have an appreciation for standard IP access-lists, let us turn our attention to their extended cousins.

Extended access-lists

A standard access-list is limited to specifying a filter via the use of a source IP address. In comparison, an extended access-list provides you with the ability to filter by source address, destination address, and upper layer protocol information, such as TCP and UDP port values. In fact, extended access-lists provide you with the ability to create very complex packet filters whose capabilities can significantly extend beyond those of a standard access-list.

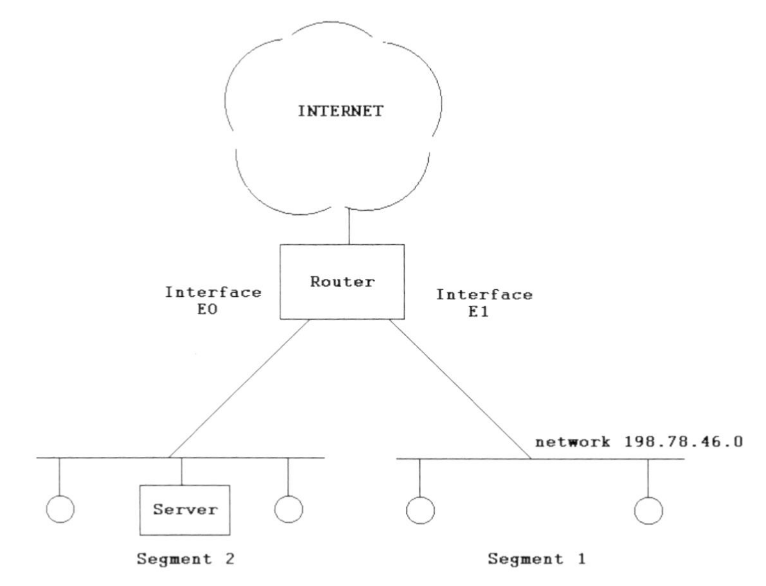

Figure 10.4 Using a router to interconnect two Ethernet segments

Extended access-lists have the following format:

access-list number {permit|deny} protocol source IP
address
source-mask destination IP address destination-mask
/[operator operand] [established]

Similar to standard access-lists, extended access-lists are numbered. Extended access-lists are numbered between 100 and 199 to distinguish them from standard IP access-lists. The protocol parameter identifies a specific TCP/IP protocol, such as ip, tcp, udp, icmp, and several routing protocols that can be filtered. Examples of the latter include the Interior Gateway Routing Protocol (IGRP) and the Open Shortest Path First (OSPF) routing protocol. The arguments source IP address and destination IP address represent the source and destination IP addresses expressed in dotted decimal notation. The arguments source-mask and destination-mask represent router wildcards used in the same manner as previously described when we examined the operation of standard access-lists. To obtain the ability to specify additional information about packets for filtering, you can include the optional arguments operator and operand in your extended access-list. When used, the operator and operand can be employed to compare tcp and udp port values. Concerning tcp and udp, the argument operators can be one of the following four keywords:

lt less than
gt greater than
eq equal
neq not equal

In comparison, the argument operand represents the integer value of the destination port for the specified protocol. For the TCP protocol the additional optional keyword 'established' is supported. When specified, a match occurs if a TCP datagram has its ACK or RST field bits set, indicating that an established connection has occurred.

To illustrate the use of an extended access-list, let us assume that the router previously illustrated in Figure 10.4 will be connected to the Internet. Let us further assume that you want to enable any host on the network behind the router whose IP address is 198.78.46.0 to establish TCP connections to any host on the Internet. However, let us also assume that, with the exception of accepting electronic mail via the Simple Mail Transport Protocol (SMTP), it is organizational policy to bar any host on the Internet from establishing TCP connections to hosts on the 198.78.46.0 network. To accomplish the preceding, you must ensure that the initial request for an SMTP connection, which is made on TCP destination port 25, occurs from a port number greater than 1023, with the originator always using destination port 25 to access the mail exchanger on your organization's network and that host using a port number greater than 1023 to respond.

Based upon the preceding and assuming that the address of the mail exchanger on the 198.78.46.0 network is 198.78.46.77, the following two access-lists would be employed:

```
access-list 101 permit tcp198.78.46.00.0.0.255 0.0.0.0
255.255.255.255
access-list 102 permit tcp 0.0.0.0 255.255.255.255 198.78.46.07
0.0.0.255 established
access-list 102 permit tcp 0.0.0.0 255.255.255.255 198.78.46.07
eq25
interface serial 0
ip access-group 101
interface ethernet 0
ip access-group 102
```

In the preceding example note that access-list 101 is applied to the router's serial port and is constructed to enable any host on the 198.78.46.0 network to establish a TCP connection to the Internet. The second access-list, which is numbered 102 in the above example, is applied to the Ethernet 0 (E0) interface previously illustrated in Figure 10.4. The first statement in the 102 access-list permits any TCP packet that represents an established connect-ion to occur, while the second statement in the access-list permits TCP packets from any source address flowing to the specific network address 198.78.46.77 with a port value of 25 to flow through the interface. Thus, an inbound connection via port 25 must occur in order for the first statement in the 102 access-list to permit succeeding packets with port numbers greater than 1023 to flow through the router.

Limitations

Although access-lists provide a significant capability to filter packets, they are far from being a comprehensive security mechanism. Thus, in concluding this examination of Cisco access-lists, a few words of caution are in order concerning their limitations.

In examining access-lists, we noted that they are constructed to filter based upon network addresses. This means they are vulnerable to address impersonation or spoofing. Another key limitation associated with their use is the fact that they cannot note whether or not a packet is part of an existing upper-layer conversation or what the conversation is about. This means that a person could run a dictionary attack through the packet filtering capability of a router if their address was not barred. Similarly, a host that is allowed ftp access could issue an mget *.* command and retrieve several gigabytes of data from a server, in effect creating a denial of service attack. Due to the preceding limitations, most organizations supplement router access-lists through proxy services incorporated into a firewall, which is the subject of the next section.

10.3 USING FIREWALL PROXY SERVICES

As the use of the TCP/IP protocol suite expanded during the 1990s in tandem with the growth in the use of the Internet, organizations began to realize that a new security threat emerged as their networks were connected to the Internet. As academic, government, and commercial networks were connected to the Internet, they became subject to attack from an unlimited population of computer users located throughout the world. Router access-lists provide a mechanism to enable or disable the flow of packets through router ports based upon source and destination IP address and the type of application data expressed in the form of a port number. Organizations began to realize that by themselves router access-lists were not a sufficient barrier to prevent many types of undesirable operations against hosts residing behind a router. One solution developed to provide a higher level of security to organizational networks was the use of a firewall with a proxy service capability located behind the router, which is the focus of this section.

In this section we will first briefly review the operation of router access-lists and some of their limitations. Using this information as a base, we will then describe and discuss the operation of different types of firewall proxy services and how they can be used to obtain an enhanced level of network protection.

10.3.1 Access-list limitations

Most routers include a packet filtering capability created by the coding of one or more statements into what is referred to as an access-list and then applying the access-list to a router interface. Access-list statements include parameters that are evaluated against values in packet fields formed at layers 3 and 4 of the International Organization for Standardization (ISO) Open System Interconnection (ISO) Reference Model. In a TCP/IP protocol suite environment this means that an access-list primarily operates by examining the source and destination IP addresses in a packet and the port number contained in the packet, which identifies the application being transported in the packet and which is formed at layer 4 of the ISO Reference Model.

One of the key limitations associated with the use of access-lists is the fact that they are in effect blind with respect to the operation being performed. This results from the inability of router access-lists to look further into the contents of a packet and determine whether or not an apparently harmful operation is occurring, and if so, to either stop the operation or generate an appropriate alert message to one or more persons in the form of an audio signal, email message, pager alert, or combination of such mechanisms.

To illustrate a potential limitation of router access-lists consider the popular file transfer protocol (ftp) application used to transfer files between hosts. When using a router's access-list, you can enable or deny ftp sessions based upon the source IP address and/or destination IP address contained in each packet transporting ftp information. Suppose that your organization operates an ftp server supporting anonymous access, allowing any person connected to the Internet to access and retrieve information from the ftp server, a relatively common occurrence on the Internet. Let us further assume that your organization has a large number of files on the server available for downloading. This means that a person could either intentionally or unintentionally use the ftp mget (multiple get) command to retrieve a large number of files with one ftp command line entry. In fact, if the person accessing your organization's ftp server issued the mget command using the wildcard operator of an asterisk (*) in the filename and file extension position to form the command line entry mget *.* then this command would result in your organization's ftp server downloading every file in the directory, one after another, to the remote user. If your organization has a large number of files whose aggregate data storage represents several gigabytes of data and a low speed connection to the Internet, such as a 56-kbps, 64-kbps, or fractional T1 connection, the use of an mget *.* command could tie up the outbound use of the Internet connection for many hours and possible days. If your organization operates a World Wide Web server as well as an ftp server and provides Internet access to employees over a common access line, the use of mget on an intentional basis can be considered to represent an unsophisticated but effective denial of service (DOS) attack method. This type of attack is perfectly legal, as the person employing the mget command is performing a perfectly valid operation, even though the result of the operation could tie up your organization's connection to the Internet for hours or even days. Similarly, letting persons have the ability to download data to your organization's ftp server means that they could consider using the reverse of mget, which is the mput command. Through the use of mput with wildcards, they could set up an antiquated 286 machine and pump gigabytes of data to your ftp server, clogging the inbound portion of your organization's Internet access line. Recognizing that the need to examine application layer operations and provide organizations with the ability to control applications resulted in the development of a proxy services capability, which is included in many firewalls.

10.3.2 Proxy services

Proxy services represents a generic term associated with the use of a proxy server. The proxy server is normally implemented as a software coding module on a firewall, and supports one or more applications for which the server acts as an intermediary or proxy between the requestor and the actual server that provides the requested service. When implemented in this manner, all requests for a specific application are first examined by the proxy service operating on the proxy server. If the proxy service was previously configured to enable or disable one or more application features for a specific TCP/IP application then the proxy service examines the contents of each packet and possibly a sequence of packets and compares the contents against the proxy service configuration. If the contents of the packet or sequence of packets that denote a specific operation are permitted by the configuration of the proxy service then the service permits the packet to flow to the appropriate server. Otherwise the packet is either directly sent to the great bit bucket in the sky or possibly permitted, with the server generating a warning message and an alert or alarm message to the firewall administrator or other designated personnel.

To illustrate the use of a proxy service, let us return to our ftp server access example. A common ftp proxy service permits a firewall administrator to enable or disable different ftp commands. Using this feature, the firewall administrator can control the ability of ftp users to issue different types of ftp commands, such as mget and mput.

In a Microsoft Windows environment you can use mget in either a streaming or an interactive mode. Concerning the latter, ftp will prompt you through the use of a question mark (?) whether or not the next file should be transferred. An example of the use of mget is illustrated in Figure 10.5. Note that by simply entering a carriage return in response to the ? prompt, the next file is transferred. Thus, it is relatively easy for a hacker to write a script to stream files when using mget under Windows' interactive mode and a no-brainer under its streaming mode.

If you are familiar with the manner in which an ftp server is configured, you probably realize that the ftp server administrator is limited to assigning read and/or write permissions to directories and possibly, depending upon the operating system used, to files within a directory for either anonymous or non-anonymous users, with the latter a term used to denote persons who have an account on the server. However, there is no mechanism that this author is aware of that enables an ftp server administrator or a router administrator to selectively enable or disable individual ftp commands. Thus, an ftp proxy service provides the ftp server administrator with a significantly

```
Command Prompt - ftp 205.131.176.11
Microsoft(R) Windows NT(TM)
(C) Copyright 1985-1996 Microsoft Corp.

C:\>ftp 205.131.176.11
Connected to 205.131.176.11.
220 www Microsoft FTP Service (Version 3.0).
User (205.131.176.11:(none)): gxheld
331 Password required for gxheld.
Password:
230-For Official Government Authorized Users only.
230 User gxheld logged in.
ftp> mget *.*
mget 01enrol.htm?
200 PORT command successful.
150 Opening ASCII mode data connection for 01enrol.htm(16994 bytes).
226 Transfer complete.
16994 bytes received in 0.15 seconds (113.29 Kbytes/sec)
mget 1-12PRAC.HTM?
```

Figure 10.5 Using mget under Windows NT requires a response to each file prompt, which can be simply a carriage return

enhanced capability that can be used to configure the capability and features of ftp services that other users can access.

The capability to employ proxy services is based on the use of a firewall located between a router and network servers connected to a LAN behind the router. Thus, the type of proxy services that can be provided is only limited by the requirements of an organization and the programming capability of firewall programmers. Some of the more popular types of proxy services include remote terminal Telnet and TN3720 proxy services, Simple Mail Transport Protocol (SMTP) proxy service, HyperText Transport Protocol (HTTP) proxy service, the previously discussed ftp proxy service, and ICMP proxy service. The latter represents a special type of proxy service and deserves a degree of elaboration due to the enhanced security capability it provides against certain types of hacker attacks.

10.3.3 ICMP proxy services

The Internet Control Message Protocol (ICMP) represents a layer 3 protocol within the TCP/IP protocol suite. ICMP is used to transmit error messages as well as status queries and responses to those queries. ICMP packets are formed by the use of an Internet Protocol (IP) header that contains an appropriate numeric in its Type field that identifies the packet as an ICMP packet. Although the use of ICMP is primarily oriented towards transporting error messages between devices operating a TCP/IP protocol stack and transparent to users of a network, the protocol is also popularly used by many individuals who are probably unaware that they are using transmitting ICMP packets.

Two of the more popular types of ICMP packets are Echo Request and Echo Response, which are better known to most persons as a Ping operation or application. Depending upon the manner by which the Ping application is implemented on a specific TCP/IP protocol suite, a user typically enters the application command name Ping followed by the host name or host IP address and one or more optional parameters that govern how Ping will operate.

The use of Ping is primarily designed as a mechanism to allow a user to determine if a remote host is operational and using a TCP/IP protocol stack. Pinging a distant host with an ICMP Echo Request packet results in the distant host returning an ICMP Echo Response packet if the distant host is reachable, operational and its TCP/IP stack is functioning. The reason that the use of Ping also notes if a distant host is reachable and that a Ping timeout does not necessarily mean the distant host is not operational is that one or more communications devices in the path to the distant host could be down while the distant host is operational. However, in most cases Ping represents the first troubleshooting method to use when it appears that a host is not responding to a query.

In addition to indicating that a host is potentially reachable and operational, the use of Ping provides information concerning the round-trip delay to a remote host. This information results from the Ping application on the originator setting a clock and noting the time until a response is received or a time-out period occurs and no response is received. The time between the transmission of the Ping and the receipt of a response represents the packet round-trip, and can provide valuable information about why a time-dependent operation such as Voice over IP (VoIP) produces a reproduced voice that sounds like a famous mouse instead of a person.

When for the first time you Ping a destination using a host name, your protocol stack may have to perform an address resolution operation to determine the IP address needed for routers to correctly direct the packet to its destination, leading to an additional delay. Therefore, most implementations of Ping by default issue between three and five consecutive Echo Request packets. However, some implementations of Ping permit the user to set an option which results in the host continuously issuing Pings, one after another until the person operating the computer generating the Pings issues a CTRL-BREAK to terminate the application.

Although continuous Pinging may appear innocent, in actuality it represents a method for a hacker to initiate a denial of service attack. This is because the Pinged host must stop what it is doing, even if its only for a few milliseconds, and respond to the Ping with an Echo Response ICMP packet. If the person who sets the Ping application to continuous Pinging also sets the packet size beyond its default size of 32 or 64 bytes, depending upon implementation, that person forces the destination to respond with increased length responses, which requires the use of additional network resources. Thus, although the use of the Ping application may not bring the destination host literally to its knees, it can be configured to operate in a manner that can significantly interfere with the ability of the destination to perform its intended operations.

Another problem associated with the unrestricted use of Ping is that it can be used as a mechanism to discover hosts on a distant network as a prelude for attacking those hosts. For example, a hacker could write a script to cycle though all 254 possible addresses on a Class C IP network as a mechanism to discover which addresses are currently operational.

Based upon the preceding, many organizations may wish to control the operation of Ping and other types of ICMP messages. While many router access-lists provide administrators with the ability to filter ICMP packets based upon source and/or destination IP address and the type of ICMP message, such access-list filtering is an all-or-nothing operation. That is, a router access-list cannot selectively examine and note that a sequence of ICMP Echo Requests from the same source address occurred after a predefined number of requests flowed through the router and that subsequent requests should be prohibited. In comparison, an ICMP proxy service feature can be configured to differentiate between a single sequence of Echo Request packets and the intentional or unintentional setting of a Ping application to continuously Ping a host. Similarly, an ICMP proxy service capability can be employed to distinguish between a person who may have difficulty accessing a server and another person who is using the Ping application in an attempt to discover all hosts on your organization's network. Thus, ICMP proxy service represents an important type of proxy service whose use can enhance the security of a network.

10.3.4 Limitations

Although proxy services can provide a considerable degree of security enhancement to networks, there are certain limitations associated with their use that warrant discussion. First and foremost, a proxy service requires a detailed examination of the contents of individual packets and sequences of individual but related packets, forcing the application to look deeper into each packet. This results in an additional degree of processing occurring on each packet, which introduces a degree of delay. Secondly, a sequence of packets may have to be examined to determine if it is acceptable to enable those packets to flow to their destination. This means that one or more packets in each sequence may have to be buffered or temporarily stored until the proxy service can determine if the packets can proceed to their destination or should be sent to the great bit bucket in the sky. This also means that additional buffer storage in the proxy server or firewall will be required, and the temporary storage of packets adds to the latency of remote requests flowing to servers operated by your organization. In fact, according to tests performed by several communications testing laboratories, using proxy services from different vendor firewalls resulted in between 20% and 40% of the bandwidth of an Internet connection at the proxy server side being unused during the full utilization on the other side due to proxy server delays. This also resulted in a packet loss between 20% and 40%, resulting in additional remote transmissions to the desired site. Thus, you must consider the effect of proxy service delays and the potential need to upgrade your Internet access line against the potential enhancements to the security of your organization's network.

10.3.5 Operational example

Now that we have an appreciation for the capabilities of a proxy firewall, we will conclude this section by examining several configuration screens generated by the Interceptor firewall, a product of Technologic of Atlanta, GA.

Figure 10.6 illustrates the Interceptor's Advanced Policy Options screen, on which the cursor is shown pointed to the toggled check associated with the FTP Put command to block FTP uploads. In examining Figure 10.6 and subsequent Interceptor screen displays, you will note that they represent HTML screens displayed using a Netscape browser. The Technologic Interceptor firewall generates HTML forms to enable network managers to view, add, and modify firewall configuration data. To secure such operations, the firewall uses encryption and authentication by supporting Netscape's Secure Sockets Layer (SSL) protocol for encrypting all traffic between the firewall and a Web browser used to configure the firewall while passwords are used for authentication. This means that network managers can safely configure the firewall via the World Wide Web.

Using classes

The Technologic Interceptor firewall includes a class definition facility that provides users with a mechanism to replace address patterns, times of day, or URLs by symbolic names. Classes are initiated by selecting the Classes

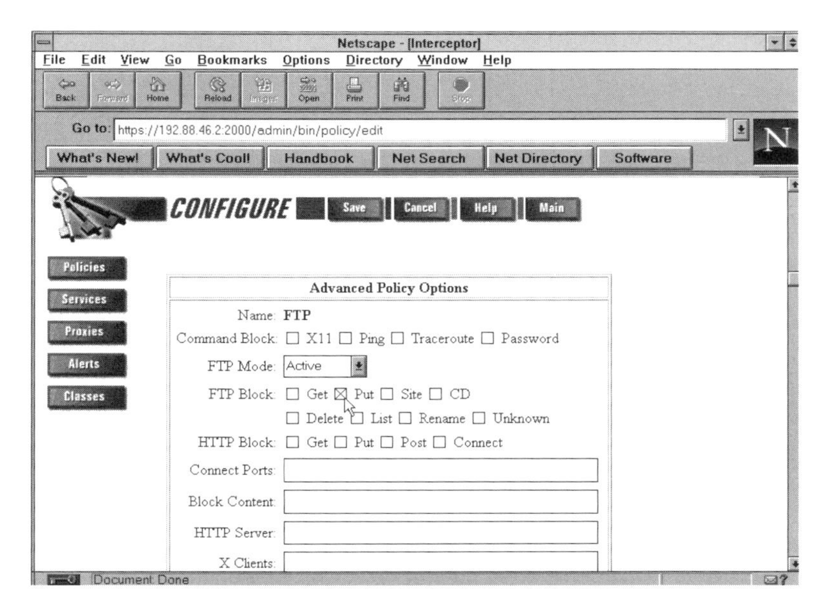

Figure 10.6 Using the Technologic Interceptor firewall configuration screen to block all FTP PUT commands

button on the left portion of the configuration screen. By using an equals sign (=) as a prefix, they are distinguished from literal patterns.

Through the use of classes, you can considerably facilitate the configuration of the firewall. For example, suppose that you want to control access from users behind the firewall to Internet services. To do so, you would first enter the IP addresses of computers that will be given permission to access common services that you wish them to use. Then you would define a class name that would be associated with the group of IP addresses and create a policy that defines the services that the members of the class are authorized to use.

Figure 10.7 illustrates the use of the Technologic Interceptor Edit Policy configuration screen to enable inbound traffic for FTP, HTTP, Telnet, and SNMP. Note that this policy uses the classname '=ALL-Internal-Hosts' in the box labeled 'From.' Although not shown, you would have first used the class configuration to enter that class name and the IP addresses that you want associated with that class. Then, this new edit policy would allow those IP addresses in the predefined class =ALL-Internal-Hosts to use FTP, HTTP, Telnet, and SMTP applications.

Alert generation

The capability of a firewall is significantly enhanced by an alert generation capability, enabling a firewall to alert a network manager or administrator to a possible attack on their network. Figure 10.8 illustrates the Technologic Interceptor Add Alert screen display, with the IP-Spoof pattern shown selected.

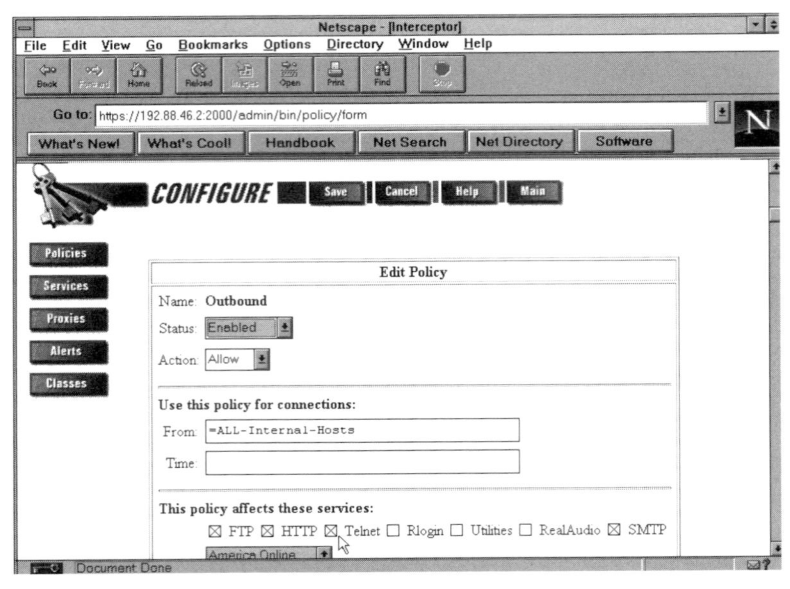

Figure 10.7 Using the Technologic Interceptor firewall to create a policy allowing outbound FTP, HTTP, Telnet, and SMTP traffic from all users in the previously defined class 'All_Internal-Hosts'

In the example shown in Figure 10.8 the IP Spoof alert is used as a mechanism to denote a connection request occurring from a host claiming to have an IP address that does not belong to it. In actuality, it can be very difficult to note the occurrence of IP spoofing. This is because, unless the firewall previously obtained information about IP addresses, such as their locations on segments whose access is obtained via different firewall ports, or notes restrictions on service by IP address, it will assume that an IP address is valid. In comparison, other patterns, such as refused connections or failed authentication, are much easier to note. For each alert you would first specify a name for the alert definition, such as IP-Spoof, for that pattern. After selecting the pattern, you can specify the days and times of day and the frequency of occurrence that, when matched, should generate an alert. The Interceptor supports two methods of alert generation: either via email or pager. If you select the use of a pager to transmit an alert, you can include a message, such as a numeric alert code, to inform the recipient of the type of alert.

Packet filtering

In concluding our brief examination of the operation of a firewall, we will examine the initiation of packet filtering. Although the packet filtering capability of firewalls functions in a manner similar to that router feature, the firewall is usually easier to configure, and provides more flexibility in enabling or disabling access based upon the set of rules that can be developed.

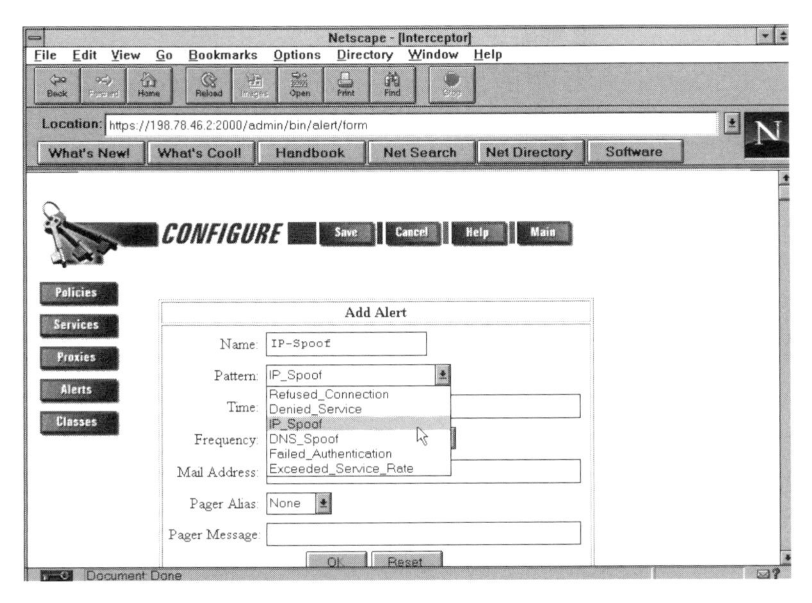

Figure 10.8 Using the Technologic Interceptor firewall Add Alert configuration screen

Figure 10.9 illustrates the Technologic Interceptor Network Services display, which lists the protocols for which this firewall accepts connections. Note that the HTTP protocol is shown selected, as we will edit that service. Also note the columns labeled 'Max' and 'Rate.' The column labeled 'Max' indicates the maximum number of simultaneous connections allowed for each service, while the column labeled 'Rate' indicates the maximum rate of new connections for each service on a per minute basis. By specifying entries for one or both columns, you can significantly control access to the network services that you provide as well as balancing the loads on heavily used services.

Figure 10.10 illustrates the Technologic Interceptor Edit Service display configuration screen. In this example, HTTP service is enabled for up to 256 connections, and a queue size of 64 has been entered, limiting TCP HTTP pending connections to that value. The Max Rate entry of 300 represents the maximum rate of new connections that will be allowed to an HTTP service. Once this rate is exceeded, the firewall will temporarily disable access to that service for a period of one minute. If you allow both internal and external access to an internal Web server, the ability to control the maximum rate of incoming connections to a particular service can be an important weapon in the war against so-called denial of service attacks. With this technique, a malicious person or group of hackers programs one or more computers to issue bogus service initiation requests using random IP addresses that more than likely do not exist. Since each access request results in a server initiating a handshake response, the response is directed to a bogus address that does not respond. The server will typically keep the connection open for 60 or 120 seconds, which represents a period of time a valid user may not be able to access the server when its connection capability is at the maximum.

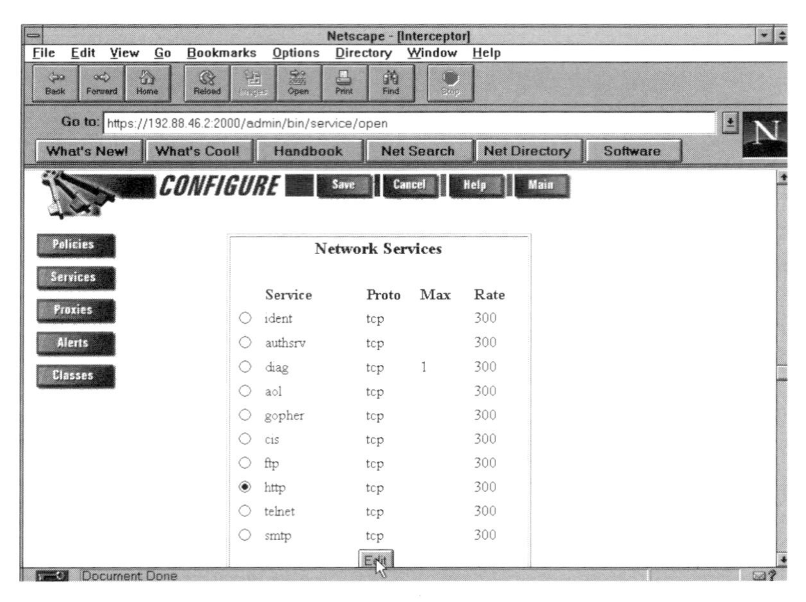

Figure 10.9 Using the Technologic Interceptor firewall configuration screen to edit HTTP network services

While there is no one uniform solution to this problem, you could use the Max Connects option to limit HTTP connections so you will always be able to let internal users access your Web server. In addition, if you specify a low Max Connects rate, you can negate some of the flooding of bogus connection attempts, which will allow some legitimate users to reach your organization's Web server.

The gap to consider

While routers and firewalls can be used to prevent unauthorized access to network hosts, they do not guarantee the security of the communications connection between client and server, or the security of the data being transported. To obtain this security, you must use some type of authentication and encryption. For example, when using Web browsers, you should consider the use of two related Internet protocols, Secure Sockets Layer (SSL), which was developed by Netscape, or the Secure Hypertext Transfer Protocol (S-HTTP), which was developed by Enterprise Integration Technologies, as well as digital certificates available from several organizations. The former support several cryptographic algorithms that use public key encryption, for which digital certificates provide authentication.

10.4 NETWORK ADDRESS TRANSLATION

As mentioned at the beginning of this chapter, we will conclude with a discussion of a feature that can be performed by both routers and firewalls.

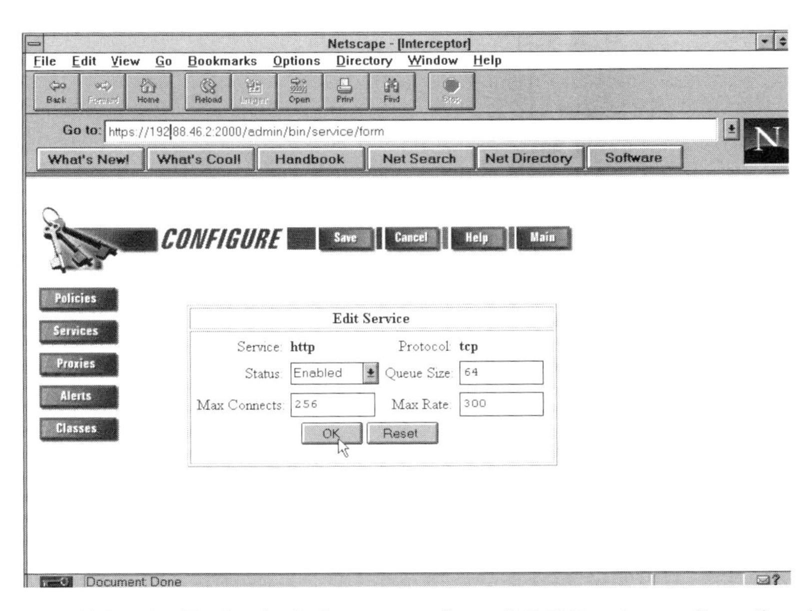

Figure 10.10 Using the Technologic Interceptor firewall Edit Service configuration display to set a series of rules to govern access to HTTP

That feature is Network Address Translation (NAT). This has evolved because of the growing scarcity of IPv4 address space. As the use of the Internet expanded, the ability of organizations to obtain registered IP addresses from their service providers became more difficult. Recognizing the fact that only a small portion of local network users would be accessing the Internet at any particular point in time, it became possible for organizations to assign each station a private IP address, typically from one of the blocks of addresses reserved in RFC 1918, which was covered in a previous chapter. Then, an address translator could be used to map or translate unregistered private IP addresses into registered addresses on the other side of the device. If an organization had 1000 stations, the mapping of 1000 private unregistered IP addresses to the 254 addresses in a Class C network enables one Class C network address to be used instead of four. However, if more than 254 users simultaneously required Internet access, some user requests must be queued until a previously used registered address becomes available. Although NAT was primarily developed as a technique to conserve difficult-to-obtain IPv4 addresses, a side-benefit of its use is the fact that it hides the addresses of stations behind the translator. This means that a direct attack upon organizational hosts is no longer possible, and resulted in NAT functionality being incorporated into firewalls in addition to its use in routers.

Regardless of the device used to perform NAT, its operation is similar. That is, as packets arrive at the device performing NAT, the private source address is translated into a public address for transmission onto the Internet. In comparison, inbound packets have their public IP address translated into their equivalent private IP address based upon the state of an IP address mapping table maintained by the device.

10.4.1 TYPES OF ADDRESS TRANSLATIONS

There are three types of NAT that devices can employ. These types or methods of address translation include static NAT, pooled NAT, and port-level NAT, with the latter also referred to as Port Address Translation (PAT).

Static NAT

Static NAT results in the permanent mapping of each host on an internal network to an address on an external network. Although static mapping does not provide a reduction in the number of IP addresses needed by an organization, after it has been configured no further action is necessary and its simple table lookup minimizes delay.

Pooled NAT

When a pooled NAT technique is used, a pool of addresses on the external network is used for the dynamic assignment of IP addresses in place of private addresses on the internal network. Although pooled NAT enables users to conserve the use of public IP addresses, its use can adversely affect certain types of applications. For example, SNMP managers track devices based upon the device IP address and an object identifier. Because pooled NAT means that network addresses will more than likely change over time, this means that devices in front of the translating device cannot be configured to reliably transmit traps to devices behind the translating device. One possible solution to this problem is to permanently map an SNMP manager to an IP address while all other devices share the remaining addresses in the address pool. Of course, the device that supports pooled NAT must also be capable of permitting support for static mapping.

Port Address Translation

A third type of address translation results in the mapping of internal addresses to a single IP address on the external network. To accomplish this, the address translator assigns different port numbers to TCP and UDP source port fields. The port numbers used for mapping are those above 1023, providing 64 512 (=65 535 − 1023) simultaneous TCP/IP or UDP/IP connections on a single IP address. Because mapping occurs to a single IP address through the use of different port numbers, this technique is referred to as Port Address Translation (PAT). The use of PAT results in all traffic transmitted onto the public network appearing to come from a single IP address.

Regardless of the method of NAT used, its use hides an organization's actual IP addresses. When incorporated into a firewall, NAT represents a technique that forces direct IP address attacks to the firewall, which is hopefully hardened to withstand such attacks.

Appendix A

THE SNMP MANAGEMENT INFORMATION BASE (MIB-II)

This appendix lists the object identifiers included in the 10 groups that form the SNMP Management Information Base (MIB-II). The groups included in this appendix are as follows:

A.1 The System Group
A.2 The Interfaces Group
A.3 The Address Translation Group
A.4 The IP Group
A.5 The ICMP Group
A.6 The TCP Group
A.7 The UDP Group
A.8 The EGP Group
A.9 The Transmission Group
A.10 The SNMP Group

A.1 THE SYSTEM GROUP

```
-- Implementation of the System group is mandatory for all
-- systems. If an agent is not configured to have a value
-- for any of these variables, a string of length 0 is
-- returned.
```

sysDescr OBJECT-TYPE

```
  SYNTAX DisplayString (SIZE (0..255))
  ACCESS read-only
  STATUS mandatory
```

```
  DESCRIPTION
    "A textual description of the entity. This value should include
    the full name and version identification of the system's hardware
    type, software operating system, and networking software. It is
    mandatory that this only contain printable ASCII characters."
  ::={ system 1 }

sysObjectID OBJECT-TYPE
  SYNTAX OBJECT IDENTIFIER
  ACCESS read-only
  STATUS mandatory
  DESCRIPTION
    "The vendor's authoritative identification of the network manage-
    ment subsystem contained in the entity. This value is allocated
    within the SMI enterprises subtree (1.3.6.1.4.1) and provides an
    easy and unambiguous means for determining 'what kind of box' is
    being managed. For example, if vendor 'Flintstones, Inc.' was
    assigned the subtree 1.3.6.1.4.1.4242, it could assign the identifier
    1.3.6.1.4.1.4242.1.1 to its 'Fred Router'."
  ::={ system 2 }

sysUpTime OBJECT-TYPE
  SYNTAX TimeTicks
  ACCESS read-only
  STATUS mandatory
  DESCRIPTION
    "The time (in hundredths of a second) since the network manage-
    ment portion of the system was last re-initialized."
  ::={ system 3 }

sysContact OBJECT-TYPE
  SYNTAX DisplayString (SIZE (0..255))
  ACCESS read-write
  STATUS mandatory
  DESCRIPTION
    "The textual identification of the contact person for this managed
    node, together with information on how to contact this person."
  ::={ system 4 }

sysName OBJECT-TYPE
  SYNTAX DisplayString (SIZE (0..255))
  ACCESS read-write
  STATUS mandatory
  DESCRIPTION
    "An administratively assigned name for this managed node. By
    convention, this is the node's fully qualified domain name."
  ::={ system 5 }
```

```
sysLocation OBJECT-TYPE
  SYNTAX DisplayString (SIZE (0..255))
  ACCESS read-write
  STATUS mandatory
  DESCRIPTION
    "The physical location of this node (e.g., 'telephone closet, 3rd
    floor')."
  ::={ system 6 }

sysServices OBJECT-TYPE
  SYNTAX INTEGER (0..127)
  ACCESS read-only
  STATUS mandatory
  DESCRIPTION
    "A value that indicates the set of services that this entity primarily
    offers.

    The value is a sum. This sum initially takes the value zero, Then, for
    each layer, L, in the range 1 through 7, that this node performs
    transactions for, 2 raised to (L-1) is added to the sum. For example,
    a node that performs primarily routing functions would have a value
    of 4 (2^(3-1)). In contrast, a node that is a host offering application
    services would have a value of 72 (2^(4-1) + 2^(7-1)). Note that in
    the context of the Internet suite of protocols, values should be
    calculated accordingly:

      layer functionality
        1 physical (e.g., repeaters)
        2 datalink/subnetwork (e.g., bridges)
        3 internet (e.g., IP gateways)
        4 end-to-end (e.g., IP hosts)
        7 applications (e.g., mail relays)

      For systems including OSI protocols, layers 5 and 6 may also
      be counted."
  ::={ system 7 }
```

A.2 THE INTERFACES GROUP

```
  -- Implementation of the Interfaces group is mandatory for
  -- all systems.

  ifNumber OBJECT-TYPE
    SYNTAX INTEGER
    ACCESS read-only
    STATUS mandatory
    DESCRIPTION
```

```
        "The number of network interfaces (regardless of their current
        state) present on this system."
      ::={ interfaces 1 }

-- the Interfaces table

-- The Interfaces table contains information on the entity's
-- interfaces. Each interface is thought of as being
-- attached to a 'subnetwork.' Note that this term should
-- not be confused with 'subnet,' which refers to an
-- addressing partitioning scheme used in the Internet suite
-- of protocols.

ifTable OBJECT-TYPE
    SYNTAX SEQUENCE OF IfEntry
    ACCESS not-accessible
    STATUS mandatory
    DESCRIPTION
      "A list of interface entries. The number of entries is given by the
      value of ifNumber."
    ::={ interfaces 2 }

ifEntry OBJECT-TYPE
    SYNTAX IfEntry
    ACCESS not-accessible
    STATUS mandatory
    DESCRIPTION
      "An interface entry containing objects at the subnetwork layer and
      below for a particular interface."
    INDEX { ifIndex }
    ::={ ifTable 1 }

IfEntry ::=
    SEQUENCE {
      ifIndex
        INTEGER,
      ifDescr
        DisplayString,
      ifType
        INTEGER,
      ifMtu
        INTEGER,
      ifSpeed
        Gauge,
      ifPhysAddress
        PhysAddress,
      ifAdminStatus
        INTEGER,
```

```
    ifOperStatus
        INTEGER,
    ifLastChange
        TimeTicks,
    ifInOctets
        Counter,
    ifInUcastPkts
        Counter,
    ifInNUcastPkts
        Counter,
    ifInDiscards
        Counter,
    ifInErrors
        Counter,
    ifInUnknownProtos
        Counter,
    ifOutOctets
        Counter,
    ifOutUcastPkts
        Counter,
    ifOutNUcastPkts
        Counter,
    ifOutDiscards
        Counter,
    ifOutErrors
        Counter,
    ifOutQLen
        Gauge,
    ifSpecific
        OBJECT IDENTIFIER
}
```

ifIndex OBJECT-TYPE

```
  SYNTAX INTEGER
  ACCESS read-only
  STATUS mandatory
  DESCRIPTION
    "A unique value for each interface. Its value ranges between 1 and
    the value of ifNumber. The value for each interface must remain
    constant, at least from one re-initialization of the entity's network
    management system to the next re-initialization."
  ::={ ifEntry 1 }
```

ifDescr OBJECT-TYPE

```
  SYNTAX DisplayString (SIZE (0..255))
  ACCESS read-only
  STATUS mandatory
  DESCRIPTION
```

```
    "A textual string containing information about the interface. This
    string should include the name of the manufacturer, the product
    name and the version of the hardware interface."
  ::={ ifEntry 2 }


ifType OBJECT-TYPE
  SYNTAX INTEGER {
    other(1),                        - - none of the following
    regular1822(2),
    hdh1822(3),
    ddn-x25(4),
    rfc877-x25(5),
    ethernet-csmacd(6),
    iso88023-csmacd(7),
    iso88024-tokenBus(8),
    iso88025-tokenRing(9),
    iso88026-man(10),
    starLan(11),
    proteon-10Mbit(12),
    proteon-80Mbit(13),
    hyperchannel(14),
    fddi(15),
    lapb(16),
    sdlc(17),
    ds1(18),                         - - T-1
    e1(19),                          - - european equiv. of T-1
    basicISDN(20),
    primaryISDN(21),                 - - proprietary serial
    propPointToPointSerial(22),
    ppp(23),
    softwareLoopback(24),
    eon(25),                         - - CLNP over IP [11]
    ethernet-3Mbit(26)
    nsip(27),                        - - XNS over IP
    slip(28),                        - - generic SLIP
    ultra(29),                       - - ULTRA technologies
    ds3(30),                         - - T-3
    sip(31),                         - - SMDS
    frame-relay(32)
  }
  ACCESS read-only
  STATUS mandatory
  DESCRIPTION
    "The type of interface, distinguished according to the physical/link
    protocol(s) immediately 'below' the network layer in the protocol
    stack."
  ::={ ifEntry 3 }
```

```
ifMtu OBJECT-TYPE
  SYNTAX INTEGER
  ACCESS read-only
  STATUS mandatory
  DESCRIPTION
    "The size of the largest datagram that can be sent/received on the
    interface, specified in octets. For interfaces that are used for
    transmitting network datagrams, this is the size of the largest
    network datagram that can be sent on the interface."
  ::={ ifEntry 4 }

ifSpeed OBJECT-TYPE
  SYNTAX Gauge
  ACCESS read-only
  STATUS mandatory
  DESCRIPTION
    "An estimate of the interface's current bandwidth in bits per second.
    For interfaces that do not vary in bandwidth or for those where no
    accurate estimation can be made, this object should contain the
    nominal bandwidth."
  ::={ ifEntry 5 }

ifPhysAddress OBJECT-TYPE
  SYNTAX PhysAddress
  ACCESS read-only
  STATUS mandatory
  DESCRIPTION
    "The interface's address at the protocol layer immediately 'below' the
    network layer in the protocol stack. For interfaces that do not have
    such an address (e.g., a serial line), this object should contain an
    octet string of zero length."
  ::={ ifEntry 6 }

ifAdminStatus OBJECT-TYPE
  SYNTAX INTEGER {
            up(1),          -- ready to pass packets
            down(2),
            testing(3)      -- in some test mode
         }
  ACCESS read-write
  STATUS mandatory
  DESCRIPTION
    "The desired state of the interface. The testing(3) state indicates that
    no operational packets can be passed."
  ::={ ifEntry 7 }

ifOperStatus OBJECT-TYPE
  SYNTAX INTEGER {
```

```
                up(1),              -- ready to pass packets
                down(2),
                testing(3)          -- in some test mode
            }
    ACCESS read-only
    STATUS mandatory
    DESCRIPTION
      "The current operational state of the interface. The testing(3) state
      indicates that no operational packets can be passed."
    ::={ ifEntry 8 }
```

ifLastChange OBJECT-TYPE

```
    SYNTAX TimeTicks
    ACCESS read-only
    STATUS mandatory
    DESCRIPTION
      "The value of sysUpTime at the time the interface entered its current
      operational state. If the current state was entered prior to the last
      re-initialization of the local network management subsystem, then
      this object contains a zero value."
    ::={ ifEntry 9 }
```

ifInOctets OBJECT-TYPE

```
    SYNTAX Counter
    ACCESS read-only
    STATUS mandatory
    DESCRIPTION
      "The total number of octets received on the interface, including
      framing characters."
    ::={ ifEntry 10 }
```

ifInUcastPkts OBJECT-TYPE

```
    SYNTAX Counter
    ACCESS read-only
    STATUS mandatory
    DESCRIPTION
      "The number of subnetwork-unicast packets delivered to a higher
      layer protocol."
    ::={ ifEntry 11 }
```

ifInNUcastPkts OBJECT-TYPE

```
    SYNTAX Counter
    ACCESS read-only
    STATUS mandatory
    DESCRIPTION
      "The number of non-unicast (i.e., subnetwork-broadcast or sub-
      network-multicast) packets delivered to a higher layer protocol."
    ::={ ifEntry 12 }
```

```
ifInDiscards OBJECT-TYPE
  SYNTAX Counter
  ACCESS read-only
  STATUS mandatory
  DESCRIPTION
    "The number of inbound packets that were chosen to be discarded
    even though no errors had been detected to prevent their being
    deliverable to a higher layer protocol. One possible reason for
    discarding such a packet could be to free up buffer space."
  ::={ ifEntry 13 }

ifInErrors OBJECT-TYPE
  SYNTAX Counter
  ACCESS read-only
  STATUS mandatory
  DESCRIPTION
    "The number of inbound packets that contained errors preventing
    them from being deliverable to a higher layer protocol."
  ::={ ifEntry 14 }

ifInUnknownProtos OBJECT-TYPE
  SYNTAX Counter
  ACCESS read-only
  STATUS mandatory
  DESCRIPTION
    "The number of packets received via the interface that were
    discarded because of an unknown or unsupported protocol."
  ::={ ifEntry 15 }

ifOutOctets OBJECT-TYPE
  SYNTAX Counter
  ACCESS read-only
  STATUS mandatory
  DESCRIPTION
    "The total number of octets transmitted out of the interface,
    including framing characters."
  ::={ ifEntry 16 }

ifOutUcastPkts OBJECT-TYPE
  SYNTAX Counter
  ACCESS read-only
  STATUS mandatory
  DESCRIPTION
    "The total number of packets that higher level protocols requested
    be transmitted to a subnetwork-unicast address, including those
    that were discarded or not sent."
  ::={ ifEntry 17 }
```

```
ifOutNUcastPkts OBJECT-TYPE
  SYNTAX Counter
  ACCESS read-only
  STATUS mandatory
  DESCRIPTION
    "The total number of packets that higher-level protocols requested
    be transmitted to a non-unicast (i.e., a subnetwork-broadcast or
    subnetwork-multicast) address, including those that were dis-
    carded or not sent."
  ::={ ifEntry 18 }

ifOutDiscards OBJECT-TYPE
  SYNTAX Counter
  ACCESS read-only
  STATUS mandatory
  DESCRIPTION
    "The number of outbound packets that were chosen to be discarded
    even though no errors had been detected to prevent their being
    transmitted. One possible reason for discarding such a packet could
    be to free up buffer space."
  ::={ ifEntry 19 }

ifOutErrors OBJECT-TYPE
  SYNTAX Counter
  ACCESS read-only
  STATUS mandatory
  DESCRIPTION
    "The number of outbound packets that could not be transmitted
    because of errors."
  ::={ ifEntry 20 }

ifOutQLen OBJECT-TYPE
  SYNTAX Gauge
  ACCESS read-only
  STATUS mandatory
  DESCRIPTION
    "The length of the output packet queue (in packets)."
  ::={ ifEntry 21 }

ifSpecific OBJECT-TYPE
  SYNTAX OBJECT IDENTIFIER
  ACCESS read-only
  STATUS mandatory
  DESCRIPTION
    "A reference to MIB definitions specific to the particular media being
    used to realize the interface. For example, if the interface is realized
    by an Ethernet, then the value of this object refers to a document
    defining objects specific to ethernet. If this information is not
```

```
      present, its value should be set to the OBJECT IDENTIFIER { 0 0 },
      which is a syntatically valid object identifier, and any conformant
      implementation of ASN.1 and BER must be able to generate and
      recognize this value."
    ::={ ifEntry 22 }
```

A.3 THE ADDRESS TRANSLATION GROUP

```
    -- Implementation of the Address Translation group is
    -- mandatory for all systems. Note, however, that this group
    -- is deprecated by MIB-II. That is, it is being included
    -- solely for compatibility with MIB-I nodes, and will most
    -- likely be excluded from MIB-III nodes. From MIB-II
    -- onwards, each network protocol group contains its own
    -- address translation tables.

    -- The Address Translation group contains one table, which is
    -- the union across all interfaces of the translation tables
    -- for converting a NetworkAddress (e.g., an IP address)
    -- into a subnetwork-specific address. For lack of a better
    -- term, this document refers to such a subnetwork-specific
    -- address as a 'physical' address.

    -- Examples of such translation tables are as follows: for broadcast
    -- media where ARP is in use, the translation table is
    -- equivalent to the ARP cache; or, on an X.25 network where
    -- non-algorithmic translation to X.121 addresses is
    -- required, the translation table contains the
    -- NetworkAddress to X.121 address equivalences.

  atTable OBJECT-TYPE
    SYNTAX SEQUENCE OF AtEntry
    ACCESS not-accessible
    STATUS deprecated
    DESCRIPTION
      "The Address Translation tables contain the NetworkAddress to
      'physical' address equivalences. Some interfaces do not use
      translation tables for determining address equivalences (e.g., DDN-
      X.25 has an algorithmic method); if all interfaces are of this type,
      then the Address Translation table is empty, i.e., has zero entries."
    ::={ at 1 }

  atEntry OBJECT-TYPE
    SYNTAX AtEntry
    ACCESS not-accessible
    STATUS deprecated
    DESCRIPTION
```

```
    "Each entry contains one NetworkAddress to 'physical' address
    equivalence."
  INDEX    { atIfIndex,
             atNetAddress }
  ::={ atTable 1 }

AtEntry ::=
  SEQUENCE {
    atIfIndex
  INTEGER,

    atPhysAddress
      PhysAddress,
    atNetAddress
      NetworkAddress
  }
```

atIfIndex OBJECT-TYPE

```
  SYNTAX INTEGER
  ACCESS read-write
  STATUS deprecated
  DESCRIPTION
    "The interface on which this entry's equivalence is effective. The
    interface identified by a particular value of this index is the same
    interface as identified by the same value of ifIndex."
  ::={ atEntry 1 }
```

atPhysAddress OBJECT-TYPE

```
  SYNTAX PhysAddress
  ACCESS read-write
  STATUS deprecated
  DESCRIPTION
    "The media-dependent 'physical' address. Setting this object to a
    null string (one of zero length) has the effect of invaliding the
    corresponding entry in the atTable object. That is, it effectively
    dissasociates the interface identified with said entry from the
    mapping identified with said entry. It is an implementation-specific
    matter as to whether the agent removes an invalidated entry from
    the table. Accordingly, management stations must be prepared to
    receive tabular information from agents that corresponds to entries
    not currently in use. Proper interpretation of such entries requires
    examination of the relevant atPhysAddress object."
  ::={ atEntry 2 }
```

atNetAddress OBJECT-TYPE

```
  SYNTAX NetworkAddress
  ACCESS read-write
  STATUS deprecated
```

```
  DESCRIPTION
    "The NetworkAddress (e.g., the IP address) corresponding to the
    media-dependent 'physical' address."
  ::={ atEntry 3 }
```

A.4 THE IP GROUP

```
-- Implementation of the IP group is mandatory for all
-- systems.

ipForwarding OBJECT-TYPE
  SYNTAX INTEGER {
    forwarding(1),          -- acting as a gateway
    not-forwarding(2)       -- NOT acting as a gateway
          }
  ACCESS read-write
  STATUS mandatory
  DESCRIPTION
     "The indication of whether this entity is acting as an IP gateway in
     respect to the forwarding of datagrams received by, but not
     addressed to, this entity. IP gateways forward datagrams. IP hosts
     do not (except those Source-Routed via the host).

     Note that for some managed nodes this object may take on only a
     subset of the values possible. Accordingly, it is appropriate for an
     agent to return a 'badValue' response if a management station
     attempts to change this object to an inappropriate value."
  ::={ ip 1 }

ipDefaultTTL OBJECT-TYPE
  SYNTAX INTEGER
  ACCESS read-write
  STATUS mandatory
  DESCRIPTION
    "The default value inserted into the Time-To-Live field of the IP
    header of datagrams originated at this entity, whenever a TTL value
    is not supplied by the transport layer protocol."
  ::={ ip 2 }

ipInReceives OBJECT-TYPE
  SYNTAX Counter
  ACCESS read-only
  STATUS mandatory
  DESCRIPTION
    "The total number of input datagrams received from interfaces,
    including those received in error."
  ::={ ip 3 }
```

ipInHdrErrors OBJECT-TYPE
SYNTAX Counter
ACCESS read-only
STATUS mandatory
DESCRIPTION
"The number of input datagrams discarded due to errors in their IP headers, including bad checksums, version number mismatch, other format errors, time-to-live exceeded, errors discovered in processing their IP options, etc."
::={ ip 4 }

ipInAddrErrors OBJECT-TYPE
SYNTAX Counter
ACCESS read-only
STATUS mandatory
DESCRIPTION
"The number of input datagrams discarded because the IP address in their IP header's destination field was not a valid address to be received at this entity. This count includes invalid addresses (e.g., 0.0.0.0) and addresses of unsupported Classes (e.g., Class E). For entities that are not IP Gateways and therefore do not forward datagrams, this counter includes datagrams discarded because the destination address was not a local address."
::={ ip 5 }

ipForwDatagrams OBJECT-TYPE
SYNTAX Counter
ACCESS read-only
STATUS mandatory
DESCRIPTION
"The number of input datagrams for which this entity was not their final IP destination, as a result of which an attempt was made to find a route to forward them to that final destination. In entities that do not act as IP Gateways, this counter will include only those packets that were Source-Routed via this entity, and the Source-Route option processing was successful."
::={ ip 6 }

ipInUnknownProtos OBJECT-TYPE
SYNTAX Counter
ACCESS read-only
STATUS mandatory
DESCRIPTION
"The number of locally addressed datagrams received successfully but discarded because of an unknown or unsupported protocol."
::={ ip 7 }

```
ipInDiscards OBJECT-TYPE
  SYNTAX Counter
  ACCESS read-only
  STATUS mandatory
  DESCRIPTION
    "The number of input IP datagrams for which no problems were
    encountered to prevent their continued processing, but which were
    discarded (e.g., for lack of buffer space). Note that this counter does
    not include any datagrams discarded while awaiting re-assembly."
  ::={ ip 8 }

ipInDelivers OBJECT-TYPE
  SYNTAX Counter
  ACCESS read-only
  STATUS mandatory
  DESCRIPTION
    "The total number of input datagrams successfully delivered to IP
    user-protocols (including ICMP)."
  ::={ ip 9 }

ipOutRequests OBJECT-TYPE
  SYNTAX    Count
  ACCESS    read-only
  STATUS    mandatory
  DESCRIPTION
    "The total number of IP datagrams which local IP user-protocols
    (including ICMP) supplied to IP in requests for transmission. Note
    that this counter does not include any datagrams counted in
    ipForwDatagrams."
  ::={ ip 10 }

ipOutDiscards OBJECT-TYPE
  SYNTAX Counter
  ACCESS read-only
  STATUS mandatory
  DESCRIPTION
    "The number of output IP datagrams for which no problem was
    encountered to prevent their transmission to their destination, but
    which were discarded (e.g., for lack of buffer space). Note that this
    counter would include datagrams counted in ipForwDatagrams if
    any such packets met this (discretionary) discard criterion."
  ::={ ip 11 }

ipOutNoRoutes OBJECT-TYPE
  SYNTAX Counter
  ACCESS read-only
  STATUS mandatory
  DESCRIPTION
```

"The number of IP datagrams discarded because no route could be found to transmit them to their destination. Note that this counter includes any packets counted in ipForwDatagrams that meet this 'no-route' criterion. Note that this includes any datagrams that a host cannot route because all of its default gateways are down."
::={ ip 12 }

ipReasmTimeout OBJECT-TYPE
SYNTAX INTEGER
ACCESS read-only
STATUS mandatory
DESCRIPTION
"The maximum number of seconds for which received fragments are held while they are awaiting reassembly at this entity."
::={ ip 13 }

ipReasmReqds OBJECT-TYPE
SYNTAX Counter
ACCESS read-only
STATUS mandatory
DESCRIPTION
"The number of IP fragments received that needed to be reassembled at this entity."
::={ ip 14 }

ipReasmOKs OBJECT-TYPE
SYNTAX Counter
ACCESS read-only
STATUS mandatory
DESCRIPTION
"The number of IP datagrams successfully reassembled."
::={ ip 15 }

ipReasmFails OBJECT-TYPE
SYNTAX Counter
ACCESS read-only
STATUS mandatory
DESCRIPTION
"The number of failures detected by the IP reassembly algorithm (for whatever reason: timed out, errors, etc.). Note that this is not necessarily a count of discarded IP fragments, since some algorithms (notably the algorithm in RFC 815) can lose track of the number of fragments by combining them as they are received."
::={ ip 16 }

ipFragOKs OBJECT-TYPE
SYNTAX Counter
ACCESS read-only

```
    STATUS mandatory
    DESCRIPTION
      "The number of IP datagrams that have been successfully
      fragmented at this entity."
    ::={ ip 17 }

ipFragFails OBJECT-TYPE
    SYNTAX Counter
    ACCESS read-only
    STATUS mandatory
    DESCRIPTION
      "The number of IP datagrams that have been discarded because
      they needed to be fragmented at this entity but could not be, e.g.,
      because their Don't Fragment flag was set."
    ::={ ip 18 }

ipFragCreates OBJECT-TYPE
    SYNTAX Counter
    ACCESS read-only
    STATUS mandatory
    DESCRIPTION
      "The number of IP datagram fragments that have been generated as
      a result of fragmentation at this entity."
    ::={ ip 19 }

-- the IP address table

-- The IP address table contains this entity's IP addressing
-- information.

ipAddrTable OBJECT-TYPE
    SYNTAX SEQUENCE OF IpAddrEntry
    ACCESS not-accessible
    STATUS mandatory
    DESCRIPTION
      "The table of addressing information relevant to this entity's IP
      addresses."
    ::={ ip 20 }

ipAddrEntry OBJECT-TYPE
    SYNTAX IpAddrEntry
    ACCESS not-accessible
    STATUS mandatory
    DESCRIPTION
      "The addressing information for one of this entity's IP addresses."
    INDEX { ipAdEntAddr }
    ::={ ipAddrTable 1 }
```

```
IpAddrEntry ::=
  SEQUENCE {
    ipAdEntAddr
      IpAddress,
    ipAdEntIfIndex
      INTEGER,
    ipAdEntNetMask
      IpAddress,
    ipAdEntBcastAddr
      INTEGER,
    ipAdEntReasmMaxSize
      INTEGER (0..65535)
  }

ipAdEntAddr OBJECT-TYPE
  SYNTAX IpAddress
  ACCESS read-only
  STATUS mandatory
  DESCRIPTION
    "The IP address to which this entry's addressing information
    pertains."
  ::={ ipAddrEntry 1 }

ipAdEntIfIndex OBJECT-TYPE
  SYNTAX INTEGER
  ACCESS read-only
  STATUS mandatory
  DESCRIPTION
    "The index value that uniquely identifies the interface to which this
    entry is applicable. The interface identified by a particular value of
    this index is the same interface as identified by the same value of
    ifIndex."
  ::={ ipAddrEntry 2 }

ipAdEntNetMask OBJECT-TYPE
  SYNTAX IpAddress
  ACCESS read-only
  STATUS mandatory
  DESCRIPTION
    "The subnet mask associated with the IP address of this entry. The
    value of the mask is an IP address with all the network bits set to 1
    and all the hosts bits set to 0."
  ::={ ipAddrEntry 3 }

ipAdEntBcastAddr OBJECT-TYPE
  SYNTAX INTEGER
  ACCESS read-only
  STATUS mandatory
```

```
    DESCRIPTION
      "The value of the least significant bit in the IP Broadcast address
      used for sending datagrams on the (logical) interface associated with
      the IP address of this entry. For example, when the Internet
      standard all-ones broadcast address is used, the value will be 1.
      This value applies to both the subnet and network broadcasts
      addresses used by the entity on this (logical) interface."
    ::={ ipAddrEntry 4 }

ipAdEntReasmMaxSize OBJECT-TYPE
    SYNTAX INTEGER (0..65535)
    ACCESS read-only
    STATUS mandatory
    DESCRIPTION
      "The size of the largest IP datagram that this entity can re-assemble
      from incoming IP fragmented datagrams received on this interface."
    ::={ ipAddrEntry 5 }

-- the IP routing table

-- The IP routing table contains an entry for each route
-- presently known to this entity.

ipRouteTable OBJECT-TYPE
    SYNTAX SEQUENCE OF IpRouteEntry
    ACCESS not-accessible
    STATUS mandatory
    DESCRIPTION
      "This entity's IP Routing table."
    ::={ ip 21 }

ipRouteEntry OBJECT-TYPE
    SYNTAX IpRouteEntry
    ACCESS not-accessible
    STATUS mandatory
    DESCRIPTION
      "A route to a particular destination."
    INDEX { ipRouteDest }
    ::={ ipRouteTable 1 }

IpRouteEntry ::=
    SEQUENCE {
      ipRouteDest
        IpAddress,
      ipRouteIfIndex
        INTEGER,
      ipRouteMetric1
        INTEGER,
```

```
    ipRouteMetric2
        INTEGER,
    ipRouteMetric3
        INTEGER,
    ipRouteMetric4
        INTEGER,
    ipRouteNextHop
        IpAddress,
    ipRouteType
        INTEGER,
    ipRouteProto
        INTEGER,
    ipRouteAge
        INTEGER,
    ipRouteMask
        IpAddress,
    ipRouteMetric5
        INTEGER,
    ipRouteInfo
        OBJECT IDENTIFIER
}
```

ipRouteDest OBJECT-TYPE

```
  SYNTAX IpAddress
  ACCESS read-write
  STATUS mandatory
  DESCRIPTION
    "The destination IP address of this route. An entry with a value of
    0.0.0.0 is considered a default route. Multiple routes to a single
    destination can appear in the table, but access to such multiple
    entries is dependent on the table-access mechanisms defined by the
    network management protocol in use."
  ::= { ipRouteEntry 1 }
```

ipRouteIfIndex OBJECT-TYPE

```
  SYNTAX INTEGER
  ACCESS read-write
  STATUS mandatory
  DESCRIPTION
    "The index value that uniquely identifies the local interface through
    which the next hop of this route should be reached. The interface
    identified by a particular value of this index is the same interface as
    identified by the same value of ifIndex."
  ::= { ipRouteEntry 2 }
```

ipRouteMetric1 OBJECT-TYPE

```
  SYNTAX INTEGER
  ACCESS read-write
```

```
  STATUS mandatory
  DESCRIPTION
    "The primary routing metric for this route. The semantics of this
    metric are determined by the routing-protocol specified in the
    route's ipRouteProto value. If this metric is not used, its value
    should be set to -1."
  ::={ ipRouteEntry 3 }
```

ipRouteMetric2 OBJECT-TYPE

```
  SYNTAX INTEGER
  ACCESS read-write
  STATUS mandatory
  DESCRIPTION
    "An alternate routing metric for this route. The semantics of this
    metric are determined by the routing-protocol specified in the
    route's ipRouteProto value. If this metric is not used, its value
    should be set to -1."
  ::={ ipRouteEntry 4 }
```

ipRouteMetric3 OBJECT-TYPE

```
  SYNTAX INTEGER
  ACCESS read-write
  STATUS mandatory
  DESCRIPTION
    "An alternate routing metric for this route. The semantics of this
    metric are determined by the routing-protocol specified in the
    route's ipRouteProto value. If this metric is not used, its value
    should be set to -1."
  ::={ ipRouteEntry 5 }
```

ipRouteMetric4 OBJECT-TYPE

```
  SYNTAX INTEGER
  ACCESS read-write
  STATUS mandatory
  DESCRIPTION
    "An alternate routing metric for this route. The semantics of this
    metric are determined by the routing-protocol specified in the
    route's ipRouteProto value. If this metric is not used, its value
    should be set to -1."
  ::={ ipRouteEntry 6 }
```

ipRouteNextHop OBJECT-TYPE

```
  SYNTAX IpAddress
  ACCESS read-write
  STATUS mandatory
  DESCRIPTION
```

```
    "The IP address of the next hop of this route. (In the case of a route
    bound to an interface that is realized via a broadcast media, the
    value of this field is the agent's IP address on that interface.)"
  ::={ ipRouteEntry 7 }

ipRouteType OBJECT-TYPE
  SYNTAX INTEGER {
              other(1),             - - none of the following

              invalid(2),           - - an invalidated route

                                    - - route to directly
              direct(3),            - - connected (sub-)network

                                    - - route to a non-local
              indirect(4)           - - host/network/sub-network
  }
  ACCESS read-write
  STATUS mandatory
  DESCRIPTION
    "The type of route. Note that the values direct(3) and indirect(4) refer
    to the notion of direct and indirect routing in the IP architecture.

    Setting this object to the value invalid(2) has the effect of
    invalidating the corresponding entry in the ipRouteTable object.
    That is, it effectively dissasociates the destination identified with
    said entry from the route identified with said entry. It is an
    implementation-specific matter as to whether the agent removes an
    invalidated entry from the table. Accordingly, management stations
    must be prepared to receive tabular information from agents that
    corresponds to entries not currently in use. Proper interpretation of
    such entries requires examination of the relevant ipRouteType
    object."
  ::={ ipRouteEntry 8 }

ipRouteProto OBJECT-TYPE
  SYNTAX INTEGER {
        other(1),         - - none of the following

                          - - non-protocol information,
                          - - e.g., manually configured
        local(2),         - - entries

                          - - set via a network
        netmgmt(3),       - - management protocol

                          - - obtained via ICMP,
        icmp(4),          - - e.g., Redirect
```

```
                              - - the remaining values are
                              - - all gateway routing
                              - - protocols
        egp(5),
        ggp(6),
        hello(7),
        rip(8),
        is-is(9),
        es-is(10),
        ciscoIgrp(11),
        bbnSpfIgp(12),
        ospf(13),
        bgp(14)
      }
    ACCESS read-only
    STATUS mandatory
    DESCRIPTION
      "The routing mechanism via which this route was learned. Inclusion
      of values for gateway routing protocols is not intended to imply that
      hosts should support those protocols."
    ::={ ipRouteEntry 9 }
```

ipRouteAge OBJECT-TYPE

```
    SYNTAX INTEGER
    ACCESS read-write
    STATUS mandatory
    DESCRIPTION
      "The number of seconds since this route was last updated or
      otherwise determined to be correct. Note that no semantics of 'too
      old' can be implied except through knowledge of the routing protocol
      by which the route was learned."
    ::={ ipRouteEntry 10 }
```

ipRouteMask OBJECT-TYPE

```
    SYNTAX IpAddress
    ACCESS read-write
    STATUS mandatory
    DESCRIPTION
      "Indicate the mask to be logical-ANDed with the destination address
      before being compared with the value in the ipRouteDest field. For
      those systems that do not support arbitrary subnet masks, an agent
      constructs the value of the ipRouteMask by determining whether
      the value of the correspondent ipRouteDest field belong to a class A,
      B, or C network, and then using one of the following:

        mask              network
        255.0.0.0         class-A
        255.255.0.0       class-B
```

```
        255.255.255.0   class-C

      If the value of the ipRouteDest is 0.0.0.0 (a default route), then the
      mask value is also 0.0.0.0. It should be noted that all IP routing
      subsystems implicitly use this mechanism."
    ::={ ipRouteEntry 11 }

ipRouteMetric5 OBJECT-TYPE
    SYNTAX INTEGER
    ACCESS read-write
    STATUS mandatory
    DESCRIPTION
      "An alternate routing metric for this route. The semantics of this
      metric are determined by the routing-protocol specified in the
      route's ipRouteProto value. If this metric is not used, its value
      should be set to -1."
    ::={ ipRouteEntry 12 }

ipRouteInfo OBJECT-TYPE
    SYNTAX OBJECT IDENTIFIER
    ACCESS read-only
    STATUS mandatory
    DESCRIPTION
      "A reference to MIB definitions specific to the particular routing
      protocol that is responsible for this route, as determined by the
      value specified in the route's ipRouteProto value. If this information
      is not present, its value should be set to the OBJECT IDENTIFIER {
      0 0 }, which is a syntatically valid object identifier, and any
      conformant implementation of ASN.1 and BER must be able to
      generate and recognize this value."
    ::={ ipRouteEntry 13 }


-- the IP Address Translation table

-- The IP address translation table contain the IpAddress to
-- 'physical' address equivalences. Some interfaces do not
-- use translation tables for determining address
-- equivalences (e.g., DDN-X.25 has an algorithmic method);
-- if all interfaces are of this type, then the Address
-- Translation table is empty, i.e., has zero entries.

ipNetToMediaTable OBJECT-TYPE
    SYNTAX SEQUENCE OF IpNetToMediaEntry
    ACCESS not-accessible
    STATUS mandatory
    DESCRIPTION
```

```
    "The IP Address Translation table used for mapping from IP
    addresses to physical addresses."
  ::={ ip 22 }

ipNetToMediaEntry OBJECT-TYPE
  SYNTAX IpNetToMediaEntry
  ACCESS not-accessible
  STATUS mandatory
  DESCRIPTION
    "Each entry contains one IpAddress to 'physical' address equiva-
    lence."
  INDEX { ipNetToMediaIfIndex,
          ipNetToMediaNetAddress }
  ::={ ipNetToMediaTable 1 }

IpNetToMediaEntry ::=
  SEQUENCE {
    ipNetToMediaIfIndex
      INTEGER,
    ipNetToMediaPhysAddress
      PhysAddress,
    ipNetToMediaNetAddress
      IpAddress,
    ipNetToMediaType
      INTEGER
  }

ipNetToMediaIfIndex OBJECT-TYPE
  SYNTAX INTEGER
  ACCESS read-write
  STATUS mandatory
  DESCRIPTION
    "The interface on which this entry's equivalence is effective. The
    interface identified by a particular value of this index is the same
    interface as identified by the same value of ifIndex."
  ::={ ipNetToMediaEntry 1 }

ipNetToMediaPhysAddress OBJECT-TYPE
  SYNTAX PhysAddress
  ACCESS read-write
  STATUS mandatory
  DESCRIPTION
    "The media-dependent 'physical' address."
  ::={ ipNetToMediaEntry 2 }

ipNetToMediaNetAddress OBJECT-TYPE
  SYNTAX IpAddress
  ACCESS read-write
```

```
  STATUS mandatory
  DESCRIPTION
    "The IpAddress corresponding to the media-dependent 'physical'
    address."
  ::={ ipNetToMediaEntry 3 }

ipNetToMediaType OBJECT-TYPE
  SYNTAX   INTEGER {
              other(1),      - - none of the following
              invalid(2),    - - an invalidated mapping
              dynamic(3),
              static(4)
          }
  ACCESS   read-write
  STATUS   mandatory
  DESCRIPTION
    "The type of mapping.

    Setting this object to the value invalid(2) has the effect of
    invalidating the corresponding entry in the ipNetToMediaTable.
    That is, it effectively dissasociates the interface identified with said
    entry from the mapping identified with said entry. It is an
    implementation-specific matter as to whether the agent removes an
    invalidated entry from the table. Accordingly, management stations
    must be prepared to receive tabular information from agents that
    corresponds to entries not currently in use. Proper interpretation of
    such entries requires examination of the relevant ipNetToMedia-
    Type object."
  ::={ ipNetToMediaEntry 4 }


- - additional IP objects

ipRoutingDiscards OBJECT-TYPE
  SYNTAX Counter
  ACCESS read-only
  STATUS mandatory
  DESCRIPTION
    "The number of routing entries that were chosen to be discarded
    even though they are valid. One possible reason for discarding such
    an entry could be to free-up buffer space for other routing entries."
  ::={ ip 23 }
```

A.5 THE ICMP GROUP

```
- - Implementation of the ICMP group is mandatory for all
- - systems.
```

```
icmpInMsgs OBJECT-TYPE
  SYNTAX Counter
  ACCESS read-only
  STATUS mandatory
  DESCRIPTION
    "The total number of ICMP messages that the entity received. Note
    that this counter includes all those counted by icmpInErrors."
  ::={ icmp 1 }

icmpInErrors OBJECT-TYPE
  SYNTAX Counter
  ACCESS read-only
  STATUS mandatory
  DESCRIPTION
    "The number of ICMP messages that the entity received but
    determined as having ICMP-specific errors (bad ICMP checksums,
    bad length, etc.)."
  ::={ icmp 2 }

icmpInDestUnreachs OBJECT-TYPE
  SYNTAX Counter
  ACCESS read-only
  STATUS mandatory
  DESCRIPTION
    "The number of ICMP Destination Unreachable messages received."
  ::={ icmp 3 }

icmpInTimeExcds OBJECT-TYPE
  SYNTAX Counter
  ACCESS read-only
  STATUS mandatory
  DESCRIPTION
    "The number of ICMP Time Exceeded messages received."
  ::={ icmp 4 }

icmpInParmProbs OBJECT-TYPE
  SYNTAX Counter
  ACCESS read-only
  STATUS mandatory
  DESCRIPTION
    "The number of ICMP Parameter Problem messages received."
  ::={ icmp 5 }

icmpInSrcQuenchs OBJECT-TYPE
  SYNTAX Counter
  ACCESS read-only
  STATUS mandatory
  DESCRIPTION
```

```
    "The number of ICMP Source Quench messages received."
  ::={ icmp 6 }

icmpInRedirects OBJECT-TYPE
  SYNTAX Counter
  ACCESS read-only
  STATUS mandatory
  DESCRIPTION
    "The number of ICMP Redirect messages received."
  ::={ icmp 7 }

icmpInEchos OBJECT-TYPE
  SYNTAX Counter
  ACCESS read-only
  STATUS mandatory
  DESCRIPTION
    "The number of ICMP Echo (request) messages received."
  ::={ icmp 8 }

icmpInEchoReps OBJECT-TYPE
  SYNTAX Counter
  ACCESS read-only
  STATUS mandatory
  DESCRIPTION
    "The number of ICMP Echo Reply messages received."
  ::={ icmp 9 }

icmpInTimestamps OBJECT-TYPE
  SYNTAX Counter
  ACCESS read-only
  STATUS mandatory
  DESCRIPTION
    "The number of ICMP Timestamp (request) messages received."
  ::={ icmp 10 }

icmpInTimestampReps OBJECT-TYPE
  SYNTAX Counter
  ACCESS read-only
  STATUS mandatory
  DESCRIPTION
    "The number of ICMP Timestamp Reply messages received."
  ::={ icmp 11 }

icmpInAddrMasks OBJECT-TYPE
  SYNTAX Counter
  ACCESS read-only
  STATUS mandatory
  DESCRIPTION
```

"The number of ICMP Address Mask Request messages received."
::={ icmp 12 }

icmpInAddrMaskReps OBJECT-TYPE
SYNTAX Counter
ACCESS read-only
STATUS mandatory
DESCRIPTION
"The number of ICMP Address Mask Reply messages received."
::={ icmp 13 }

icmpOutMsgs OBJECT-TYPE
SYNTAX Counter
ACCESS read-only
STATUS mandatory
DESCRIPTION
"The total number of ICMP messages that this entity attempted to send. Note that this counter includes all those counted by icmpOutErrors."
::={ icmp 14 }

icmpOutErrors OBJECT-TYPE
SYNTAX Counter
ACCESS read-only
STATUS mandatory
DESCRIPTION
"The number of ICMP messages that this entity did not send due to problems discovered within ICMP such as a lack of buffers. This value should not include errors discovered outside the ICMP layer, such as the inability of IP to route the resultant datagram. In some implementations there may be no types of error that contribute to this counter's value."
::={ icmp 15 }

icmpOutDestUnreachs OBJECT-TYPE
SYNTAX Counter
ACCESS read-only
STATUS mandatory
DESCRIPTION
"The number of ICMP Destination Unreachable messages sent."
::={ icmp 16 }

icmpOutTimeExcds OBJECT-TYPE
SYNTAX Counter
ACCESS read-only
STATUS mandatory
DESCRIPTION
"The number of ICMP Time Exceeded messages sent."
::={ icmp 17 }

```
icmpOutParmProbs OBJECT-TYPE
  SYNTAX Counter
  ACCESS read-only
  STATUS mandatory
  DESCRIPTION
    "The number of ICMP Parameter Problem messages sent."
  ::={ icmp 18 }

icmpOutSrcQuenchs OBJECT-TYPE
  SYNTAX Counter
  ACCESS read-only
  STATUS mandatory
  DESCRIPTION
    "The number of ICMP Source Quench messages sent."
  ::={ icmp 19 }

icmpOutRedirects OBJECT-TYPE
  SYNTAX Counter
  ACCESS read-only
  STATUS mandatory
  DESCRIPTION
    "The number of ICMP Redirect messages sent. For a host, this object
    will always be zero, since hosts do not send redirects."
  ::={ icmp 20 }

icmpOutEchos OBJECT-TYPE
  SYNTAX Counter
  ACCESS read-only
  STATUS mandatory
  DESCRIPTION
    "The number of ICMP Echo (request) messages sent."
  ::={ icmp 21 }

icmpOutEchoReps OBJECT-TYPE
  SYNTAX Counter
  ACCESS read-only
  STATUS mandatory
  DESCRIPTION
    "The number of ICMP Echo Reply messages sent."
  ::={ icmp 22 }

icmpOutTimestamps OBJECT-TYPE
  SYNTAX Counter
  ACCESS read-only
  STATUS mandatory
  DESCRIPTION
    "The number of ICMP Timestamp (request) messages sent."
  ::={ icmp 23 }
```

```
icmpOutTimestampReps OBJECT-TYPE
  SYNTAX Counter
  ACCESS read-only
  STATUS mandatory
  DESCRIPTION
    "The number of ICMP Timestamp Reply messages sent."
  ::={ icmp 24 }

icmpOutAddrMasks OBJECT-TYPE
  SYNTAX Counter
  ACCESS read-only
  STATUS mandatory
  DESCRIPTION
    "The number of ICMP Address Mask Request messages sent."
  ::={ icmp 25 }

icmpOutAddrMaskReps OBJECT-TYPE
  SYNTAX Counter
  ACCESS read-only
  STATUS mandatory
  DESCRIPTION
    "The number of ICMP Address Mask Reply messages sent."
  ::={ icmp 26 }
```

A.6 THE TCP GROUP

```
-- Implementation of the TCP group is mandatory for all
-- systems that implement the TCP.

-- Note that instances of object types that represent
-- information about a particular TCP connection are
-- transient; they persist only as long as the connection
-- in question.

tcpRtoAlgorithm OBJECT-TYPE
  SYNTAX   INTEGER {
               other(1),          -- none of the following

               constant(2),       -- a constant rto
               rsre(3),           -- MIL-STD-1778, Appendix B
               vanj(4)            -- Van Jacobson's algorithm [10]
           }
  ACCESS read-only
  STATUS mandatory
  DESCRIPTION
    "The algorithm used to determine the timeout value used for
    retransmitting unacknowledged octets."
  ::={ tcp 1 }
```

```
tcpRtoMin OBJECT-TYPE
  SYNTAX INTEGER
  ACCESS read-only
  STATUS mandatory
  DESCRIPTION
    "The minimum value permitted by a TCP implementation for the
    retransmission timeout, measured in milliseconds. More refined
    semantics for objects of this type depend upon the algorithm used to
    determine the retransmission timeout. In particular, when the
    timeout algorithm is rsre(3), an object of this type has the semantics
    of the LBOUND quantity described in RFC 793."

  ::={ tcp 2 }

tcpRtoMax OBJECT-TYPE
  SYNTAX INTEGER
  ACCESS read-only
  STATUS mandatory
  DESCRIPTION
    "The maximum value permitted by a TCP implementation for the
    retransmission timeout, measured in milliseconds. More refined
    semantics for objects of this type depend upon the algorithm used to
    determine the retransmission timeout. In particular, when the
    timeout algorithm is rsre(3), an object of this type has the semantics
    of the UBOUND quantity described in RFC 793."
  ::={ tcp 3 }

tcpMaxConn OBJECT-TYPE
  SYNTAX INTEGER
  ACCESS read-only
  STATUS mandatory
  DESCRIPTION
    "The limit on the total number of TCP connections the entity can
    support. In entities where the maximum number of connections is
    dynamic, this object should contain the value -1."
  ::={ tcp 4 }

tcpActiveOpens OBJECT-TYPE
  SYNTAX Counter
  ACCESS read-only
  STATUS mandatory
  DESCRIPTION
    "The number of times TCP connections have made a direct
    transition to the SYN-SENT state from the CLOSED state."
  ::={ tcp 5 }

tcpPassiveOpens OBJECT-TYPE
  SYNTAX Counter
```

```
    ACCESS read-only
    STATUS mandatory
    DESCRIPTION
      "The number of times TCP connections have made a direct
      transition to the SYN-RCVD state from the LISTEN state."
    ::={ tcp 6 }
```

tcpAttemptFails OBJECT-TYPE

```
    SYNTAX Counter
    ACCESS read-only
    STATUS mandatory
    DESCRIPTION
      "The number of times TCP connections have made a direct
      transition to the CLOSED state from either the SYN-SENT state or
      the SYN-RCVD state, plus the number of times TCP connections
      have made a direct transition to the LISTEN state from the SYN-
      RCVD state."
    ::={ tcp 7 }
```

tcpEstabResets OBJECT-TYPE

```
    SYNTAX Counter
    ACCESS read-only
    STATUS mandatory
    DESCRIPTION
      "The number of times TCP connections have made a direct
      transition to the CLOSED state from either the ESTABLISHED state
      or the CLOSE-WAIT state."
    ::={ tcp 8 }
```

tcpCurrEstab OBJECT-TYPE

```
    SYNTAX Gauge
    ACCESS read-only
    STATUS mandatory
    DESCRIPTION
      "The number of TCP connections for which the current state is
      either ESTABLISHED or CLOSE-WAIT."
    ::={ tcp 9 }
```

tcpInSegs OBJECT-TYPE

```
    SYNTAX Counter
    ACCESS read-only
    STATUS mandatory
    DESCRIPTION
      "The total number of segments received, including those received in
      error. This count includes segments received on currently estab-
      lished connections."
    ::={ tcp 10 }
```

```
tcpOutSegs OBJECT-TYPE
  SYNTAX Counter
  ACCESS read-only
  STATUS mandatory
  DESCRIPTION
    "The total number of segments sent, including those on current
    connections but excluding those containing only retransmitted
    octets."
  ::={ tcp 11 }

tcpRetransSegs OBJECT-TYPE
  SYNTAX Counter
  ACCESS read-only
  STATUS mandatory
  DESCRIPTION
    "The total number of segments retransmitted—that is, the number
    of TCP segments transmitted containing one or more previously
    transmitted octets."
  ::={ tcp 12 }


-- the TCP Connection table

-- The TCP Connection table contains information about this
-- entity's existing TCP connections.

tcpConnTable OBJECT-TYPE
  SYNTAX SEQUENCE OF TcpConnEntry
  ACCESS not-accessible
  STATUS mandatory
  DESCRIPTION
    "A table containing TCP connection-specific information."
  ::={ tcp 13 }

tcpConnEntry OBJECT-TYPE
  SYNTAX TcpConnEntry
  ACCESS not-accessible
  STATUS mandatory
  DESCRIPTION
    "Information about a particular current TCP connection. An object
    of this type is transient, in that it ceases to exist when (or soon after)
    the connection makes the transition to the CLOSED state."
  INDEX   { tcpConnLocalAddress,
            tcpConnLocalPort,
            tcpConnRemAddress,
            tcpConnRemPort }
  ::={ tcpConnTable 1 }
```

```
TcpConnEntry ::=
  SEQUENCE {
    tcpConnState
       INTEGER,
    tcpConnLocalAddress
      IpAddress,
    tcpConnLocalPort
      INTEGER (0..65535),
    tcpConnRemAddress
     IpAddress,
    tcpConnRemPort
      INTEGER (0..65535)
  }

tcpConnState OBJECT-TYPE
  SYNTAX    INTEGER {
                closed(1),
                listen(2),
                synSent(3),
                synReceived(4),
                established(5),
                finWait1(6),
                finWait2(7),
                closeWait(8),
                lastAck(9),
                closing(10),
                timeWait(11),
                deleteTCB(12)
            }
  ACCESS read-write
  STATUS mandatory
  DESCRIPTION
    "The state of this TCP connection.

    The only value that may be set by a management station is
    deleteTCB(12). Accordingly, it is appropriate for an agent to return a
    'badValue' response if a management station attempts to set this
    object to any other value.

    If a management station sets this object to the value deleteTCB(12),
    then this has the effect of deleting the TCB (as defined in RFC 793) of
    the corresponding connection on the managed node, resulting in
    immediate termination of the connection.

    As an implementation-specific option, a RST segment may be sent
    from the managed node to the other TCP endpoint (note, however,
    that RST segments are not sent reliably)."
  ::={ tcpConnEntry 1 }
```

```
tcpConnLocalAddress OBJECT-TYPE
  SYNTAX IpAddress
  ACCESS read-only
  STATUS mandatory
  DESCRIPTION
    "The local IP address for this TCP connection. In the case of a
    connection in the listen state which is willing to accept connections
    for any IP interface associated with the node, the value 0.0.0.0 is
    used."
  ::={ tcpConnEntry 2 }

tcpConnLocalPort OBJECT-TYPE
  SYNTAX INTEGER (0..65535)
  ACCESS read-only
  STATUS mandatory
  DESCRIPTION
    "The local port number for this TCP connection."
  ::={ tcpConnEntry 3 }

tcpConnRemAddress OBJECT-TYPE
  SYNTAX IpAddress
  ACCESS read-only
  STATUS mandatory
  DESCRIPTION
    "The remote IP address for this TCP connection."
  ::={ tcpConnEntry 4 }

tcpConnRemPort OBJECT-TYPE
  SYNTAX INTEGER (0..65535)
  ACCESS read-only
  STATUS mandatory
  DESCRIPTION
    "The remote port number for this TCP connection."
  ::={ tcpConnEntry 5 }


-- Additional TCP objects

tcpInErrs OBJECT-TYPE
  SYNTAX Counter
  ACCESS read-only
  STATUS mandatory
  DESCRIPTION
    "The total number of segments received in error (e.g., bad TCP
    checksums)."
  ::={ tcp 14 }
```

tcpOutRsts OBJECT-TYPE
SYNTAX Counter
ACCESS read-only
STATUS mandatory
DESCRIPTION
"The number of TCP segments sent containing the RST flag."
::={ tcp 15 }

A.7 THE UDP GROUP

- - Implementation of the UDP group is mandatory for all
- - systems that implement the UDP.

udpInDatagrams OBJECT-TYPE
SYNTAX Counter
ACCESS read-only
STATUS mandatory
DESCRIPTION
"The total number of UDP datagrams delivered to UDP users."
::={ udp 1 }

udpNoPorts OBJECT-TYPE
SYNTAX Counter
ACCESS read-only
STATUS mandatory
DESCRIPTION
"The total number of received UDP datagrams for which there was no application at the destination port."
::={ udp 2 }

udpInErrors OBJECT-TYPE
SYNTAX Counter
ACCESS read-only
STATUS mandatory
DESCRIPTION
"The number of received UDP datagrams that could not be delivered for reasons other than the lack of an application at the destination port."
::={ udp 3 }

udpOutDatagrams OBJECT-TYPE
SYNTAX Counter
ACCESS read-only
STATUS mandatory
DESCRIPTION
"The total number of UDP datagrams sent from this entity."
::={ udp 4 }

```
-- the UDP Listener table

-- The UDP Listener table contains information about this
-- entity's UDP end-points on which a local application is
-- currently accepting datagrams.

udpTable OBJECT-TYPE
  SYNTAX SEQUENCE OF UdpEntry
  ACCESS not-accessible
  STATUS mandatory
  DESCRIPTION
    "A table containing UDP listener information."
  ::={ udp 5 }

udpEntry OBJECT-TYPE
  SYNTAX UdpEntry
  ACCESS not-accessible
  STATUS mandatory
  DESCRIPTION
    "Information about a particular current UDP listener."
  INDEX { udpLocalAddress, udpLocalPort }
  ::={ udpTable 1 }

UdpEntry ::=
  SEQUENCE {
    udpLocalAddress
      IpAddress,
    udpLocalPort
      INTEGER (0..65535)
  }

udpLocalAddress OBJECT-TYPE
  SYNTAX IpAddress
  ACCESS read-only
  STATUS mandatory
  DESCRIPTION
    "The local IP address for this UDP listener. In the case of a UDP
    listener that is willing to accept datagrams for any IP interface
    associated with the node, the value 0.0.0.0 is used."
  ::={ udpEntry 1 }

udpLocalPort OBJECT-TYPE
  SYNTAX INTEGER (0..65535)
  ACCESS read-only
  STATUS mandatory
  DESCRIPTION
    "The local port number for this UDP listener."
  ::={ udpEntry 2 }
```

A.8 THE EGP GROUP

```
-- Implementation of the EGP group is mandatory for all
-- systems that implement the EGP.

egpInMsgs OBJECT-TYPE
  SYNTAX Counter
  ACCESS read-only
  STATUS mandatory
  DESCRIPTION
    "The number of EGP messages received without error."
  ::={ egp 1 }

egpInErrors OBJECT-TYPE
  SYNTAX Counter
  ACCESS read-only
  STATUS mandatory
  DESCRIPTION
    "The number of EGP messages received that proved to be in error."
  ::={ egp 2 }

egpOutMsgs OBJECT-TYPE
  SYNTAX Counter
  ACCESS read-only
  STATUS mandatory
  DESCRIPTION
    "The total number of locally generated EGP messages."
  ::={ egp 3 }

egpOutErrors OBJECT-TYPE
  SYNTAX Counter
  ACCESS read-only
  STATUS mandatory
  DESCRIPTION
    "The number of locally generated EGP messages not sent due to
    resource limitations within an EGP entity."
  ::={ egp 4 }


-- the EGP Neighbor table

-- The EGP Neighbor table contains information about this
-- entity's EGP neighbors.

egpNeighTable OBJECT-TYPE
  SYNTAX SEQUENCE OF EgpNeighEntry
  ACCESS not-accessible
```

```
  STATUS mandatory
  DESCRIPTION
    "The EGP neighbor table."
  ::={ egp 5 }
```

egpNeighEntry OBJECT-TYPE

```
  SYNTAX EgpNeighEntry
  ACCESS not-accessible
  STATUS mandatory
  DESCRIPTION
    "Information about this entity's relationship with a particular EGP
    neighbor."
  INDEX { egpNeighAddr }
  ::={ egpNeighTable 1 }
```

EgpNeighEntry ::=

```
  SEQUENCE {
    egpNeighState
      INTEGER,
    egpNeighAddr
      IpAddress,
    egpNeighAs
      INTEGER,
    egpNeighInMsgs
      Counter,
    egpNeighInErrs
      Counter,
    egpNeighOutMsgs
      Counter,
    egpNeighOutErrs
      Counter,
    egpNeighInErrMsgs
      Counter,
    egpNeighOutErrMsgs
      Counter,
    egpNeighStateUps
      Counter,
    egpNeighStateDowns
      Counter,
    egpNeighIntervalHello
      INTEGER,
    egpNeighIntervalPoll
      INTEGER,
    egpNeighMode
      INTEGER,
    egpNeighEventTrigger
      INTEGER
  }
```

```
egpNeighState OBJECT-TYPE
  SYNTAX INTEGER {
              idle(1),
              acquisition(2),
              down(3),
              up(4),
              cease(5)
           }
  ACCESS read-only
  STATUS mandatory
  DESCRIPTION
    "The EGP state of the local system with respect to this entry's EGP
    neighbor. Each EGP state is represented by a value that is one
    greater than the numerical value associated with said state in RFC
    904."
  ::={ egpNeighEntry 1 }

egpNeighAddr OBJECT-TYPE
  SYNTAX IpAddress
  ACCESS read-only
  STATUS mandatory
  DESCRIPTION
    "The IP address of this entry's EGP neighbor."
  ::={ egpNeighEntry 2 }

egpNeighAs OBJECT-TYPE
  SYNTAX INTEGER
  ACCESS read-only
  STATUS mandatory
  DESCRIPTION
    "The autonomous system of this EGP peer. Zero should be specified
    if the autonomous system number of the neighbor is not yet
    known."
  ::={ egpNeighEntry 3 }

egpNeighInMsgs OBJECT-TYPE
  SYNTAX Counter
  ACCESS read-only
  STATUS mandatory
  DESCRIPTION
    "The number of EGP messages received without error from this EGP
    peer."
  ::={ egpNeighEntry 4 }

egpNeighInErrs OBJECT-TYPE
  SYNTAX Counter
  ACCESS read-only
  STATUS mandatory
```

```
    DESCRIPTION
      "The number of EGP messages received from this EGP peer that
      proved to be in error (e.g., bad EGP checksum)."
    ::={ egpNeighEntry 5 }
```

egpNeighOutMsgs OBJECT-TYPE

```
    SYNTAX Counter
    ACCESS read-only
    STATUS mandatory
    DESCRIPTION
      "The number of locally generated EGP messages to this EGP peer."
    ::={ egpNeighEntry 6 }
```

egpNeighOutErrs OBJECT-TYPE

```
    SYNTAX Counter
    ACCESS read-only
    STATUS mandatory
    DESCRIPTION
      "The number of locally generated EGP messages not sent to this
      EGP peer due to resource limitations within an EGP entity."
    ::={ egpNeighEntry 7 }
```

egpNeighInErrMsgs OBJECT-TYPE

```
    SYNTAX Counter
    ACCESS read-only
    STATUS mandatory
    DESCRIPTION
      "The number of EGP-defined error messages received from this EGP
      peer."
    ::={ egpNeighEntry 8 }
```

egpNeighOutErrMsgs OBJECT-TYPE

```
    SYNTAX Counter
    ACCESS read-only
    STATUS mandatory
    DESCRIPTION
      "The number of EGP-defined error messages sent to this EGP peer."
    ::={ egpNeighEntry 9 }
```

egpNeighStateUps OBJECT-TYPE

```
    SYNTAX Counter
    ACCESS read-only
    STATUS mandatory
    DESCRIPTION
      "The number of EGP state transitions to the UP state with this EGP
      peer."
    ::={ egpNeighEntry 10 }
```

```
egpNeighStateDowns OBJECT-TYPE
  SYNTAX Counter
  ACCESS read-only
  STATUS mandatory
  DESCRIPTION
    "The number of EGP state transitions from the UP state to any other
    state with this EGP peer."
  ::={ egpNeighEntry 11 }

egpNeighIntervalHello OBJECT-TYPE
  SYNTAX INTEGER
  ACCESS read-only
  STATUS mandatory
  DESCRIPTION
    "The interval between EGP Hello command retransmissions (in
    hundredths of a second). This represents the t1 timer as defined in
    RFC 904."
  ::={ egpNeighEntry 12 }

egpNeighIntervalPoll OBJECT-TYPE
  SYNTAX INTEGER
  ACCESS read-only
  STATUS mandatory
  DESCRIPTION
    "The interval between EGP poll command retransmissions (in
    hundredths of a second). This represents the t3 timer as defined in
    RFC 904."
  ::={ egpNeighEntry 13 }

egpNeighMode OBJECT-TYPE
  SYNTAX INTEGER { active(1), passive(2) }
  ACCESS read-only
  STATUS mandatory
  DESCRIPTION
    "The polling mode of this EGP entity, either passive or active."
  ::={ egpNeighEntry 14 }

egpNeighEventTrigger OBJECT-TYPE
  SYNTAX INTEGER { start(1), stop(2) }
  ACCESS read-write
  STATUS mandatory
  DESCRIPTION
    "A control variable used to trigger operator-initiated Start and Stop
    events. When read, this variable always returns the most recent
    value that egpNeighEventTrigger was set to. If it has not been set
    since the last initialization of the network management subsystem
    on the node, it returns a value of 'stop'.
```

```
      When set, this variable causes a Start or Stop event on the specified
      neighbor, as specified on pages 8–10 of RFC 904. Briefly, a Start
      event causes an Idle peer to begin neighbor acquisition and a non-
      Idle peer to reinitiate neighbor acquisition. A stop event causes a
      non-Idle peer to return to the Idle state until a Start event occurs,
      either via egpNeighEventTrigger or otherwise."
  ::={ egpNeighEntry 15 }


-- Additional EGP objects

egpAs OBJECT-TYPE
  SYNTAX INTEGER
  ACCESS read-only
  STATUS mandatory
  DESCRIPTION
    "The autonomous system number of this EGP entity."
  ::={ egp 6 }
```

A.9 THE TRANSMISSION GROUP

```
-- Based on the transmission media underlying each interface
-- on a system, the corresponding portion of the
-- Transmission group is mandatory for that system.

-- When Internet-standard definitions for managing
-- transmission media are defined, the transmission group is
-- used to provide a prefix for the names of those objects.

-- Typically, such definitions reside in the experimental
-- portion of the MIB until they are "proven", then as a
-- part of the Internet standardization process, the
-- definitions are accordingly elevated and a new object
-- identifier, under the transmission group is defined. By
-- convention, the name assigned is:
--
-- type OBJECT IDENTIFIER ::={ transmission number }
--
-- where 'type' is the symbolic value used for the media in
-- the ifType column of the ifTable object, and 'number' is
-- the actual integer value corresponding to the symbol.
```

A.10 THE SNMP GROUP

```
-- Implementation of the SNMP group is mandatory for all
-- systems that support an SNMP protocol entity. Some of
```

```
- - the objects defined below will be zero-valued in those
- - SNMP implementations that are optimized to support only
- - those functions specific to either a management agent or
- - a management station. In particular, it should be
- - observed that the objects below refer to an SNMP entity,
- - and there may be several SNMP entities residing on a
- - managed node (e.g., if the node is hosting acting as
- - a management station).

snmpInPkts OBJECT-TYPE
  SYNTAX Counter
  ACCESS read-only
  STATUS mandatory
  DESCRIPTION
    "The total number of Messages delivered to the SNMP entity from
    the transport service."
  ::={ snmp 1 }

snmpOutPkts OBJECT-TYPE
  SYNTAX Counter
  ACCESS read-only
  STATUS mandatory
  DESCRIPTION
    "The total number of SNMP Messages which were passed from the
    SNMP protocol entity to the transport service."
  ::={ snmp 2 }

snmpInBadVersions OBJECT-TYPE
  SYNTAX Counter
  ACCESS read-only
  STATUS mandatory
  DESCRIPTION
    "The total number of SNMP Messages which were delivered to the
    SNMP protocol entity and were for an unsupported SNMP version."
  ::={ snmp 3 }

snmpInBadCommunityNames OBJECT-TYPE
  SYNTAX Counter
  ACCESS read-only
  STATUS mandatory
  DESCRIPTION
    "The total number of SNMP Messages delivered to the SNMP
    protocol entity that used a SNMP community name not known to
    said entity."
  ::={ snmp 4 }

snmpInBadCommunityUses OBJECT-TYPE
  SYNTAX Counter
```

```
  ACCESS read-only
  STATUS mandatory
  DESCRIPTION
    "The total number of SNMP Messages delivered to the SNMP
    protocol entity that represented an SNMP operation that was not
    allowed by the SNMP community named in the Message."
  ::={ snmp 5 }
```

snmpInASNParseErrs OBJECT-TYPE

```
  SYNTAX Counter
  ACCESS read-only
  STATUS mandatory
  DESCRIPTION
    "The total number of ASN.1 or BER errors encountered by the SNMP
    protocol entity when decoding received SNMP Messages."
  ::={ snmp 6 }
```

```
- - { snmp 7 } is not used
```

snmpInTooBigs OBJECT-TYPE

```
  SYNTAX Counter
  ACCESS read-only
  STATUS mandatory
  DESCRIPTION
    "The total number of SNMP PDUs that were delivered to the SNMP
    protocol entity and for which the value of the error-status field is
    'tooBig'."
  ::={ snmp 8 }
```

snmpInNoSuchNames OBJECT-TYPE

```
  SYNTAX Counter
  ACCESS read-only
  STATUS mandatory
  DESCRIPTION
    "The total number of SNMP PDUs that were delivered to the SNMP
    protocol entity and for which the value of the error-status field is
    'noSuchName'."
  ::={ snmp 9 }
```

snmpInBadValues OBJECT-TYPE

```
  SYNTAX Counter
  ACCESS read-only
  STATUS mandatory
  DESCRIPTION
    "The total number of SNMP PDUs that were delivered to the SNMP
    protocol entity and for which the value of the error-status field is
    'badValue'."
  ::={ snmp 10 }
```

snmpInReadOnlys OBJECT-TYPE
SYNTAX Counter
ACCESS read-only
STATUS mandatory
DESCRIPTION
"The total number of valid SNMP PDUs that were delivered to the SNMP protocol entity and for which the value of the error-status field is 'readOnly'. It should be noted that it is a protocol error to generate an SNMP PDU that contains the value 'readOnly' in the error-status field, as such this object is provided as a means of detecting incorrect implementations of the SNMP."
::={ snmp 11 }

snmpInGenErrs OBJECT-TYPE
SYNTAX Counter
ACCESS read-only
STATUS mandatory
DESCRIPTION
"The total number of SNMP PDUs that were delivered to the SNMP protocol entity and for which the value of the error-status field is 'genErr'."
::={ snmp 12 }

snmpInTotalReqVars OBJECT-TYPE
SYNTAX Counter
ACCESS read-only
STATUS mandatory
DESCRIPTION
"The total number of MIB objects that have been retrieved successfully by the SNMP protocol entity as the result of receiving valid SNMP Get-Request and Get-Next PDUs."
::={ snmp 13 }

snmpInTotalSetVars OBJECT-TYPE
SYNTAX Counter
ACCESS read-only
STATUS mandatory
DESCRIPTION
"The total number of MIB objects that have been altered successfully by the SNMP protocol entity as the result of receiving valid SNMP Set-Request PDUs."
::={ snmp 14 }

snmpInGetRequests OBJECT-TYPE
SYNTAX Counter
ACCESS read-only
STATUS mandatory
DESCRIPTION

"The total number of SNMP Get-Request PDUs that have been accepted and processed by the SNMP protocol entity."
::={ snmp 15 }

snmpInGetNexts OBJECT-TYPE
SYNTAX Counter
ACCESS read-only
STATUS mandatory
DESCRIPTION
"The total number of SNMP Get-Next PDUs that have been accepted and processed by the SNMP protocol entity."
::={ snmp 16 }

snmpInSetRequests OBJECT-TYPE
SYNTAX Counter
ACCESS read-only
STATUS mandatory
DESCRIPTION
"The total number of SNMP Set-Request PDUs that have been accepted and processed by the SNMP protocol entity."
::={ snmp 17 }

snmpInGetResponses OBJECT-TYPE
SYNTAX Counter
ACCESS read-only
STATUS mandatory
DESCRIPTION
"The total number of SNMP Get-Response PDUs that have been accepted and processed by the SNMP protocol entity."
::={ snmp 18 }

snmpInTraps OBJECT-TYPE
SYNTAX Counter
ACCESS read-only
STATUS mandatory
DESCRIPTION
"The total number of SNMP Trap PDUs that have been accepted and processed by the SNMP protocol entity."
::={ snmp 19 }

snmpOutTooBigs OBJECT-TYPE
SYNTAX Counter
ACCESS read-only
STATUS mandatory
DESCRIPTION
"The total number of SNMP PDUs that were generated by the SNMP protocol entity and for which the value of the error-status field is 'tooBig.' "
::={ snmp 20 }

```
snmpOutNoSuchNames OBJECT-TYPE
  SYNTAX Counter
  ACCESS read-only
  STATUS mandatory
  DESCRIPTION
    "The total number of SNMP PDUs that were generated by the SNMP
    protocol entity and for which the value of the error-status is
    'noSuchName'."
  ::={ snmp 21 }

snmpOutBadValues OBJECT-TYPE
  SYNTAX Counter
  ACCESS read-only
  STATUS mandatory
  DESCRIPTION
    "The total number of SNMP PDUs that were generated by the SNMP
    protocol entity and for which the value of the error-status field is
    'badValue'."
  ::={ snmp 22 }

-- { snmp 23 } is not used

snmpOutGenErrs OBJECT-TYPE
  SYNTAX Counter
  ACCESS read-only
  STATUS mandatory
  DESCRIPTION
    "The total number of SNMP PDUs that were generated by the SNMP
    protocol entity and for which the value of the error-status field is
    'genErr'."
  ::={ snmp 24 }

snmpOutGetRequests OBJECT-TYPE
  SYNTAX Counter
  ACCESS read-only
  STATUS mandatory
  DESCRIPTION
    "The total number of SNMP Get-Request PDUs that have been
    generated by the SNMP protocol entity."
  ::={ snmp 25 }

snmpOutGetNexts OBJECT-TYPE
  SYNTAX Counter
  ACCESS read-only
  STATUS mandatory
  DESCRIPTION
    "The total number of SNMP Get-Next PDUs that have been
    generated by the SNMP protocol entity."
  ::={ snmp 26 }
```

snmpOutSetRequests OBJECT-TYPE
SYNTAX Counter
ACCESS read-only
STATUS mandatory
DESCRIPTION
"The total number of SNMP Set-Request PDUs that have been generated by the SNMP protocol entity."
::={ snmp 27 }

snmpOutGetResponses OBJECT-TYPE
SYNTAX Counter
ACCESS read-only
STATUS mandatory
DESCRIPTION
"The total number of SNMP Get-Response PDUs that have been generated by the SNMP protocol entity."
::={ snmp 28 }

snmpOutTraps OBJECT-TYPE
SYNTAX Counter
ACCESS read-only
STATUS mandatory
DESCRIPTION
"The total number of SNMP Trap PDUs that have been generated by the SNMP protocol entity."
::={ snmp 29 }

snmpEnableAuthenTraps OBJECT-TYPE
SYNTAX INTEGER { enabled(1), disabled(2) }
ACCESS read-write
STATUS mandatory
DESCRIPTION
"Indicates whether the SNMP agent process is permitted to generate authentication-failure traps. The value of this object overrides any configuration information; as such, it provides a means whereby all authentication-failure traps may be disabled.

Note that it is strongly recommended that this object be stored in non-volatile memory so that it remains constant between re-initializations of the network management system."
::={ snmp 30 }

Appendix B

DEMONSTRATION SOFTWARE

Demonstration diagnostic testing and SNMP/RMON communications programs are available from AG Software and Triticom, two firms specializing in the development of software products that considerably facilitate the work of network managers and LAN administrators.

The EtherPeek for Windows Demo software can be downloaded directly from AG's website. Readers can find them at http://www.aggroup.com.

Triticom's demonstration programs can be accessed through their website: http://www.triticom.com. In the Triticom directory you will find several subdirectories, with each subdirectory containing one or more Triticom demonstration programs. One subdirectory named SimpleView contains a demonstration copy of the Triticom SimpleView Windows SNMP Network Management Program. A second subdirectory labeled DemoPack contains demonstration copies of its LANdecoder, EtherVision, TokenVision, and RMONster and Interconnection programs.

INDEX

Note: Figures and Tables are indicated (in this index) by *italic page numbers*

Index compiled by Paul Nash